ALLE · ZEIT · WACH · 1842

Topics in Boundary Element Research

Edited by C. A. Brebbia

Volume 3: Computational Aspects

With 126 Figures and 28 Tables

Springer-Verlag Berlin Heidelberg New York
London Paris Tokyo

Editor:
Dr. Carlos A. Brebbia

Computational Mechanics Institute
Wessex Institute of Technology
52 Henstead Road
Southampton SO1 2DD
England

ISBN-13:978-3-642-82665-8 e-ISBN-13:978-3-642-82663-4
DOI: 10.1007/978-3-642-82663-4

Typesetting: Graphischer Betrieb Konrad Triltsch, Würzburg

2161/3020-5 4 3 2 1 0

Contributors

M. Doblaré	University of Zaragoza, Spain	(Chap. 4)
T. V. Hromadka II	University of California, U.S.A.	(Chap. 7)
Y. Iso	Kyoto University, Japan	(Chap. 1)
M. J. A. O'Callaghan	University College, Cork, Eire	(Chap. 5)
K. Onishi	Fukuoka University, Japan	(Chap. 1)
H. Pina	Instituto Superior Técnico, Lisbon, Portugal	(Chap. 3)
P. M. Quinlan	University College, Cork, Eire	(Chap. 5)
H. Schippers	National Aerospace Laboratory, Amsterdam, The Netherlands	(Chap. 6)
S. Takahashi	Science University of Tokyo, Japan	(Chap. 1)
M. Tanaka	Shinshu University, Japan	(Chap. 2)
J. C. F. Telles	Federal University of Rio de Janeiro, Brazil	(Chap. 9)
L. C. Wrobel	Federal University of Rio de Janeiro, Brazil	(Chap. 8)

Introduction to the Series "Topics in Boundary Element Research"

The continuing interest in the application of Boundary Element Methods in engineering has generated a series of books and numerous scientific papers, not least those regularly presented at the International Conferences on Boundary Elements which have been held under my direction since 1978. Most recently a new journal, "Engineering Analysis", has been launched which concentrates on new developments in this important area. In spite of all this activity, the need exists for a serial publication in which the most recent advances in the method are documented in a more complete form than is usually the case in papers presented at conferences or scientific gatherings. This unfulfilled need prompted me to launch the present series.

Each volume in this series will comprise chapters describing new applications of the method. The emphases will be on contributions which are self-contained and explain a particular topic in sufficient detail for the analytical engineer or scientist to be able to understand the theory and in due course to write the relevant computer software. All chapters are written by scientists who are actively involved in Boundary Element research, the internationally best known names being balanced with those of new researchers who have recently made significant contributions in this area.

Another objective of the series is to report work for direct application by the practising engineer. Furthermore, I feel that it is important to include sections which discuss the modelling strategies and presentation of results as well as theoretical chapters. The relationship between Boundary Element analysis codes and computer aided design packages will be discussed in subsequent volumes to achieve the right perspective on the application of Boundary Elements. It is all too easy when dealing with these types of analytical techniques to forget that they exist within the framework of the final engineering product.

It is my intention that the series should be open to all those researchers who have made significant contributions to the advancement of the new method. In this regard I shall be happy to receive any suggestions that such members of the scientific community may wish to make, in an effort to produce a publication that is indispensable to all concerned with the advancement of Boundary Elements.

Carlos A. Brebbia
Editor

Preface

Past volumes of this series have concentrated on the theoretical and the more formal aspects of the boundary element method. The present book instead stresses the computational aspects of the technique and its applications with the objective of facilitating the implementation of BEM in the engineering industry and its better understanding in the teaching and research environments.

The book starts by discussing the topics of convergence of solutions, application to nonlinear problems and numerical integration. This is followed by a long chapter on the computational aspects of the method, discussing the different numerical schemes and the way in which influence functions can be computed. Three separate chapters deal with important techniques which are related to classical boundary elements, namely the edge method, multigrid schemes and the complex variable boundary element approach.

The last two chapters are of special interest as they present and explain in detail two FORTRAN codes which have numerous applications in engineering, i.e. a code for the solution of potential problems and another for elastostatics. Each subroutine in the programs is listed and explained. The codes follow the same format as the ones in the classical book "The Boundary Element Method for Engineers" (by C. A. Brebbia, Computational Mechanics Publications, first published in 1978) but are more advanced in terms of elements and capabilities. In particular the new listings deal with symmetry, linear elements for the two dimensional elasticity, some mixed type of boundary conditions and the treatment of infinite regions.

The importance of this volume is that it attempts to bridge the gap between theory and applications and provide a valuable help to those engineers or scientists who are interested in using the new technique but are not conversant with its computational aspects. The chapters and the codes published here should give them a solid starting point for applying boundary elements to solve their problems. By doing so this volume attempts to help raise the level of awareness of boundary elements amongst the scientific and engineering communities.

Southampton, December 1986

Carlos A. Brebbia
Editor

Contents

Chapter 1

Numerical Convergence of Boundary Solutions in Transient Heat Conduction Problems

by Y. Iso, S. Takahashi, and K. Onishi

Abstract

Mathematical theorems on uniform convergence of the boundary solution and stability of the computing scheme are proved for the approximation of a singular boundary integral equation using a time-dependent fundamental solution for two-dimensional isotropic heat conduction problems with non-isothermal boundary conditions. Discussions are extended to efficient numerical techniques developed for the transient boundary solution of problems having singular points and for the solution of external problems. The zoning technique and the double node technique are used together with linear boundary elements. Specific examples involving re-entrant corners and the external region of a circle are considered to show high accuracy of the boundary element solution.

1.1 Introduction

The direct boundary element method using the time-dependent heat potential was introduced by Rizzo and Shippy [1], Chang et al. [2], and Shaw [3] to approximate the boundary solution of transient thermal diffusion problems. Wrobel and Brebbia [4] improved the numerical method for the computer application to the heat transfer problems in engineerings. Two nice reference books were published recently by Banerjee and Butterfield [5], and by Brebbia et al. [6] for further developments of this method.

Purpose of this paper is to show: (1) mathematical theorems on the convergence of direct boundary solutions, and (2) numerical solutions near singular points and for unbounded domains. The convergence is proved mathematically for the problem with flux-type boundary conditions and for the problem with boundary conditions of the linear radiation type.

Mathematical theorem on uniform convergence and stability for the boundary solution of two-dimensional transient linear heat conduction problems with the pure Neumann condition was given by Onishi [7], Onishi and Kuroki [8]. Recently, the first and the second author improved the proof of the theorem. More recently, Iso [9] extended the theorem to the problem with linear radiation boundary conditions.

1.2 Boundary Element Approximations

1.2.1 Motivation

Let Ω be a two-dimensional bounded convex domain with the smooth boundary denoted by Γ. The domain may be regarded to be a heat-conducting medium. With the rectangular coordinate system, we denote the coordinates of the spatial points by $\boldsymbol{P}(x, y)$.

We consider the following initial-boundary value problem for heat-conducting equation in the finite time interval $0 < t \leqq T$:

(i) field equation with the unit thermal diffusivity:

$$\left(\frac{\partial}{\partial t} - \nabla^2\right) u(\boldsymbol{P}, t) = 0 \quad \text{in } \Omega \times (0, T] \tag{2.1}$$

(ii) initial condition:

$$u(\boldsymbol{P}, 0) = u_0(\boldsymbol{P}) \quad \text{on } \bar{\Omega} \tag{2.2}$$

(iii) boundary conditions:

$$u(\boldsymbol{P}, t) = \bar{u}_1(\boldsymbol{P}) \qquad \text{on } \Gamma_u \times (0, T] \tag{2.3}$$

$$q(\boldsymbol{P}, t) = \bar{q}(\boldsymbol{P}) \qquad \text{on } \Gamma_q \times (0, T] \tag{2.4}$$

$$q(\boldsymbol{P}, t) = h\,\{u(\boldsymbol{P}, t) - \bar{u}_2(\boldsymbol{P})\} \quad \text{on } \Gamma_r \times (0, T]. \tag{2.5}$$

Here, q is the boundary heat flux defined by $q = -\partial u/\partial n$ in which $\boldsymbol{n}$ is the unit outward normal vector at the boundary, ∇^2 is two-dimensional Laplacian, $\overline{\Gamma_u \cup \Gamma_q \cup \Gamma_r} = \Gamma$ in which some of the boundary parts Γ_u, Γ_q, Γ_r may be null sets. Furthermore, we assume that u_0, $\bar{u}_1$, $\bar{q}$, and $\bar{u}_2$ are given functions and that u_0 is the smooth function on $\bar{\Omega}$, and $\bar{u}_1$, $\bar{q}$, $\bar{u}_2$ are also the smooth functions on $\bar{\Gamma}_u$, $\bar{\Gamma}_q$, $\bar{\Gamma}_r$, respectively.

We intend to construct the solution of the problem (2.1)–(2.5) by boundary element methods.

1.2.2 Boundary Integral Equations

The time-dependent fundamental solution v^* of the adjoint equation of (2.1), which has its singularity at $(\boldsymbol{P}, \tau)$, is given by

$$v^*(\boldsymbol{Q}, t\colon \boldsymbol{P}, \tau) = \left(\frac{1}{\sqrt{4\pi(\tau - t)}}\right)^2 \exp\left(-\frac{\overline{PQ}^2}{4(\tau - t)}\right) H(\tau - t) \tag{2.6}$$

for $0 < \tau \leqq T$. Here, $\boldsymbol{Q}$ is a spatial point and has its coordinates $\boldsymbol{Q}(\xi, \eta)$, $\overline{PQ}$ means the distance between two points $\boldsymbol{P}$ and $\boldsymbol{Q}$, and H is the Heaviside function, i.e., $H(x) = 0$ for $x < 0$ and $H(x) = 1$ for $x \geqq 0$.

Using the fundamental solution (2.6), we can transform Eq. (2.1) into the following integral equation:

$$\begin{aligned} u(\boldsymbol{P},\tau) - \int_{\Gamma}\int_{0}^{\tau} u(\boldsymbol{Q},t)\, q^*(\boldsymbol{Q},t:\boldsymbol{P},\tau)\, d\Gamma_Q\, dt &= -\int_{\Gamma}\int_{0}^{\tau} q(\boldsymbol{Q},t)\, v^*(\boldsymbol{Q},t:\boldsymbol{P},\tau)\, d\Gamma_Q\, dt \\ &\quad + \int_{\Omega} u_0(\boldsymbol{Q})\, v^*(\boldsymbol{Q},0:\boldsymbol{P},\tau)\, d\Omega_Q \end{aligned} \tag{2.7}$$

for $\boldsymbol{P} \in \mathring{\Omega}$, where q^* is defined by $q^* = -\partial v^*(\boldsymbol{Q},t:\boldsymbol{P},\tau)/\partial n_Q$, and $\partial/\partial n_Q$, $d\Gamma_Q$, $d\Omega_Q$ are considered for the variables ξ, η.

According to Eq. (2.7), we can find the temperature $u(\boldsymbol{P},t)$ at inner points $\boldsymbol{P}$ of the domain Ω, if we know the values of $u(\boldsymbol{Q},t)$ and $q(\boldsymbol{Q},t)$ on the boundary Γ. Therefore our aim is to find the values u and q on the boundary.

First of all, we make $\boldsymbol{P}$ tend to the boundary point $\boldsymbol{P}_0$ in Eq. (2.7). As $\boldsymbol{Q}$ is on Γ, $v^*(\boldsymbol{Q},t:\boldsymbol{P},\tau)$ and $q^*(\boldsymbol{Q},t:\boldsymbol{P},\tau)$ have their singularities at $(\boldsymbol{Q},t)=(\boldsymbol{P},\tau)$. But it is well known that the singularity of v^* is as much as absolutely integrable on Γ (Friedman [10]). Only the singularity of q^* must be taken into consideration.

We introduce principal value of integrals in connection with the singular integration of q^*. There are many definitions for it, such as those given by Friedman [10], by Onishi and Kuroki [11], which are essentially equivalent. Here, according to Iso [8], we introduce it in the following sense. We adopt this definition, because we are conscious of the discretization of the boundary integrals.

Let S be the boundary of the cylindrical domain:

$$S = \{(\boldsymbol{Q},t) \mid \boldsymbol{Q} \in \Gamma, 0 \leqq t \leqq T\}$$

and let B_δ be a "rectangle" on S as shown in Fig. 2.1 such as

$$B_\delta = \{(\boldsymbol{Q},t) \in S \mid \tau - \delta_t \leqq t \leqq \tau,\ \overline{\boldsymbol{Q}\boldsymbol{P}_0} \leqq \delta_\xi\}$$

where $\boldsymbol{P}_0 \in \Gamma$. Let S_δ be $\Gamma \times [0,\tau] \backslash B_\delta$. We now consider the following limit:

$$\lim_{\boldsymbol{P}\to\boldsymbol{P}_0} \int_{\Gamma}\int_{0}^{\tau} u(\boldsymbol{Q},t)\, q^*(\boldsymbol{Q},t:\boldsymbol{P},\tau)\, d\Gamma_Q\, dt \tag{2.8}$$

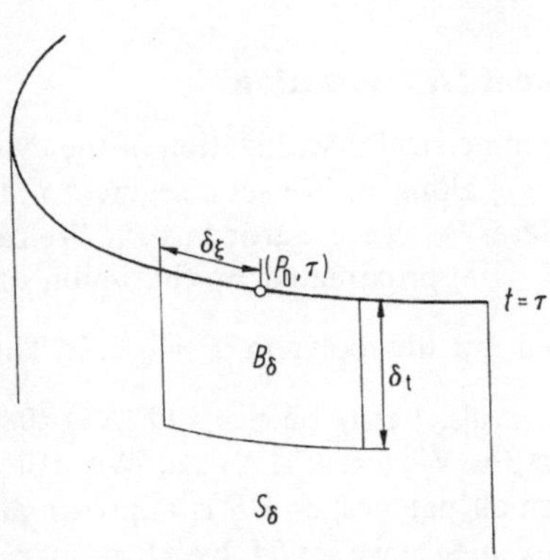

Fig. 2.1. Geometry of the integration domain

where u is a smooth function on S, and $\boldsymbol{P}$ tends to $\boldsymbol{P}_0$ in the normal direction to the boundary. We divide this integral as follows:

$$\int_\Gamma \int_0^\tau u(\boldsymbol{Q},t)\, q^*(\boldsymbol{Q},t:\boldsymbol{P},\tau)\, d\Gamma_Q\, dt = \iint_{S_\delta} u(\boldsymbol{Q},t)\, q^*(\boldsymbol{Q},t:\boldsymbol{P},\tau)\, d\Gamma_Q\, dt$$
$$+ \iint_{B_\delta} \{u(\boldsymbol{Q},t) - u(\boldsymbol{P}_0,\tau)\}\, q^*(\boldsymbol{Q},t:\boldsymbol{P},\tau)\, d\Gamma_Q\, dt + u(\boldsymbol{P}_0,\tau) \iint_{B_\delta} q^*\, d\Gamma_Q\, dt\,. \quad (2.9)$$

When $\boldsymbol{P}$ tends to $\boldsymbol{P}_0$ and δ_ξ tends to 0, the third term of the right hand side of Eq. (2.9) tends to $\frac{1}{2}u(\boldsymbol{P}_0,\tau)$, namely:

$$\lim_{\delta_\xi \to 0} \lim_{\boldsymbol{P} \to \boldsymbol{P}_0} u(\boldsymbol{P}_0,\tau) \iint_{B_\delta} q^*(\boldsymbol{Q},t:\boldsymbol{P},\tau)\, d\Gamma_Q\, dt = \tfrac{1}{2}u(\boldsymbol{P}_0,\tau)\,.$$

At this time, the first and the second term of the right hand side of (2.9) tend to the finite values. Here, we give the following definition to the principal value of the integral.

Definition 2.1. The principal value of the integral (2.8):

$$\text{p.v.} \int_\Gamma \int_0^\tau u(\boldsymbol{Q},t)\, q^*(\boldsymbol{Q},t:\boldsymbol{P}_0,\tau)\, d\Gamma_Q\, dt$$

is defined by the following limit:

$$\lim_{\delta_\xi \to 0} \lim_{\boldsymbol{P} \to \boldsymbol{P}_0} \left\{ \iint_{S_\delta} u(\boldsymbol{Q},t)\, q^*(\boldsymbol{Q},t:\boldsymbol{P},\tau)\, d\Gamma_Q\, dt \right.$$
$$\left. + \iint_{B_\delta} (u(\boldsymbol{Q},t) - u(\boldsymbol{P}_0,\tau))\, q^*(\boldsymbol{Q},t:\boldsymbol{P},\tau)\, d\Gamma_Q\, dt \right\}$$

where $\boldsymbol{P}$ tends to $\boldsymbol{P}_0$ in the normal direction to the boundary.

From this definition, we get the following boundary integral equation:

$$\tfrac{1}{2}u(\boldsymbol{P},\tau) - \text{p.v.} \int_\Gamma \int_0^\tau u(\boldsymbol{Q},t)\, q^*(\boldsymbol{Q},t:\boldsymbol{P},\tau)\, d\Gamma_Q\, dt$$
$$= - \int_\Gamma \int_0^\tau q(\boldsymbol{Q},t)\, v^*(\boldsymbol{Q},t:\boldsymbol{P},\tau)\, d\Gamma_Q\, dt + \int_\Omega u_0(\boldsymbol{Q})\, v^*(\boldsymbol{Q},0:\boldsymbol{P},\tau)\, d\Omega_Q \quad (2.10)$$

for the point $\boldsymbol{P} \in \Gamma$.

1.2.3 Boundary Element Discretization

We now consider the geometrical discretization of the cylindrical domain. We take N_Γ nodal points $\{P_j\}_{j=1,N_\Gamma}$ along Γ. We set a segment Γ_i by the small line segment P_iP_{i+1} $(1 \leqq i \leqq N_\Gamma)$. Here, $\boldsymbol{P}_{N_\Gamma+1}$ is regarded as $\boldsymbol{P}_1$. We call Γ_i a boundary element. Hence the boundary Γ is approximated by the union of the boundary elements, i.e., Γ is approximated by the polygon $\tilde{\Gamma} = \bigcup_{i=1}^{N_\Gamma} \Gamma_i$. The time interval $[0, T]$ is divided into N_T subintervals. Let t_k be $t_k = k(T/N_T)$ $(0 \leqq k \leqq N_T)$. Moreover, let $S_j^k = \Gamma_j \times [t_k, t_{k+1}]$, $(1 \leqq j \leqq N_\Gamma, 0 \leqq k \leqq N_{T-1})$. We call S_j^k a patch. Hence S is divided into $N_\Gamma \times N_T$ small patches, and S is approximated by the union of these $N_\Gamma \times N_T$ patches. In convenience we set $\Delta\Gamma$ by $\Delta\Gamma = \max_{1 \leqq i \leqq N_\Gamma} |\Gamma_i|$, and ΔT by T/N_T.

Furthermore we divide Ω into N_Ω cells $\{\Omega_e\}_{e=1,N_\Omega}$, where Ω_e is a triangle. We divide Ω in such a way as the nodal points of the cells coincide with the nodal points of the boundary elements taken on Γ.

1.2.4 Interpolation of Functions

In this section, we shall give two manners for interpolating boundary functions. One is called "constant in time and linear in $\tilde{\Gamma}$ interpolation", and the other is called "bilinear Lagrange interpolation".

Let $g(\boldsymbol{P}, t)$ be a continuous function on S. We introduce function $g^I(\tilde{\boldsymbol{P}}, t)$, which is defined on $\tilde{\Gamma} \times [0, T]$ and approximates $g(\boldsymbol{P}, t)$ by the following way:

$$g^I(\boldsymbol{P}, t)|_{\tilde{S}_i^k} = g(\boldsymbol{P}_i, t_{k+1}) L_{2,i} + g(\boldsymbol{P}_{i+1}, t_{k+1}) L_{1,i}. \tag{2.11}$$

Here, $\tilde{\boldsymbol{P}}$ represents the point on $\tilde{\Gamma}$. Furthermore $L_{1,i}$ and $L_{2,i}$ are linear functions on $\tilde{\Gamma}$ such as:

$$L_{1,i}(\boldsymbol{P}_i) = 0, \quad L_{1,i}(\boldsymbol{P}_{i+1}) = 1,$$
$$L_{2,i}(\boldsymbol{P}_i) = 1, \quad L_{2,i}(\boldsymbol{P}_{i+1}) = 0.$$

Hence $g(\boldsymbol{P}, t)$ is approximated as a piecewise constant function in time and piecewiese linear function on $\tilde{\Gamma}$.

Next we introduce the bilinear Lagrange interpolation. Let N_j $(1 \leqq j \leqq 4)$ be Lagrange interpolation function on S_i^k. We set $\hat{g}(\tilde{\boldsymbol{P}}, t)$ on $\tilde{\Gamma}$ as follows:

$$\hat{g}(\tilde{\boldsymbol{P}}, t)|_{\tilde{S}_i^k} = g(\boldsymbol{P}_i, t_k) N_1 + g(\boldsymbol{P}_{i+1}, t_k) N_2 + g(\boldsymbol{P}_i, t_{k+1}) N_3 + g(\boldsymbol{P}_{i+1}, t_{k+1}) N_4. \tag{2.12}$$

We will discretize the equation using the constant in time and linear in $\tilde{\Gamma}$ interpolation. For only the error estimate of the same order can be proved for both of them mathematically, though the number of arithmatic operations involved is more for the bilinear Lagrange interpolation. But from the practical point of view, the latter interpolation yiels often more accurate results than the former. The details can be found in Onishi and Kuroki [11].

Furthermore we interpolate functions defined on $\bar{\Omega}$. The conventional finite element approximation is used. Let ϕ_i^e $(1 \leqq i \leqq 3)$ be finite element linear base functions on the triangle Ω_e, called roof functions. Let ψ be a smooth function on $\bar{\Omega}$. The interpolation function on Ω_e is to be defined as follows:

$$\psi^I|_{\bar{\Omega}_e} = \psi(Q_1)\phi_1^e + \psi(Q_2)\phi_2^e + \psi(Q_3)\phi_3^e$$

where Q_1, Q_2, Q_3 are verteces of the triangle Ω_e.

Under these preparation, we shall discretize the equation in the next subsection.

1.2.5 Discretization of the Equations

As the discrete problem for Eqs. (2.10) and (2.3)–(2.5), we consider the following system of linear equations (2.13)–(2.17):

$$\begin{aligned} \tfrac{1}{2} u_i^1 = &- \sum_{1 \leqq j \leqq N_\Gamma} \int_{S_j^0} q^I(\tilde{\boldsymbol{Q}}, t)\, v^*(\tilde{\boldsymbol{Q}}, t: \boldsymbol{P}_i, t_1)\, d\Gamma_{\tilde{Q}}\, dt \\ &+ \sum_{1 \leqq e \leqq N_\Omega} \int_{\Omega_e} u_0^I(\boldsymbol{Q})\, v^*(\boldsymbol{Q}, 0: \boldsymbol{P}_i, t_1)\, d\Omega_Q, \quad (1 \leqq i \leqq N_\Gamma) \end{aligned} \tag{2.13}$$

and

$$\frac{1}{2}u_i^k - \sum_{\substack{0 \leq l < k-2 \\ 1 \leq j \leq N_\Gamma}} \int_{S_j^l} u^l(\tilde{Q}, t) \{\nabla v^*(\tilde{Q}, t: \boldsymbol{P}_i, t_k)\} \cdot \boldsymbol{n}_{\Gamma_j} d\Gamma_{\tilde{Q}} dt$$

$$= - \sum_{\substack{0 \leq l \leq k-1 \\ 1 \leq j \leq N_\Gamma}} \int_{S_j^l} q^l(\tilde{Q}, t) v^*(\tilde{Q}, t: P_i, t_k) d\Gamma_{\tilde{Q}} dt$$

$$+ \sum_{1 \leq e \leq N_\Omega} \int_{\Omega_e} u_0^l(Q) v^*(Q, 0: \boldsymbol{P}_i, t_k) d\Omega_Q, \quad (1 \leq i \leq N_\Gamma, 2 \leq k \leq N_T). \quad (2.14)$$

Here $\boldsymbol{\nabla}$ is the two-dimensional nabla and $\boldsymbol{n}_{\Gamma_j}$ represents a unit outward normal vector of Γ_j. Furthermore, $\tilde{Q}$, etc., represent variable points on $\tilde{\Gamma}$. Additionally we consider the interpolations:

$$u^l = \bar{u}_1^l(\tilde{\boldsymbol{P}}), \qquad \tilde{\Gamma}_u \times [0, T] \quad (2.15)$$

$$q^l = \bar{q}^l(\tilde{\boldsymbol{P}}), \qquad \tilde{\Gamma}_q \times [0, T] \quad (2.16)$$

$$q^l = h\{u^l(\tilde{\boldsymbol{P}}, t) - \bar{u}_2^l(\tilde{\boldsymbol{P}})\}, \quad \tilde{\Gamma}_r \times [0, T]. \quad (2.17)$$

Here $\tilde{\Gamma}_u$, $\tilde{\Gamma}_q$, and $\tilde{\Gamma}_r$ are approximations of Γ_u, Γ_q, and Γ_r respectively. We are to approximate Eq. (2.10) by (2.13) and (2.14), and to approximate the boundary conditions (2.3)–(2.5) by (2.15)–(2.17).

1.2.6 Problem with Boundary Condition of the Linear Radiation Type

In Sect. 3, we will show the theorem on convergence and stability for the boundary element method, especially for the problem with boundary condition of the linear radiation type. The theorem is an extension of Onishi's results [7], and was proved rigorously by Iso [9]. For the preparation for the proof in the next section, we rewrite the boundary element scheme by the following way. At the beginning, it is necessary to declare precisely the boundary condition to be considered. In place of Eqs. (2.3)–(2.5), we impose the boundary condition in the general form:

$$q(\boldsymbol{P}, t) + m(\boldsymbol{P}, t)\, u(\boldsymbol{P}, t) = f(\boldsymbol{P}, t) \quad \text{on } \Gamma \times (0, T] \quad (2.18)$$

where $m(\boldsymbol{P}, t)$ and $f(\boldsymbol{P}, t)$ are smooth function on $\Gamma \times [0, T]$.

We can eliminate q in Eq. (2.10) by substituting Eq. (2.18). The equation becomes:

$$\frac{1}{2}u(\boldsymbol{P}, \tau) - \text{p.v.} \int_\Gamma \int_0^\tau u(Q, t)\, q^*(Q, t: \boldsymbol{P}, \tau)\, d\Gamma_Q dt$$

$$- \int_\Gamma \int_0^\tau u(Q, t)\, m(Q, t)\, v^*(Q, t: \boldsymbol{P}, \tau)\, d\Gamma_Q dt$$

$$= \int_\Gamma \int_0^\tau f(Q, t)\, v^*(Q, t: \boldsymbol{P}, \tau)\, d\Gamma_Q dt + \int_\Omega u^0(Q)\, v^*(Q, 0: \boldsymbol{P}, \tau)\, d\Omega_Q. \quad (2.19)$$

We rewrite the discrete problem of Eqs. (2.13)−(2.17) by the following equations:

$$\frac{1}{2}u_i^1 - \sum_{1\leq j\leq N_\Gamma} \int_{S_j^0} (u_j^1 L_{2,j} + u_{j+1}^1 L_{1,j})\, m^I(\tilde{Q}, t)\, v^*(\tilde{Q}, t: \boldsymbol{P}_i, t_1)\, d\Gamma_{\tilde{Q}}\, dt$$
$$= \sum_{1\leq j\leq N_\Gamma} \int_{S_j^0} f^I(Q, t)\, v^*(Q, t: \boldsymbol{P}_i, t_1)\, d\Gamma_{\tilde{Q}}\, dt$$
$$+ \sum_{1\leq e\leq N_\Omega} \int_{\Omega_e} u_0^I(Q)\, v^*(Q, 0: \boldsymbol{P}_i, t_1)\, d\Omega_Q, \quad (1 \leq i \leq N_\Gamma) \tag{2.20}$$

and

$$\frac{1}{2}u_i^k - \sum_{\substack{0\leq l\leq k-2\\ 1\leq j\leq N_\Gamma}} \int_{S_j^l} (u_j^{l+1} L_{2,j} + u_{j+1}^{l+1} L_{1,j})\, \{\nabla v^*(\tilde{Q}, t: \boldsymbol{P}_i, t_k)\} \cdot \boldsymbol{n}_{\Gamma_j}\, d\Gamma_{\tilde{Q}}\, dt$$
$$- \sum_{\substack{0\leq l\leq k-1\\ 1\leq j\leq N_\Gamma}} \int_{S_j^l} (u_j^{l+1} L_{2,j} + u_{j+1}^{l+1} L_{1,j})\, m^I(\tilde{Q}, t)\, v^*(\tilde{Q}, t: \boldsymbol{P}_i, t_k)\, d\Gamma_Q\, dt$$
$$= \sum_{\substack{0\leq l\leq k-1\\ 1\leq j\leq N_\Gamma}} \int_{S_j^l} f^I(\tilde{Q}, t)\, v^*(\tilde{Q}, t: \boldsymbol{P}_i, t_k)\, d\Gamma_{\tilde{Q}}\, dt$$
$$+ \sum_{1\leq e\leq N_\Omega} \int_{\Omega_e} u_0^I(Q)\, v^*(Q, 0: \boldsymbol{P}_i, t_k)\, d\Omega_Q, \quad (1 \leq i \leq N_\Gamma, 2 \leq k \leq N_T). \tag{2.21}$$

In convenience, we write these in the matrix form:

$$\left\{\begin{pmatrix} \frac{1}{2} & & 0 \\ & \ddots & \\ 0 & & \frac{1}{2} \end{pmatrix} + (a_{ij}^k)\right\} \begin{pmatrix} u_1^k \\ \vdots \\ u_{N_\Gamma}^k \end{pmatrix} + \sum_{l=1}^{k-1} \{(h_{ij}^l) + (a_{ij}^l)\} \begin{pmatrix} u_1^l \\ \vdots \\ u_{N_\Gamma}^l \end{pmatrix} = \begin{pmatrix} G_1^k \\ \vdots \\ G_{N_\Gamma}^k \end{pmatrix}, \quad (1 \leq k \leq N_T). \tag{2.22}$$

Here, the components of the matrices are defined as follows:

$$a_{ij}^l = - \int_{S_{j-1}^{l-1}} L_{1,j-1}\, m^I(\tilde{Q}, t)\, v^*(\tilde{Q}, t: \boldsymbol{P}_i, t_k)\, d\Gamma_{\tilde{Q}}\, dt$$
$$- \int_{S_j^{l-1}} L_{2,j}\, m^I(\tilde{Q}, t)\, v^*(\tilde{Q}, t: \boldsymbol{P}_i, t_k)\, d\Gamma_{\tilde{Q}}\, dt\,, \quad (1 \leq l \leq k, 1 \leq i, j \leq N_\Gamma), \tag{2.23}$$

$$h_{ij}^l = - \int_{S_{j-1}^{l-1}} L_{1,j-1} \{\nabla v^*(\tilde{Q}, t: \boldsymbol{P}_i, t_k)\} \cdot \boldsymbol{n}_{\Gamma_{j-1}}\, d\Gamma_{\tilde{Q}}\, dt$$
$$- \int_{S_j^{l-1}} L_{2,j} \{\nabla v^*(\tilde{Q}, t: \boldsymbol{P}_i, t_k)\} \cdot \boldsymbol{n}_{\Gamma_j}\, d\Gamma_{\tilde{Q}}\, dt\,, \quad (1 \leq l \leq k-1, 1 \leq i, j \leq N_\Gamma), \tag{2.24}$$

and

$$G_i^k = \sum_{\substack{0\leq l\leq k-1\\ 1\leq j\leq N_\Gamma}} \int_{S_j^l} f^I(\tilde{Q}, t)\, v^*(\tilde{Q}, t: \boldsymbol{P}_i, t_k)\, d\Gamma_{\tilde{Q}}\, dt$$
$$+ \sum_{1\leq e\leq N_\Omega} \int_{\Omega_e} u_0^I(Q)\, v^*(Q, 0: \boldsymbol{P}_i, t_k)\, d\Omega_Q, \quad (1 \leq i \leq N_\Gamma)\,. \tag{2.25}$$

We solve Eq. (2.22) recursively and get the boundary solution vector $\{u_i^k\}$.

1.3 Convergence and Stability

1.3.1 Preliminary Estimates

We consider Eq. (2.1) with the initial condition (2.2) and the boundary condition (2.18). As the discretization for these equations, we adopt Eqs. (2.20) and (2.21). We will show solvability of the discrete problem, convergence, and stability. To show them mathematically, we have to prepare some estimates.

Lemma 3.1. The following estimate holds for $1 \leqq i, j \leqq N_\Gamma$, $1 \leqq k \leqq N_T$:

$$|a_{ij}^k| \leqq C\sqrt{\Delta T}. \tag{3.1}$$

Here C is a positive constant and is independent of $\Delta\Gamma$ and ΔT.

To prove this Lemma, we notice that, according to Eq. (2.23)

$$\begin{aligned} |a_{ij}^k| &\leqq \Bigg| \int_{S_{j-1}^{k-1}} L_{1,j-1}\, m^I(\tilde{\boldsymbol{Q}}, t)\, v^*(\tilde{\boldsymbol{Q}}, t: \boldsymbol{P}_i, t_k)\, d\Gamma_{\tilde{Q}}\, dt \\ &\quad + \int_{S_j^{k-1}} L_{2,j}\, m^I(\tilde{\boldsymbol{Q}}, t)\, v^*(\tilde{\boldsymbol{Q}}, t: \boldsymbol{P}_i, t_k)\, d\Gamma_{\tilde{Q}}\, dt \Bigg| \\ &\leqq M \Bigg| \int_{S_{j-1}^{k-1}} v^*(\tilde{\boldsymbol{Q}}, t: \boldsymbol{P}_i, t_k)\, d\Gamma_{\tilde{Q}}\, dt + \int_{S_j^{k-1}} v^*(\tilde{\boldsymbol{Q}}, t: \boldsymbol{P}_i, t_k)\, d\Gamma_{\tilde{Q}}\, dt \Bigg| \end{aligned}$$

Here we put

$$M = \sup_{\substack{0 \leqq t \leqq T \\ P \in \Gamma}} |m(\boldsymbol{P}, t)|.$$

We denote $\tilde{\boldsymbol{Q}} = (\tilde{\xi}_1, \tilde{\xi}_2) \in \Gamma_j$ having two end nodes $\boldsymbol{P}_j = (x_1^j, x_2^j)$, $\boldsymbol{P}_{j+1} = (x_1^{j+1}, x_2^{j+1})$ in terms of the parameter s $(0 \leqq s \leqq 1)$ as:

$$\tilde{\xi}_1 = x_1^j + (x_1^{j+1} - x_1^j)\, s$$

and

$$\tilde{\xi}_2 = x_2^j + (x_2^{j+1} - x_2^j)\, s.$$

It is sufficient for us to consider the case that $\boldsymbol{P}_i$ coincides with the origin (0, 0). For this case, we have:

$$v^*(\tilde{\boldsymbol{Q}}, t: \boldsymbol{P}_i, t_k) = \frac{1}{4\pi(t_k - t)} \exp\left[-\frac{1}{4(t_k - t)} \sum_{q=1}^{2} \{x_q^j + (x_q^{j+1} - x_q^j)\, s\}^2\right].$$

Hence we know:

$$\begin{aligned} &\int_{S_j^{k-1}} v^*(\tilde{\boldsymbol{Q}}, t: \boldsymbol{P}_i, t_k)\, d\Gamma_{\tilde{Q}}\, dt \\ &= \int_{t_k - \Delta T}^{t_k} \int_0^1 \frac{1}{4\pi(t_k - t)} \exp\left[-\frac{1}{4(t_k - t)} \sum_{q=1}^{2} \{x_q^j + (x_q^{j+1} - x_q^j)\, s\}^2\right] |\tilde{\Gamma}_j|\, ds\, dt \\ &\leqq C \int_{t_k - \Delta T}^{t_k} \int_\alpha^\beta \frac{1}{4\pi(t_k - t)} \exp\left[-\frac{(s')^2}{4(t_k - t)}\right] ds'\, dt \end{aligned}$$

$$= C \int_{t_k - \Delta T}^{t_k} \int_{\frac{\alpha}{2\sqrt{t_k - t}}}^{\frac{\beta}{2\sqrt{t_k - t}}} \frac{1}{2\pi\sqrt{t_k - t}} e^{-\eta^2} d\eta\, dt$$

$$\leqq \frac{C}{2\sqrt{\pi}} \sqrt{\Delta T}$$

in which C, α, β are constants and C is dependent on neither $\Delta\Gamma$ nor ΔT. This completes the proof of the Lemma.

Solvability of the discrete problem follows from Lemma 3.1 immediately.

Lemma 3.2. If $\Delta\Gamma$ and ΔT satisfy $\Delta T/(\Delta\Gamma)^2 \leqq \varepsilon$ for some positive constant ε, which is dependent on neither $\Delta\Gamma$ nor ΔT, then the system of linear equations (2.22) is uniquely solvable for $1 \leqq k \leqq N_T$.

To prove this Lemma, it is sufficient to show diagonal dominance for the coefficient matrix:

$$\begin{pmatrix} \frac{1}{2} & & 0 \\ & \ddots & \\ 0 & & \frac{1}{2} \end{pmatrix} + (a_{ij}^k)$$

of the first term of the left hand side of the system of equations (2.22). If we take ε sufficiently small, so that the following inequalities hold:

$$\sum_{j \neq i} |a_{ij}^k| \leqq N_\Gamma \cdot C\sqrt{\Delta T} \leqq N_\Gamma \cdot C \cdot \Delta\Gamma\,(\sqrt{\Delta T}/\Delta\Gamma) \leqq C\,|\Gamma|\,\sqrt{\varepsilon} < 1/2$$

then the coefficient matrix becomes diagonally dominant. Here $|\Gamma|$ denotes the measure of the boundary Γ.

Before considering convergence and stability, we must pay attentions to the following fact. The discrete problems of Eqs. (2.20) and (2.21) are not the approximations to Eq. (2.19) but to the following equation:

$$\begin{aligned} &\tfrac{1}{2} u(\boldsymbol{P}, \tau) - \int_\Gamma \int_0^{\tau - \Delta T} u(\boldsymbol{Q}, t)\, q^*(\boldsymbol{Q}, t : \boldsymbol{P}, \tau)\, d\Gamma_Q\, dt \\ &\quad - \int_\Gamma \int_0^\tau u(\boldsymbol{Q}, t)\, m(\boldsymbol{Q}, t)\, v^*(\boldsymbol{Q}, t : \boldsymbol{P}, \tau)\, d\Gamma_Q\, dt \\ &\quad = \int_\Gamma \int_0^\tau f(\boldsymbol{Q}, t)\, v^*(\boldsymbol{Q}, t : \boldsymbol{P}, \tau)\, d\Gamma_Q\, dt + \int_\Omega u_0(\boldsymbol{Q})\, v^*(\boldsymbol{Q}, 0 : \boldsymbol{P}, \tau)\, d\Omega_Q . \end{aligned} \tag{3.2}$$

This discretization may seem to be rough. In the first term of the left hand side of Eqs. (2.20) and (2.21), the coefficient 1/2 itself is an approximate value.

According to Def. 2.1 and Eq. (2.9), we can get the following error estimate.

Lemma 3.3. Let $u(\boldsymbol{P}, t)$ be a smooth function. If $\Delta\Gamma$ and ΔT satisfy $\Delta T/(\Delta\Gamma)^2 \leqq \varepsilon$ for some positive constant ε and if we fix δ_ξ and δ_t by $\delta_\xi = \Delta\Gamma$ and $\delta_t = \Delta T$, then the following estimate holds:

$$\begin{aligned} &\Big| \tfrac{1}{2} u(\boldsymbol{P}_0, \tau) - \lim_{P \to P_0} u(\boldsymbol{P}_0, \tau) \iint_{B_\delta} q^*(\boldsymbol{Q}, t : \boldsymbol{P}, \tau)\, d\Gamma_Q\, dt \\ &\qquad + \lim_{P \to P_0} \iint_{B_\delta} \{u(\boldsymbol{Q}, t) - u(\boldsymbol{P}_0, \tau)\}\, q^*(\boldsymbol{Q}, t : \boldsymbol{P}, \tau)\, d\Gamma_Q\, dt \Big| \leqq C \cdot \Delta\Gamma . \end{aligned} \tag{3.3}$$

Here C is a positive constant and it is dependent on neither $\Delta\Gamma$ nor ΔT.

We can show, on the other hand, the following estimate for the principal value of the singular integral.

Lemma 3.4. For a smooth function $u(P, t)$, the following estimate holds:

$$\left| \text{p.v.} \int_{\tau-\Delta T}^{\tau} \int_{\Gamma} u(Q, t)\, q^*(Q, t: P, \tau)\, d\Gamma_Q\, dt \right| \leqq C\sqrt{\Delta T}. \tag{3.4}$$

Here C is a positive constant and is dependent on neither $\Delta\Gamma$ nor ΔT.

Proof:

$$\left| \text{p.v.} \int_{\tau-\Delta T}^{\tau} \int_{\Gamma} u(Q, t)\, q^*(Q, t: P, \tau)\, d\Gamma_Q\, dt \right|$$

$$\leqq \int_{\tau-\Delta T}^{\tau} \int_{\Gamma \cap \{Q \mid |Q-P| \geqq \delta\}} |u(Q, t)\, q^*(Q, t: P, \tau)|\, d\Gamma_Q\, dt$$

$$+ \text{p.v.} \int_{\tau-\Delta T}^{\tau} \int_{\Gamma \cap \{Q \mid |Q-P| \leqq \delta\}} |u(Q, t)\, q^*(Q, t: P, \tau)|\, d\Gamma_Q\, dt.$$

Here, δ is a positive constant which is sufficiently small. We represent the set $\Gamma \cap \{Q \mid |Q-P| \leqq \delta\}$ by introducing the local coordinate in such a way that the point Q belonging to the set is presented as follows:

$$Q = (x_1, x_2) + (\eta_1(s), \eta_2(s))$$

$$n(s) = (-\eta_2'(s), \eta_1'(s)), \quad \eta(0) = 0$$

$$\{\eta_1'(s)\}^2 + \{\eta_2'(s)\}^2 = 1 \quad \text{for} \quad \alpha \leqq s \leqq \beta, \alpha < 0, \beta > 0.$$

Using this coordinate, we can obtain:

$$\left| \text{p.v.} \int_{\tau-\Delta T}^{\tau} \int_{\Gamma} u(Q, t)\, q^*(Q, t: P, \tau)\, d\Gamma_Q\, dt \right|$$

$$\leqq C_1 \int_0^{\Delta T} \int_{\Gamma \cap \{Q \mid |Q-P| \geqq \delta\}} \frac{|Q-P|}{8\pi t^2} \exp\left[-\frac{|Q-P|^2}{4t}\right] d\Gamma_Q\, dt$$

$$+ C_2 \int_0^{\Delta T} \int_\alpha^\beta \frac{|\eta(s) \cdot \eta(s)|}{8\pi t^2} \exp\left[-\frac{|\eta(s)|^2}{4t}\right] ds\, dt$$

$$\leqq C_1 \int_0^{\Delta T} \int_{\Gamma \cap \{Q \mid |Q-P| \geqq \delta\}} \frac{4^{3/2}}{\pi} \frac{1}{\sqrt{t}} \frac{1}{|Q-P|^2} \frac{|Q-P|^3}{(4t)^{3/2}} \exp\left[-\frac{|Q-P|^2}{4t}\right] d\Gamma_Q\, dt$$

$$+ C_2' \int_0^{\Delta T} \int_\alpha^\beta \frac{|Q(s)|^2}{(4t)^2} \exp\left[-\frac{|Q(s)|^2}{4t}\right] ds\, dt \leqq C_1' \sqrt{\Delta T} + C_2'' \sqrt{\Delta T}.$$

This completes the proof.

Considering these two estimates (3.3) and (3.4), we can adopt well Eqs. (2.20) and (2.21).

1.3.2 Theorem on Convergence and Stability

From the solvability as discussed in the last section, we can obtain the sequence $\{u_j^k\}_{j=1,N_\Gamma}^{k=1,N_T}$ by solving the system of Eq. (2.22). We can then construct a numerical solution function $u^N(\boldsymbol{P}, t)$ by interpolating $\{u_j^k\}$ by the same way as in Sect. 2.4. On the other hand, we denote $u^I(\boldsymbol{P}, t)$ by the interpolating function of the exact solution function on Γ. Then, we can prove the following theorem, which guarantees convergence of the numerical solution.

Theorem 3.1. If $\Delta\Gamma$ and ΔT satisfy $\Delta T/(\Delta\Gamma)^2 \leqq \varepsilon$ for some positive constant ε, then the following estimate holds:

$$\| u^N - u^I \|_\infty \leqq C \cdot \Delta\Gamma .$$

Here C is a positive constant and is dependent on neither $\Delta\Gamma$ nor ΔT.

To prove this theorem, we put e_i^k by

$$e_i^k = u(\boldsymbol{P}_i, t_k) - u_i^k$$

and we denote by $e^I(\tilde{\boldsymbol{P}}, t)$ the interpolating function of $\{e_i^k\}$. We substract Eq. (2.21) from (2.19) to obtain:

$$\tfrac{1}{2} e^I(\boldsymbol{P}_i, t_k) - \int_0^{t_k-\Delta T} \int_{\tilde{\Gamma}} e^I(\tilde{\boldsymbol{Q}}, t)\, q^*(\tilde{\boldsymbol{Q}}, t : \boldsymbol{P}_i, t_k)\, d\Gamma_{\tilde{Q}}\, dt$$
$$- \int_0^{t_k} \int_{\tilde{\Gamma}} e^I(\tilde{\boldsymbol{Q}}, t)\, m^I(\tilde{\boldsymbol{Q}}, t)\, v^*(\tilde{\boldsymbol{Q}}, t : \boldsymbol{P}_i, t_k)\, d\Gamma_{\tilde{Q}}\, dt = g(\boldsymbol{P}_i, t_k) .$$

Here $g(\boldsymbol{P}_i, t_k)$ means the error term. According to the previous Lemmas, $g(\boldsymbol{P}_i, t_k)$ satisfies the inequality: $|g(\boldsymbol{P}_i, t_k)| \leqq C\,\Delta\Gamma$, where C is a positive constant and is dependent on neither $\Delta\Gamma$ nor ΔT. Furthermore, we put:

$$e(t) = \sup_{\tilde{\boldsymbol{P}} \in \tilde{\Gamma}} |e^I(\tilde{\boldsymbol{P}}, t)|$$

We consider the integral inequality which the error function $e(t)$ satisfies:

$$\tfrac{1}{2} e(t) \leqq C \int_0^{t_k} \frac{e(t)}{\sqrt{t_k - t}}\, dt + g(t_k) . \qquad (3.5)$$

Here $g(t_k)$ is an error which is committed in every step in the solution procedure of Eqs. (2.20) and (2.21). C is some positive constant. Next we construct a function which majorize the error function. To construct the major function, we solve the following integral equation by the successive approximation technique.

$$\tfrac{1}{2} e(\tau) = C \int_0^{\tau} e(t) \frac{dt}{\sqrt{\tau - t}} + g \quad (0 \leqq \tau \leqq T) .$$

To this end, we set $e_1(t) = g$ and calculate the sequence $e_n(t)$ by the recurrence formula:

$$e_{n+1}(\tau) = C \int_0^{\tau} e_n(t) \frac{dt}{\sqrt{\tau - t}} + g \quad (n = 1, 2, 3, \ldots) .$$

Using the method of the majorant, we can get the following inequality:

$$e(t) \leqq \sum_{n=1}^{\infty} \frac{g}{n!!} \left(\frac{\tilde{C}T}{2} \right)^n \tag{3.6}$$

where $\tilde{C}$ is some positive constant and g is the maximum of $g(t_k)$, which satisfies:

$$0 < g \leqq C' \cdot \Delta\Gamma$$

in which C' is some positive constant. By this, we can prove the Theorem.

We notice that the estimate (3.6) guarantees the stability of the scheme.

Theorem 3.2. If $\Delta\Gamma$ and ΔT satisfy $\Delta T/(\Delta\Gamma)^2 \leqq \varepsilon$ for some positive constant ε, then the computational scheme (2.22) is stable.

1.3.3 Numerical Demonstrations

Consider the first example: to find the temperature $u = u(\boldsymbol{P}, t)$ satisfying the heat conduction Eq. (2.1) in the unit circle $\Omega = \{\boldsymbol{P}(x, y) | 0 \leqq r^2 = x^2 + y^2 < 1\}$, together with zero initial temperature and the Neumann boundary condition:

$$u(\boldsymbol{P}, 0) = 0 \quad \text{on } \bar{\Omega} \tag{3.7}$$

$$q(\boldsymbol{P}, t) = \bar{q} \quad \text{on } \Gamma_u \times (0, T] \tag{3.8}$$

in which $\bar{q}$ is the given constant at the boundary surface.

The exact solution from Carslaw and Jaeger [12] is given by

$$u = 2\bar{q}\,t + \bar{q} \left\{ \frac{r^2}{2} - \frac{1}{4} - 2 \sum_{n=1}^{\infty} \exp[-\alpha_n^2 t] \frac{J_0(r\alpha_n)}{\alpha_n^2 J_0(\alpha_n)} \right\} \tag{3.9}$$

in which α_n $(n = 1, 2, 3, \ldots)$ are the positive roots of the Bessel function: $J_1(\alpha) = 0$.

We observe the boundary element solution for the problem. First we divide the boundary Γ into regular crude line segments so that there are 16 boundary nodes. The characteristic length of boundary elements is $\Delta\Gamma = 0.39$. Then we see how the calculated solutions converge to the exact solution on the boundary as the element subdivision is refined. For this purpose, we subdivide the mesh much more such as: $\Delta\Gamma = 0.39$, 0.196 and 0.0981 as shown in Fig. 3.1. For the finest mesh, the number of boundary nodes becomes 64. At a point with the coordinates (1, 0), relative errors are calculated for various subdivisions with the fixed time in-

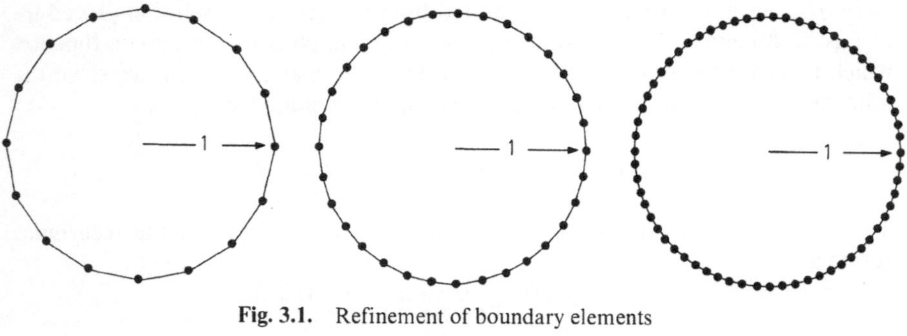

Fig. 3.1. Refinement of boundary elements

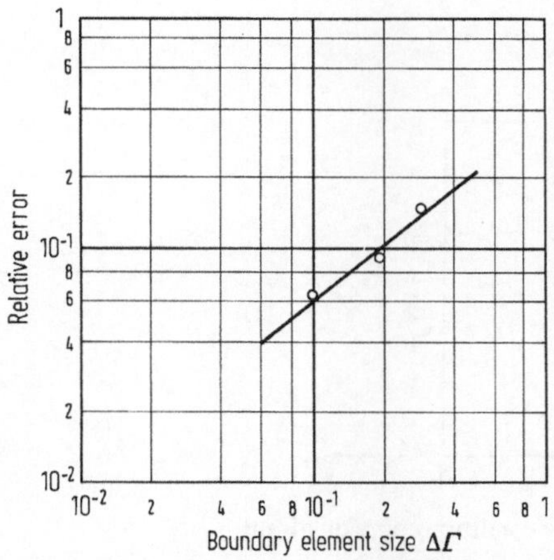

Fig. 3.2. Relative errors vs. element sizes ($x = 1$, $y = 0$, $T = 0.5$, $\Delta T = 0.05$)

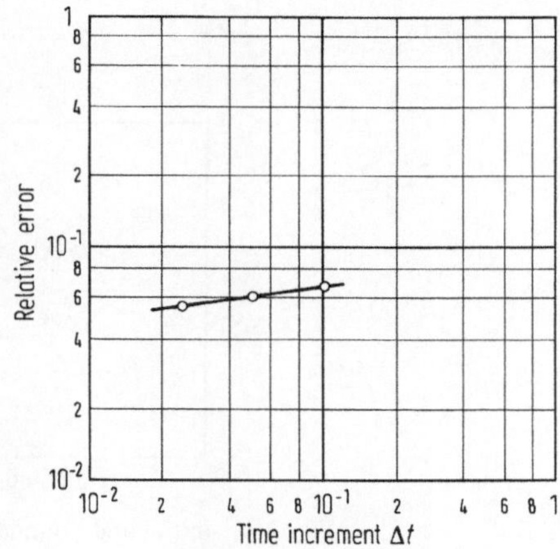

Fig. 3.3. Relative errors vs. time increments ($x = 1$, $y = 0$, $\Delta\Gamma = 0.0981$, $T = 0.5$)

crement $\Delta T = 0.05$. The slope of the regression line in Fig. 3.2 is about 1:1. This implies the first order of convergence.

So far we have kept the time increment fixed, though the boundary mesh was refined. Next, we vary the time increment, while the boundary mesh is fixed, in order to see how the calculated solution converges to the exact solution on the boundary.

Choose the finest mesh in Fig. 3.1 as the spatial subdivision. The characteristic length is $\Delta\Gamma = 0.0981$. We take $\Delta T = 0.1$ first, then we will halve it twice as $\Delta T = 0.05$, $= 0.025$. Relative errors at the point (1, 0) at the time $t = 0.5$ are plotted in Fig. 3.3 for various time increments. The low order convergence is suggested in the time increment.

Consider the second example: to find the temperature $u = u(\boldsymbol{P}, t)$ satisfying the heat conduction Eq. (2.1) in the square domain $\Omega = \{\boldsymbol{P}(x, y) \,|\, |x| < 1, |y| < 1\}$, together with the initial and boundary conditions:

$$u(\boldsymbol{P}, 0) = 1 \quad \text{on } \bar{\Omega} \tag{3.10}$$

$$u(\boldsymbol{P}, t) = 0 \quad \text{on } \Gamma_u \times (0, T]. \tag{3.11}$$

This is the very simple problem with the Dirichlet boundary condition. Recall that the mathematical proof for the convergence was not obtained for this problem in the above discussion.

Owing to the symmetry of the problem, we are sufficient to solve Eq. (2.1) on a quadrant $\Omega_{1/4} = \{(x, y) \,|\, 0 \leqq x \leqq 1, 0 \leqq y \leqq 1\}$ with appropriate initial and boundary conditions shown in Fig. 3.4. The exact solution from Carslaw and Jaeger [12] is given by

$$u = \frac{4}{\pi} \sum_{n=0}^{\infty} \frac{(-1)^n}{2n+1} \cos\left(\frac{2n+1}{2} \pi x\right) \exp\left[-(2n+1)^2 \frac{\pi^2 t}{4}\right] \times \frac{4}{\pi} \sum_{n=0}^{\infty} \frac{(-1)^n}{2n+1} \cos\left(\frac{2n+1}{2} \pi y\right) \exp\left[-(2n+1)^2 \frac{\pi^2 t}{4}\right] \tag{3.12}$$

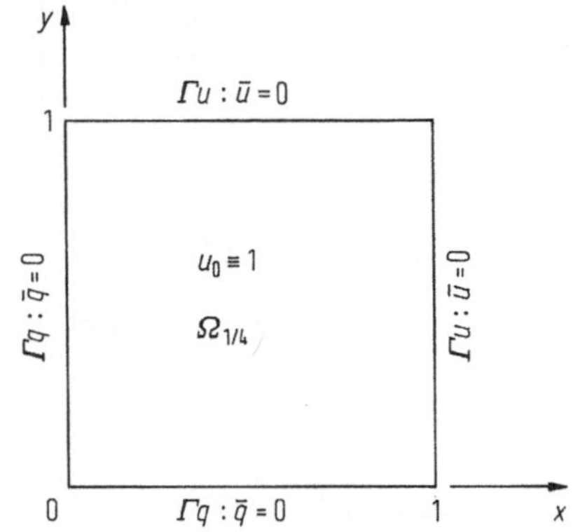

Fig. 3.4. Initial and boundary conditions on a quadrant

This infinite series is absolutely convergent for $t > 0$. When t is small, the rate of convergence is slow. The rapidly converging solution for small values of the time t is given also by Carslaw and Jaeger [12] by

$$u = \left[1 - \sum_{n=0}^{\infty} (-1)^n \left\{\operatorname{erfc}\frac{(2n+1)-x}{2\sqrt{t}} + \operatorname{erfc}\frac{(2n+1)+x}{2\sqrt{t}}\right\}\right]$$

$$\times \left[1 - \sum_{n=0}^{\infty} (-1)^n \left\{\operatorname{erfc}\frac{(2n+1)-y}{2\sqrt{t}} + \operatorname{erfc}\frac{(2n+1)+y}{2\sqrt{t}}\right\}\right] \qquad (3.13)$$

where $\operatorname{erfc}(x)$ is the complementary error function.

Now we consider the boundary element solution for the problem. First we divide the quadrant $\Omega_{1/4}$ into a regular crude net of triangles so that there are 12 boundary nodes including double nodes at corners. The characteristic length of boundary elements is $\Delta\Gamma = \frac{1}{2}$. Then we see how the calculated solutions converge to the exact solution for the temperature u and for the flux q as the element subdivision is refined. We therefore subdivide the original mesh much more such as: $\Delta\Gamma = \frac{1}{2}, \frac{1}{4}$, and $\frac{1}{8}$. For the finest mesh, the number of boundary nodes becomes 85, and the number of triangular cells is 128.

Calculated boundary temperatures along the x-axis at the constant time $t = 1.0$ are plotted for the series of subdivisions of the domain in Fig. 3.5. The time increment is fixed as $\Delta T = \frac{1}{32}$ for all computations. We can observe numerically the uniform convergence of calculated boundary solutions to the exact temperatures. At the origin of the coordinate system, relative errors are calculated for various subdivisions with the fixed time increment. The slope of the regression line in Fig. 3.6 is 1:2 for right two points. It suggests the second order of convergence. The left-most point is not colinear because we have fixed the time increment as the spatial mesh is refined.

In order to access the convergence property of the approximate fluxes, calculated heat fluxes along the boundary $x = 1.0$, $0 < y < 1.0$ are plotted for the series of subdivision of different finess in the space. The time is $t = 1.0$ using the fixed time

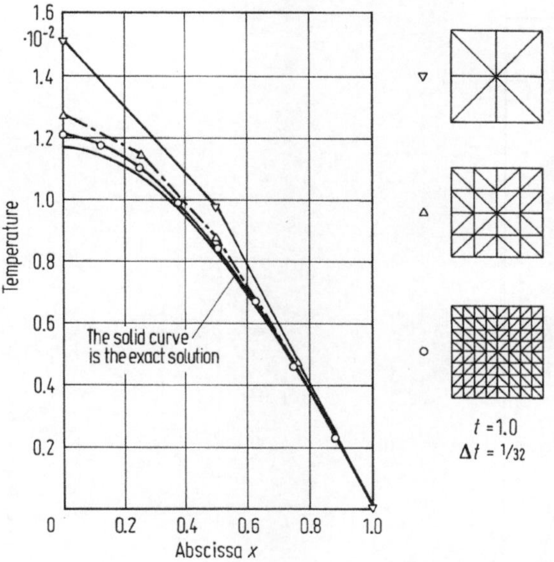

Fig. 3.5. Boundary temperatures along the x-axis for various element sizes

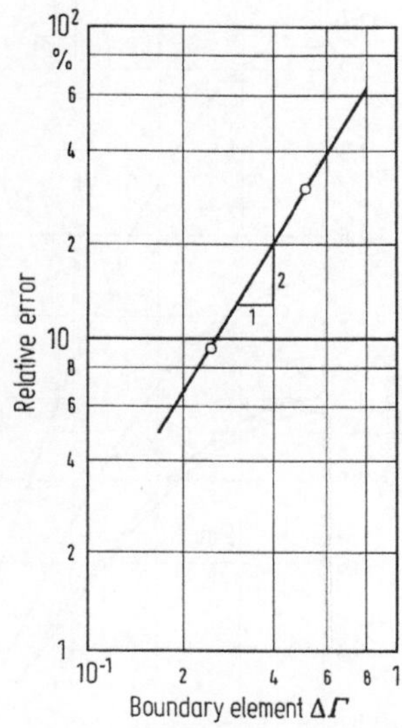

Fig. 3.6. Rate of convergence of the temperature. ($x = y = 0$)

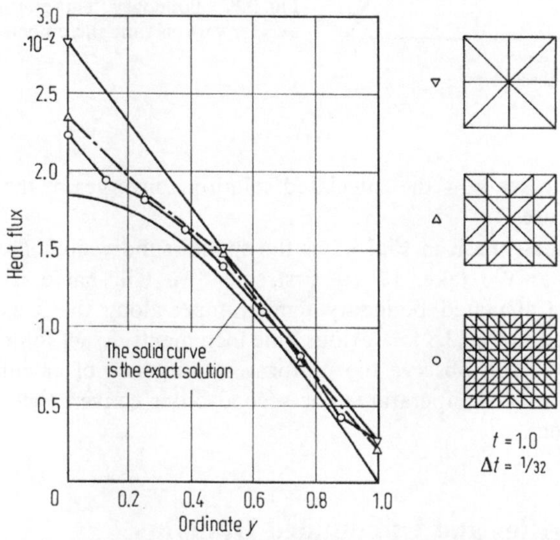

Fig. 3.7. Heat fluxes along the boundary: $x = 1, 0 \leqq y \leqq 1$

increment $\varDelta T = \frac{1}{32}$. Figure 3.7 shows that calculated boundary fluxes converge to the exact values shown by solid curve, but the uniform convergence is not indicated.

So far we have kept the time increment fixed, though the spatial mesh was refined. Next, we are going to vary the time increment, while the spatial mesh is

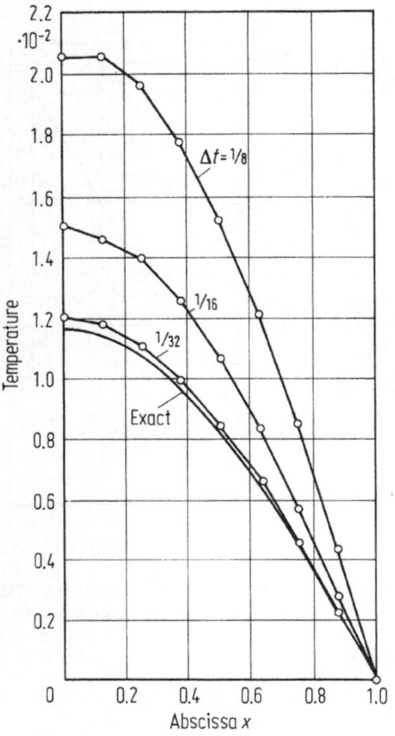

Fig. 3.8. Boundary temperatures along the x-axis for various time increments

fixed, in order to see how the calculated solution converges to the exact solution for the temperature.

Choose the finest mesh in Fig. 3.5 as the spatial subdivision. The characteristic length is $\Delta\Gamma = \frac{1}{8}$. We take $\Delta T = \frac{1}{8}$ first, then we will halve it twice such as $\Delta T = \frac{1}{16}$, $= \frac{1}{32}$. Calculated boundary temperatures along the x-axis at the time $t = 1.0$ are plotted in Fig. 3.8 for various time increments. Exact solution is drawn by a solid curve. We can observe the uniform convergence of calculated boundary solutions to the exact temperature. The second order convergence is suggested in the time increment.

1.4 Singularities and Unbounded Domains

If u_0 (Cauchy data) in Ω are either discontinuous or inconsistent with $\bar{u}_1$ (Dirichlet data) on Γ, the small-time solution has very large fluxes in some confined region called the region of infection. If the boundary changes its tangential direction by an angle $\theta > \pi$ at a corner, the derivatives of the temperature becomes unbounded at the corner. Such a corner, is called the re-entrant corner. If two media with different thermal conductivity intersect at a point, the derivatives may become unbounded at that point. Singularity at such a point is called interzonal singularity. The zoning technique is useful for these specific problems: If a domain is not convex, we subdivide it into convex subdomains (zones), if possible. We construct the boundary elements equations in each zone, and assemble the equations by taking both continuity of u (compatibility) and continuity of q (equilibrium) on the interface of zones.

1.4.1 Re-entrant Corners

Consider an example with the L-shaped domain whose initial and boundary conditions are shown in Fig. 4.1. The re-entrant corner $\boldsymbol{P}$ is thermally insulated. Bell and Crank [13] gave the analytical form of the exact solution in the neighbourhood of $\boldsymbol{P}$ as follows:

$$u = \sum_{j=0}^{\infty} \sum_{k=0}^{\infty} A_{kj} \exp\left[-\alpha_j^2 t\right] J_{k\lambda}(\alpha_j r) \cos(k\lambda\theta) + \sum_{k=0}^{\infty} c_k r^{k\lambda} \cos(k\lambda\theta) \quad (4.1)$$

in which A_{kj} and c_k are constant coefficients, α_j the zeros of some functions of Bessel functions $J_{k\lambda}$, $\lambda = \pi/\theta_0$ in which θ_0 is the angle of the corner, and (r, θ) is the local polar coordinates centred at the corner point $\boldsymbol{P}$. We can see that the temperature is bounded at the point $\boldsymbol{P}$. However, all the derivatives of the solution will be unbounded at that point.

We calculated unsteady temperature variations by the boundary element method using the subdivision shown in Fig. 4.1. Since the domain is not convex, we divide it into convex zones. The convex subdivision is not unique. One possibility is shown in the figure. The time increment was chosen as $\Delta T = \frac{1}{32}$ according to the recommendation in Onishi and Kuroki [11] for the size of space-time subdivision to retain the accuracy.

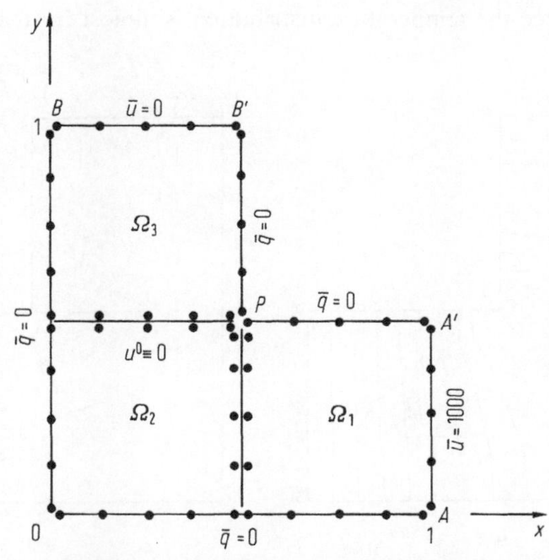

Fig. 4.1. *L*-shaped domain divided into three zones and boundary elements

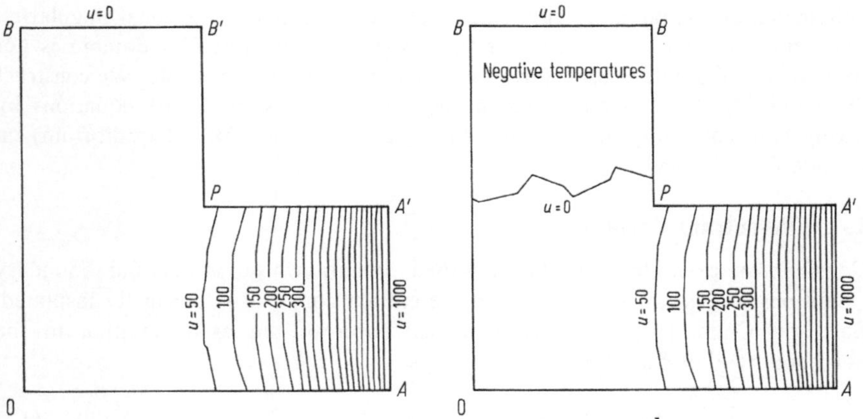

a b

Fig. 4.2 a, b. Isotherms at the time $T=\Delta T=1/32$. The zoning technique is used for case **a** and is not used for **b**

The Cauchy data are inconsistent with the Dirichlet data on the side AA'. Steep thermal gradients are expected near the side at small values of the time. Figure 4.2a shows calculated isotherms at the first time step. Calculated temperatures did not oscillate. We can observe the smooth temperature variation even in the region where the steep gradient is found. In addition, calculated temperatures are positive all over the domain. If the zoning technique is not used, a region of negative temperatures appears as shown in Fig. 4.2b.

Figure 4.3a shows calculated isotherms at the time $t=1.0$. Since the problem is antisymmetric with respect to the line OP, the exact isotherm $u=500$ in steady state must coincide with this line. The calculated isotherm $u=500$ in Fig. 4.3a is near to OP, hence the temperature distribution is almost in steady state. We can

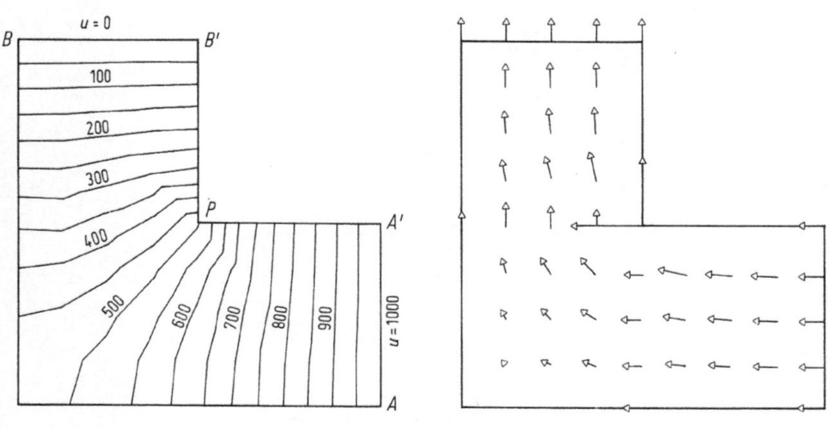

a b

Fig. 4.3 a, b. Isotherms at $T=1.0$. The zoning technique is used

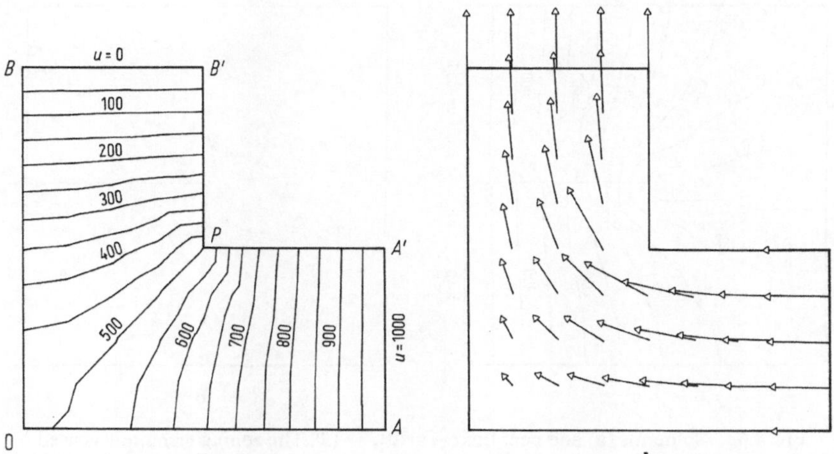

Fig. 4.4a, b. Isotherms at $T = 1.0$. The zoning technique is not used

observe very large calculated fluxes at the re-entrant corner P in Fig. 4.3b. Figure 4.4 shows calculated solution of the same problem without the zoning technique used. Isotherms coincide with the result in Fig. 4.3a. However, the calculated internal fluxes are larger, and we no longer obtain significant fluxes at the re-entrant corner.

1.4.2 Slit Boundary

In the extreme in which the boundary changes its direction by the angle $\theta = 2\pi$, the domain under consideration has a slit. The derivatives become unbounded at the tip of the slit.

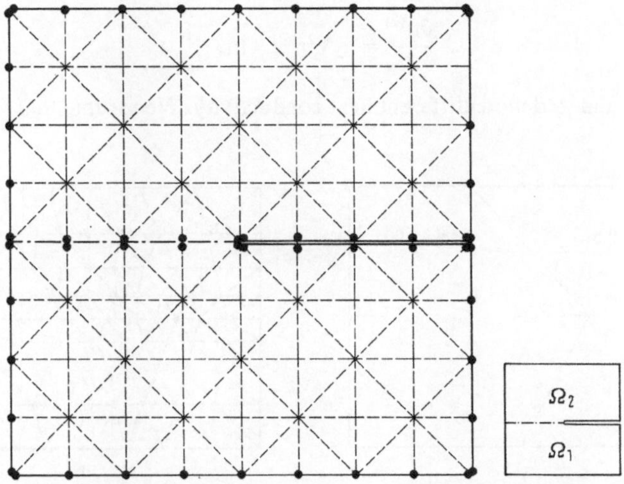

Fig. 4.5. Zones and boundary elements for cracked domain

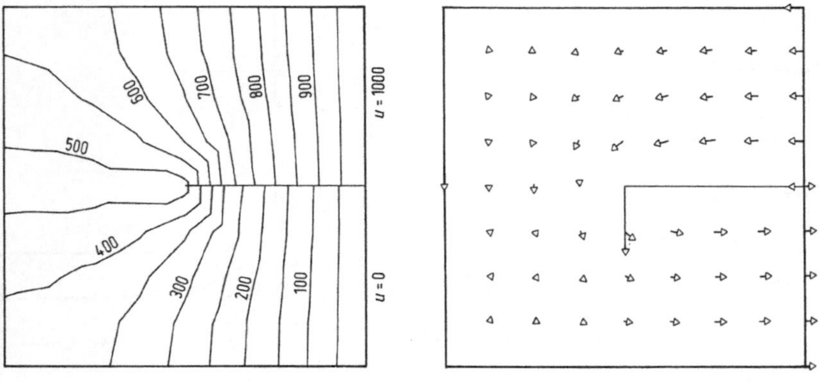

a b

Fig. 4.6. Isotherms (**a**) and heat fluxes (**b**) at $T = 1.0$. The zoning technique is used

We calculated unsteady temperature variations by the boundary element method using the subdivision shown in Fig. 4.5. We divide the nonconvex domain into convex zones. The time increment was chosen at $\Delta T = \frac{1}{32}$.

Figure 4.6a shows calculated isotherms at the time $t = 1.0$. We can observe large calculated fluxes at the tip ***P*** in Fig. 4.6b. Similar calculation may be necessary for the transient analysis of stress concentration at the crack tip in the thermoelasticity.

1.4.3 Interzonal Singularity

We consider the two-media heat conduction in an unsteady state. The heat conduction is governed in each medium by the equations:

$$\frac{\partial u^{(1)}}{\partial t} = \varkappa_1 \nabla^2 u^{(1)} \quad \text{in } \Omega^{(1)} \tag{4.2}$$

$$\frac{\partial u^{(2)}}{\partial t} = \varkappa_2 \nabla^2 u^{(2)} \quad \text{in } \Omega^{(2)} \tag{4.3}$$

in which $\varkappa_1$ and $\varkappa_2$ denote different heat conductivity. We assume that $\varkappa_1 < \varkappa_2$.

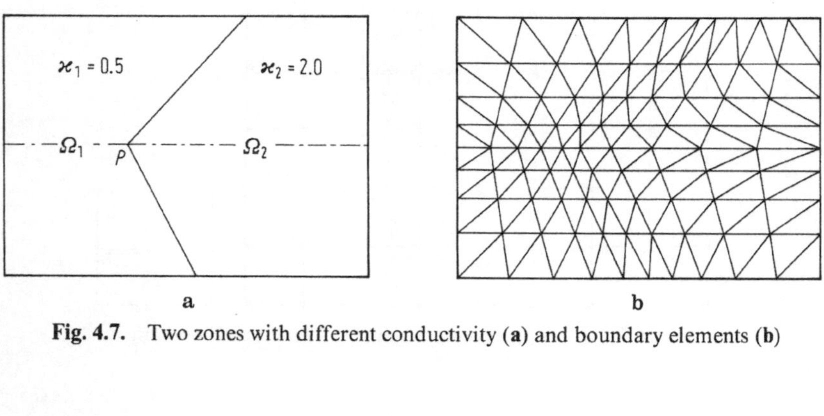

a b

Fig. 4.7. Two zones with different conductivity (**a**) and boundary elements (**b**)

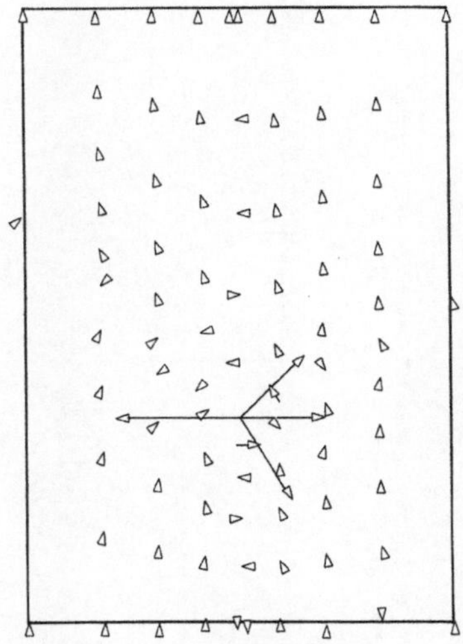

Fig. 4.8. Heat fluxes at $T = 1.0$

For a simple case shown in Fig. 4.7, the exact solution near the corner ***P*** can be expressed in polar coordinates as follows:

$$u^{(i)} = \sum_{j=0}^{\infty} \sum_{k=1}^{\infty} \exp[-\varkappa_i \alpha_j^2 t] J_{\alpha_j \lambda_k}(\alpha_j r) \{A_{jk} \cos(\lambda_k \theta) + B_{jk} \sin(\lambda_k \theta)\}$$
$$+ \sum_{k=0}^{\infty} r^{\lambda_k} \{C_k \cos(\lambda_k \theta) + C' \sin(\lambda_k \theta)\} \tag{4.4}$$

in which λ_k are zeros of the equation:

$$\varkappa_1 \tan(\lambda \delta_1/2) + \varkappa_2 \tan(\lambda \delta_2/2) = 0 \tag{4.5}$$

in which $\delta_1 = 17/(12\pi)$ and $\delta_2 = 7/(12\pi)$ (see Liggett and Liu [14]). Since the smallest positive value of λ_k, say λ_0, is less than unity, the solution has the singularity of the form $u = 0(r^{\lambda_0})$.

An example is considered for the case: $\varkappa_1 = 0.5$ and $\varkappa_2 = 2.0$. Zero initial values are assumed. The boundary on the left side is raised at the unit temperature, while the boundary on the right is kept at the zero temperature. The rest of the boundary is thermally insulated.

The boundary solution was obtained. The time increment was chosen as $\Delta T = \frac{1}{32}$. Calculated results are shown in Fig. 4.8. Each medium is subdivided into two convex zones. Large fluxes are obtained at the interzonal corner.

1.4.4 Infinite Domain

We consider a two-dimensional domain extending to infinity. Suppose that the boundary is finite. External domain of a circle of an infinite medium is a typical example.

Let us consider the problem shown in Fig. 4.9. We are interested in the external region of the unit circle. Temperatures are initially set to be zero over the domain.

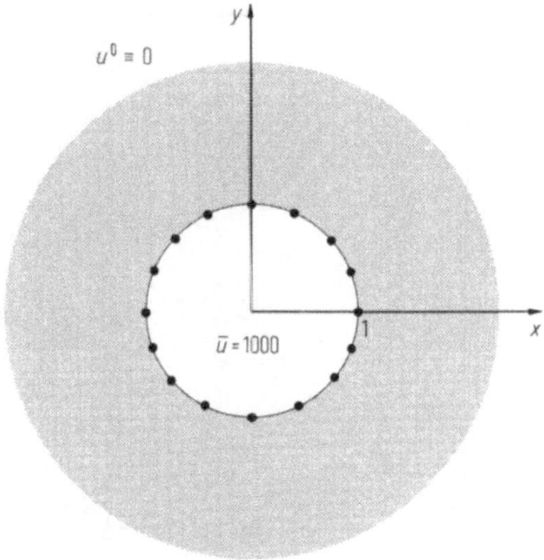

Fig. 4.9. Domain extending to infinity

The boundary along the circle is raised at the temperature 1000. The exact solution is found by Carslaw and Jaeger [12].

Boundary elements with only 16 boundary nodes is shown in Fig. 4.9. The representative size of the space mesh was $\Delta\Gamma = 2.2$, while the time increment was chosen as $\Delta T = 2.5$.

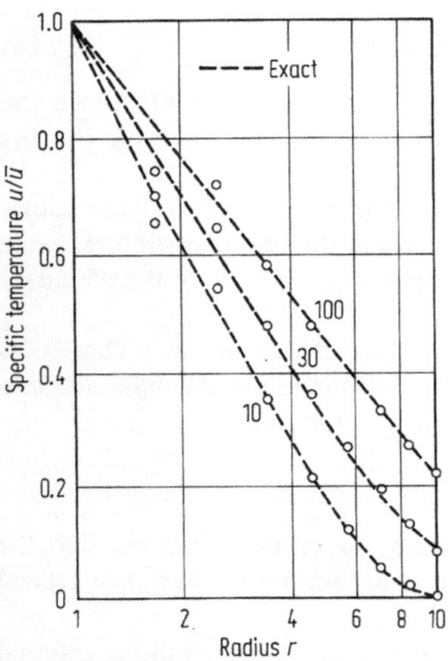

Fig. 4.10. Temperatures at various Fourier numbers

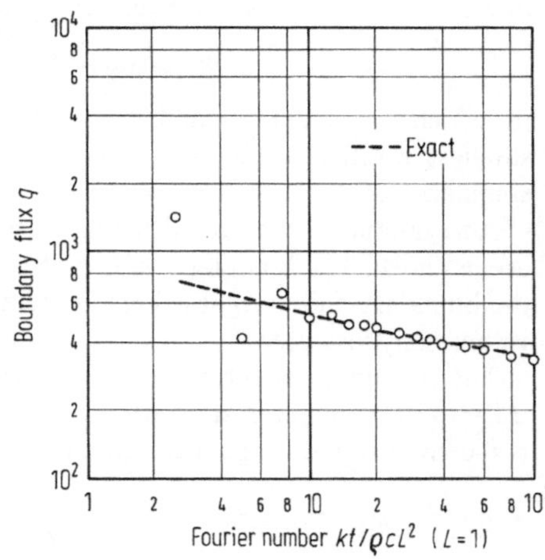

Fig. 4.11. Heat flux across the surface of the cavity

Figure 4.10 shows calculated temperature profiles in the radial direction. The calculated temperature distributions are in good agreement with the exact solutions except for the result in small radii.

In the problems with infinite domain, the heat flux across the boundary surface is the important quantity from the engineering point of view. The time variation of calculated fluxes at the internal circular cavity is shown in Fig. 4.11. The results were found to be satisfactory as compared to the exact flux.

1.5 Conclusions

Boundary element method was discussed about the solution of transient heat conduction equations. We proved convergence and stability mathematically. From the numerical experiments, the second order uniform convergence and unconditional stability of the boundary solution are obtained.

Illustrative examples showed that the boundary element method combined with the zoning technique offers non-oscillating accurate temperatures in the region of infection. Singularities were presented numerically at re-entrant corners and at the intersection of zones with different conductivity. Good results were obtained for a domain extending to infinity.

Acknowledgements

The authors wish to express their cordial thanks to Dr. C. A. Brebbia for his encouragement. This investigation is partially supported by The Japanese Ministry of Education, Grant in Aid for Scientific Research.

References

1 Rizzo, F.J. and Shippy, D.J., A method of solution for certain problems of transient heat conduction. AIAA Journal, **8,** 2004–2009, 1970
2 Chang, Y.P., Kang, C.S., and Chen, D.J., The use of fundamental Green's functions for the solution of problems of heat conduction in anisotropic media. Int. J. Heat Mass Transfer **16,** 1905–1918, 1973
3 Shaw, R.P., An integral equation approach to diffusion. Int. J. Heat Mass Transfer **17,** 693–699, 1974
4 Wrobel, L.C. and Brebbia, C.A., The boundary element method for steady state and transient heat conduction. Proc. First International Conference on Numerical Methods in *Thermal Problems,* Swansea, July 1979, Pineridge Press, Swansea, 1979
5 Banerjee, P. K. and Butterfield, R., *Boundary Element Methods in Engineering Science.* McGraw-Hill, London 1981
6 Brebbia, C.A., Telles, J.C.F., and Wrobel, L.C., *Boundary Element Techniques, Theory and Applications in Engineering.* Springer-Verlag, Berlin 1984
7 Onishi, K., Convergence in the boundary element method for heat equation. TRU Mathematics, 16, No. 2, Science University of Tokyo, 1981
8 Onishi, K. and Kuroki, T., Boundary element approximate method for transient heat conduction. *Proc. National Congress of Theoretical and Applied Mechanics,* 30, Tokyo University Press, 1981

9 Iso, Y., Convergence of boundary element solution for heat equation (submitted to J.J.A.M.)
10 Friedmann, A., *Partial Differential Equation of Parabolic Type.* Prentice-Hall, 1964
11 Onishi, K. and Kuroki, T., Boundary element method in transient heat transfer problems. The *Bulletin of the Institute for Advanced Research of Fukuoka University,* No. 55, 1981
12 Carslaw, H.S. and Jaeger, J.C., *Conduction of Heat in Solids,* Second ed., Oxford, 1959
13 Bell, G.E. and Crank, J., A method of treating boundary singularities in time-dependent problems. J. Inst. Maths. and its Appl. **12,** 37–48, 1973
14 Liggett, J.A. and Philip L.-F. Liu, *The Boundary Integral Equation Method for Porous Media Flow.* Goerge Allen & Unwin, London, 1983

Chapter 2

New Integral Equation Approach to Viscoelastic Problems

by M. Tanaka

2.1 Introduction

Computer-oriented numerical methods of solution have been successfully applied to various engineering problems. At the first stage of such developments, finite difference and finite element methods have attracted the attention of scientists and engineers. While rapid developments of these numerical methods in the last thirty years have stimulated a tremendous amount of work in computational techniques and engineering software, important research in basic physical principles such as variational techniques or method of residual was originated. It would be one of the most important consequences in the latter research that the integral equation method was re-considered through finite element techniques and a new numerical method of solution was innovated by some pioneering groups in England and also the U.S. (1). Brebbia's boundary element book (2) and the International Conference on this subject (3–8) have much contributed to recent years' rapid advances of the boundary element methods and stimulated a wide variety of boundary element applications in engineering. It can be seen that among various integral equation formulations the direct formulation is most successful and promising for engineering analysis.

As has been well known, in the boundary element method the governing differential equation of the problem under consideration is transformed into an equivalent boundary integral equation and it is solved in the same manner as in the finite element methods. In most linear problems, only the boundary surface is discretized by introducing the so called boundary elements, and hence the discretized set of equations include only nodal values on the boundary alone. This can lead us to saving time of input data formation and computation. Because of this advantage the boundary element methods have been recognized as a powerful alternative to the domain-type methods of solution such as the finite difference or finite element methods. At present some engineering software is available for analyzing a wide variety of practical problems.

As far as the boundary element methods for viscoelasticity are concerned, attempts have been made by using the Laplace transform and the correspondence principle to linear elasticity (9, 10). In this aproach, governing equations in the transformed space are solved by means of the usual boundary element method and the results obtained are inversed into an appropriate physical space. Such inversion in most cases must be made numerically and an accurate numerical inversion needs much experience in choosing appropriate values of the transformation parameter. It can be mentioned that such disadvantage has limited the applicability of the transformed boundary element method.

In this paper, new boundary element techniques are discussed for viscoelastic problems. First, the integral equations both in space and time are derived for linear viscoelasticity where isotropy and homogeneity are assumed, and their discretization is presented by using boundary elements and also time elements. Second, an incremental formulation and its numerical implementation are proposed as a more powerful alternative to the transformed method. Finally, numerical results are shown for some sample problems to demonstrate the potential usefulness of the proposed method.

2.2 Preliminaries

Here we consider a linear viscoelastic medium occupying the domain Ω with boundary Γ as shown in Fig. 1. It is assumed that the viscoelastic medium under consideration is subjected to a quasi-static time dependent deformation under an isothermal state. Using the Cartesian coordinate system O-$x_1 x_2 x_3$, the equilibrium equation of the viscoelastic problems can be expressed as

$$\sigma_{ij,j}(\boldsymbol{x}, t) + f_i(\boldsymbol{x}, t) = 0 \tag{2.1}$$

where σ_{ij} denotes the stress tensor and f_i the body force per unit volume. The summation convention for repeated indices is used in this paper, unless otherwise noticed. The arguments $(\boldsymbol{x}, t)$ imply that the variables are dependent upon both the position $\boldsymbol{x}$ and time t. The index following the comma denotes a partial derivative such that $(\cdot)_{,i} = \partial(\cdot)/\partial x_i$.

The constitutive relations of linear viscoelasticity can be expressed as follows (11)

$$\sigma_{ij}(\boldsymbol{x}, t) = G_{ijkl}(t)\, \varepsilon_{kl}(\boldsymbol{x}, 0) + \int_0^t G_{ijkl}(t-\tau) \frac{\partial \varepsilon_{kl}(\boldsymbol{x}, \tau)}{\partial \tau}\, d\tau \tag{2.2}$$

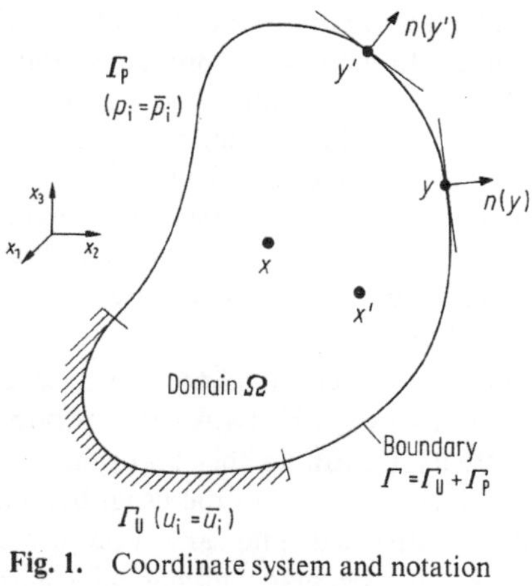

Fig. 1. Coordinate system and notation

where $\varepsilon_{kl}(x, 0)$ is the strain tensor of the point x at time $t = 0$, which is prescribed as the initial conditions of the problem. The second integral on the r.h.s. of Eq. (2.2) is well known as the hereditary integral (12) or the Stieltjes integral (11). The functions in Eq. (2.2) comprise a fourth-order tensor such that $G_{ijkl}(t) = 0$ for the time domain $t < 0$. These functions are sometimes called the relaxation functions. The symmetrical properties of stress and strain tensors lead to the following relations:

$$G_{ijkl}(t) = G_{jikl}(t) = G_{ijlk}(t)\,. \tag{2.3}$$

For an isotropic and homogeneous viscoelastic medium, the relaxation functions can be given by

$$G_{ijkl}(t) = \lambda(t)\,\delta_{ij}\,\delta_{kl} + \mu(t)(\delta_{ik}\,\delta_{jl} + \delta_{il}\,\delta_{jk}) \tag{2.4}$$

where δ_{ij} denotes Kronecker's delta. The functions $\lambda(t)$ and $\mu(t)$ have similar meanings to Lame's constants in elasticity. In the theory of viscoelasticity they are prescribed as appropriate functions of time.

If displacements and strains are infinitesimally small, it follows that

$$\varepsilon_{ij}(x, t) = \tfrac{1}{2}\left\{\frac{\partial u_i(x, t)}{\partial x_j} + \frac{\partial u_j(x, t)}{\partial x_i}\right\}. \tag{2.5}$$

Making use of Eqs. (2.2) and (2.5) for the equilibrium equation (2.1), we can derive the governing equation of viscoelasticity as follows:

$$G_{ijkl}(t)\, u_{k,lj}(x, 0) + \int_0^t G_{ijkl}(t-\tau)\,\frac{\partial u_{k,lj}(x, \tau)}{\partial \tau}\,d\tau + f_i(x, t) = 0\,. \tag{2.6}$$

The first terms of Eq. (2.6) represent the influences of the initial conditions, which can be obtained by elastic calculation. Hence, our objective is solve Eq. (2.6) under appropriate boundary conditions expressed such that

$$\begin{aligned} u_i(y, t) &= \overline{u_i(y, t)} \quad \text{on } \Gamma_u \\ p_i(y, t) \equiv \sigma_{ij}(y, t)\, n_j(y) &= \overline{p_i(y, t)} \quad \text{on } \Gamma_p \equiv \Gamma - \Gamma_u \end{aligned} \tag{2.7}$$

where the superimposed bar denotes a prescribed value.

The purpose of this paper is show how to manage the above initial-boundary value problems by means of the boundary element method with no use of the Laplace transform and the correspondence principle.

2.3 Boundary Integral Equation in Space and Time

Now we derive the boundary integral equation in space and time for the linear viscoelastic problems. To this end, we may use a generalized virtual work principle which can be obtained from a straightforward extension of Brebbia's statement [13] given for static problems into space and time. Denoting by the asterisk * the virtual variables, we obtain the extended virtual work principle for linear viscoelasticity as follows:

$$\begin{aligned}
&\int_\Gamma \int_0^t p_i(y', t-\tau) \frac{\partial u^*_{mi}(x, y'; \tau)}{\partial \tau} d\tau \, d\Gamma(y') \\
&+ \int_\Omega \int_0^t f_i(x', t-\tau) \frac{\partial u^*_{mi}(x, x'; \tau)}{\partial \tau} d\tau \, d\Omega(x') \\
&= \int_\Gamma \int_0^t u_i(y', t-\tau) \frac{\partial p^*_{mi}(x, y'; \tau)}{\partial \tau} d\tau \, d\Gamma(y') \\
&+ \int_\Omega \int_0^t u_i(x', t-\tau) \frac{\partial f^*_{mi}(x, x'; \tau)}{\partial \tau} d\tau \, d\Omega(x')
\end{aligned} \tag{3.1}$$

This statement is equivalent to the reciprocal theorem presented by Banerjee and Butterfield [14].

Provided that the virtual variables are caused by a unit force of the form

$$f^*_{mi}(x, y; t) = \delta_{mi} \, \delta(x, y) \, H(t) \tag{3.2}$$

where $\delta(\cdot)$ is the Dirac delta function and $H(t)$ the Heaviside unit step function, Eq. (3.1) yields the following integral equation:

$$\begin{aligned}
&u_i(x, t) + \int_\Gamma \left[p^*_{ij}(x, y'; t) \, u_j(y', 0) + \int_0^t p^*_{ij}(x, y'; t-\tau) \frac{\partial u_j(y', \tau)}{\partial \tau} d\tau \right] d\Gamma(y') \\
&\quad - \int_\Gamma \left[u^*_{ij}(x, y'; t) \, p_i(y', 0) + \int_0^t u^*_{ij}(x, y'; t-\tau) \frac{\partial p_j(y', \tau)}{\partial \tau} d\tau \right] d\Gamma(y') \\
&= \int_\Omega \int_0^t f_j(x', t-\tau) \frac{\partial u^*_{ij}(x, x'; \tau)}{\partial \tau} d\tau \, d\Omega(x') .
\end{aligned} \tag{3.3}$$

The virtual variables denoted by asterisk are related to the fundamental solution to the governing Eq. (2.6), which satisfies in the infinite viscoelastic medium the differential equation of the form

$$G_{ijkl}(t) \, u^*_{mk,lj}(x, y; 0) + \int_0^t G_{ijkl}(t-\tau) \frac{\partial u^*_{mk,lj}(x, y; \tau)}{\partial \tau} d\tau + \delta_{mi} \, \delta(x, y) \, H(t) = 0 . \tag{3.4}$$

The fundamental solution of this type can be obtained for particular viscoelastic models by using Kelvin's fundamental solution of linear elasticity. Kaneko, Shinokawa, Yoshida and Kawahara [15] have recently derived some useful expressions of the two-dimensional fundamental solution for a typical three-element viscoelastic model of so called Voigt type. They also show its application to boundary element computation of some sample problems.

Using the fundamental solution obtainable from Eq. (3.4) for the integral Eq. (3.3), and taking the limiting process where an internal point x approaches an arbitrary boundary point y, we can finally obtain the boundary integral equation of the following form:

$$c_{ij}(y)\,u_j(y,t)+\int_\Gamma\left[p^*_{ij}(y,y';t)\,u_j(y',0)+\int_0^t p^*_{ij}(y,y';t-\tau)\frac{\partial u_j(y',\tau)}{\partial\tau}\,d\tau\right]d\Gamma(y')$$

$$-\int_\Gamma\left[u^*_{ij}(y,y';t)\,p_j(y',0)+\int_0^t u^*_{ij}(y,y';t-\tau)\frac{\partial u_j(y',\tau)}{\partial\tau}\,d\tau\right]d\Gamma(y')$$

$$=\int_\Omega\int_0^t f_j(x,t-\tau)\frac{\partial u^*_{ij}(y,x;\tau)}{\partial\tau}\,d\tau\,d\Omega(x) \tag{3.5}$$

where c_{ij} is the same coefficient as that of the boundary integral equation in elasticity [2], and equal to $(\frac{1}{2})\,\delta_{ij}$ for a smooth boundary point.

The discretization of the boundary integral Eq. (3.5) can be progressed in the following manner. Namely, the boundary Γ is divided into a series of boundary elements, and the time domain is also subdivided into small time steps. By assuming that the boundary variables $u_i(y,t)$ and $p_i(y,t)$ are linear with respect to time t within a small time step, we can obtain the discretized set of equations as follows:

$$c_{ij}(y)\,u_j^{(\varkappa)}(y)+\sum_{\lambda=1}^{\varkappa}\int_\Gamma p^{*(\varkappa)}_{ij\lambda}(y,y')\,\Delta u_{j\lambda}(y')\,d\Gamma(y')$$

$$-\sum_{\lambda=1}^{\varkappa}\int_\Gamma u^{*(\varkappa)}_{ij\lambda}(y,y')\,\Delta p_{j\lambda}(y')\,d\Gamma(y')=F_i^{(\varkappa)}(y) \tag{3.6}$$

where

$$u_j^{(\lambda)}(y)=u_j(y,t_\lambda)$$

$$\Delta u_{j\lambda}(y)=[u_j^{(\lambda)}(y)-u_j^{(\lambda-1)}(y)]/\Delta t_\lambda$$

$$\Delta p_{j\lambda}(y)=[p_j^{(\lambda)}(y)-p_j^{(\lambda-1)}(y)]/\Delta t_\lambda$$

$$p^{*(\varkappa)}_{ij\lambda}(y,y')=\int_0^{t_\varkappa}p^*_{ij}(y,y';t_\varkappa-\tau)\,d\tau$$

$$u^{*(\varkappa)}_{ij\lambda}(y,y')=\int_0^{t_\varkappa}u^*_{ij}(y,y';t_\varkappa-\tau)\,d\tau\,. \tag{3.7}$$

It is noted that the r.h.s. of Eq. (3.6) can be computed from the integrals corresponding to the initial conditions and body forces.

A step-wise procedure can be applied to the solution of Eq. (3.6). We begin with the first time step $\varkappa=1$, and solve the boundary element equations of the following form:

$$[H]_{\varkappa=1}\{u\}_{\varkappa=1}-[G]_{\varkappa=1}\{p\}_{\varkappa=1}=\{F\}_{\lambda=0,1}\,. \tag{3.8}$$

The column vector $\{F\}_{\lambda=0,1}$ in Eq. (3.8) can be calculated from the initial conditions ($\lambda=0$) and the body forces at the time step $\lambda=1$. All the unknown boundary variables at the time step $\varkappa=1$ can be obtained by solving Eq. (3.8) under appropriate boundary conditions. Then, we can proceed to the next time step $\varkappa=2$. This step-wise solution is progressed until the final time step is reached.

2.4 Alternative Approach – Incremental Formulation

Although the integral equation formulation discussed in the last section is exact and hence can lead to accurate numerical results, the calculation of the fundamental solution is troublesome and its numerical implementation needs much computation time. Furthermore, the formulation is restricted to linear viscoelasticity. In this section, we shall propose an alternative integral equation approach which can be easily applicable even to nonlinear viscoelasticity, since the formulation is presented in terms of incremental variables.

Now, we express Eq. (2.6) in the following form:

$$\int_0^t G_{ijkl}(t-\tau)\frac{\partial u_{k,lj}(\boldsymbol{x},\tau)}{\partial \tau}\,d\tau + b_i(\boldsymbol{x},t) = 0\,, \tag{4.1}$$

where

$$b_i(\boldsymbol{x},t) = f_i(\boldsymbol{x},t) + G_{ijkl}(t)\,u_{k,lj}(\boldsymbol{x},0)\,. \tag{4.2}$$

Provided that time variations of the relaxation functions G_{ijkl} are negligible, time derivative of Eq. (4.1) yields the incremental relationship of the form

$$G_{ijkl}(t)\,\dot{u}_{k,lj}(\boldsymbol{x},t) + \dot{b}_i(\boldsymbol{x},t) = 0\,, \tag{4.3}$$

where the superimposed dot denotes time derivative and

$$\dot{b}_i(\boldsymbol{x},t) = \dot{f}_i(\boldsymbol{x},t) + \dot{G}_{ijkl}(t)\,u_{k,lj}(\boldsymbol{x},0)\,. \tag{4.4}$$

Equation (4.3) implies that its form is similar to the Navier equation in elastostatics and hence Kelvin's solution can be applied to its boundary element solution.

The fundamental solution to Eq. (4.3) can be defined for an infinite viscoelastic medium as follows:

$$G_{ijkl}(t)\,u^*_{mk,lj}(\boldsymbol{x},\boldsymbol{y};t) + \delta_{mi}\,\delta(\boldsymbol{x},\boldsymbol{y}) = 0\,. \tag{4.5}$$

For the isotropic and homogeneous viscoelasticic medium which has the relaxation functions expressed in Eq. (2.4), the fundamental solution of a Kelvin type can be given;

For the two-dimensional plane strain

$$u^*_{ij}(\boldsymbol{x},\boldsymbol{y};t) = \frac{1}{8\pi[1-\nu(t)]\,\mu(t)}\left\{[3-4\nu(t)]\ln\left(\frac{1}{r}\right)\delta_{ij} + r_{,i}r_{,j}\right\} \tag{4.6}$$

and for three-dimensional problems

$$u^*_{ij}(\boldsymbol{x},\boldsymbol{y};t) = \frac{1}{16\pi[1-\nu(t)]\,\mu(t)}\{[3-4\nu(t)]\,\delta_{ij} + r_{,i}r_{,j}\} \tag{4.7}$$

where r is the distance between the two points $\boldsymbol{x}$ and $\boldsymbol{y}$, i.e. $r = |\boldsymbol{x}-\boldsymbol{y}|$.

The traction p^*_{ij} corresponding to the fundamental solution u^*_{ij} is expressed by

$$p^*_{ij}(\boldsymbol{x},\boldsymbol{y};t) = \frac{-1}{4\alpha\pi[1-\nu(t)]\,r^{\alpha}}\{[1-2\nu(t)]\,\delta_{ij} + \beta\, r_{,i}r_{,j} - [1-2\nu(t)](r_{,i}n_j - r_{,j}n_i)\} \tag{4.8}$$

where $\alpha = 1$, $\beta = 2$ for 2-D problems, and $\alpha = 2$, $\beta = 3$ for 3-D problems. The relaxation functions $\lambda(t)$, $\mu(t)$, $\nu(t)$ and $K(t)$ are analogous to familiar moduli in elasticity, called Lame's constants and bulk modulus. The only difference is that the viscoelastic moduli are functions of time. It is noted that between these moduli holds the following relationship:

$$\nu(t) = \frac{\lambda(t)}{2[\lambda(t) + \mu(t)]}, \quad K(t) = \tfrac{1}{3}[3\lambda(t) + 2\mu(t)]. \tag{4.9}$$

A virtual work principle can be stated in the incremental form as follows:

$$\int_\Gamma \dot{p}_j(y', t)\, u^*_{ij}(x, y; t)\, d\Gamma(y') + \int_\Omega \dot{b}_j(x', t)\, u^*_{ij}(x, x'; t)\, d\Omega(x')$$
$$= \int_\Gamma \dot{u}_j(y', t)\, p^*_{ij}(x, y; t)\, d\Gamma(y') + \int_\Omega \dot{u}_j(x', t) f^*_{ij}(x, x'; t)\, d\Omega(x'). \tag{4.10}$$

The virtual variables represented by the asterisk are selected as the fundamental solutions of Kelvin type mentioned before, and hence it follows that

$$f^*_{ij}(x, x'; t) = \delta_{ij}\, \delta(x, x'). \tag{4.11}$$

In a similar manner to the elastostatic one, the statement (4.10) yields the boundary integral equation of the form

$$C_{ij}(y)\, \dot{u}_j(y) + \int_\Gamma p^*_{ij}(y, y'; t)\, \dot{u}_j(y', t)\, d\Gamma(y') - \int_\Gamma u^*_{ij}(y, y'; t)\, \dot{p}_j(y', t)\, d\Gamma(y')$$
$$= \int_\Omega u^*_{ij}(y, x; t)\, \dot{b}_j(x, t)\, d\Omega(x) \tag{4.12}$$

where c_{ij} is the same coefficient as in elastostatics and $c_{ij} = (\frac{1}{2})\, \delta_{ij}$ for a smooth boundary. One can put $c_{ij} = \delta_{ij}$, when the point y belongs to the inner domain Ω.

Now, we discuss numerical implementation of the boundary integral Eq. (4.12). To this end, the time domain is divided into small time steps. The initial conditions are taken into account and their corresponding domain integrals are numerically evaluated. Then, the incremental variables $\dot{u}_i(y, 0)$ and $\dot{p}_i(y, 0)$ can be computed from the discretized set of Eq. (4.12) by means of the usual boundary element method available in the literature. The state variables can be obtained by Euler's forward integration scheme such that

$$u_i(y, t_\lambda) = u_i(y, t_{\lambda-1}) + \dot{u}_i(y, t_{\lambda-1})\, \Delta t_\lambda$$
$$p_i(y, t_\lambda) = p_i(y, t_{\lambda-1}) + \dot{p}_i(y, t_{\lambda-1})\, \Delta t_\lambda \tag{4.13}$$

where Δt_λ denotes the interval of the λ time step. If this step-wise integration shown in Eq. (4.13) is carried out successively for all the time steps $\lambda = 1, 2, \ldots,$ state variables at arbitrary time can be obtained.

2.5 Numerical Results and Discussion

Based on the incremental formulation, a new computer program is developed. Since the formulation employs the modified Kelvin's fundamental solution the constants of which are replaced with appropriate functions of time, the usual

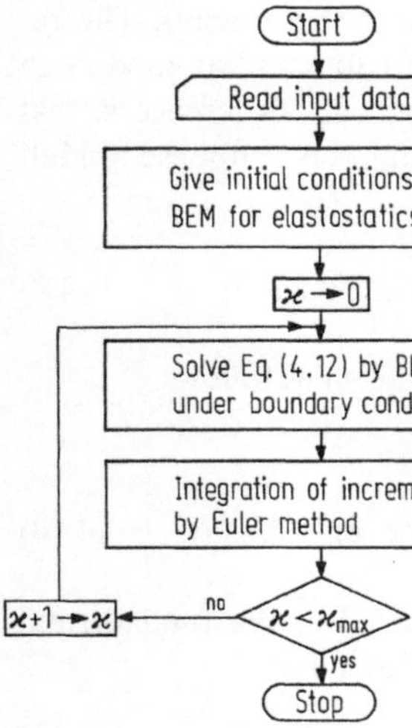

Fig. 2. Main flow of solution procedure

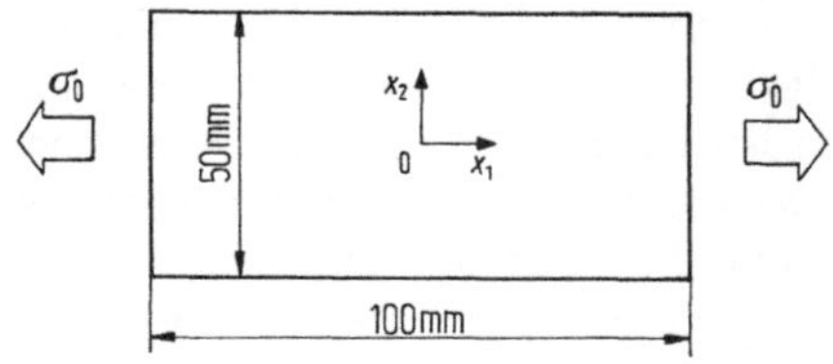

Fig. 3. Rectangular plate subjected to uniaxial tension

boundary element method can be easily extended to its numerical implementation. The main flow of the computer program is illustrated in Fig. 2. The initial conditions which are prescribed as the values of strain or displacement in Ω at time $t=0$ can be calculated by mens of the usual boundary method for elastostatics. The same BEM is repeatedly used for computation of the incremental variables at every time step. In what follows, we shall show some computational results obtained by such a newly developed solution procedure.

The plane strain state is assumed in numerical computations. The relaxation function $\mu(t)$ is assumed such that

$$\mu(t) = \mu_0 \left[a + b \exp\left(\frac{c\,t}{\tau_0} \right) \right] \tag{5.1}$$

where a, b and c are material constants, while μ_0 is equivalent to the shear modulus in elastostatics, and τ_0 is called the relaxation time. In addition, it is assumed that $K(t) = K_0$ (constant) and $\mu_0/K_0 = \frac{3}{8}$.

The first example is the creep deformation of a rectangular plate subjected to the uniaxial tension $\sigma_0 H(t)$ at both the ends of the plate as shown in Fig. 3. It is assumed that $\mu_0 = 500$ MPa, $\sigma_0 = 5$ MPa, $a = 0.5$, $b = 0.5$, $c = -0.5$ and $\tau_0 = 1$ sec. In Figs. 4a and b are shown the boundary element discretization in which thirty

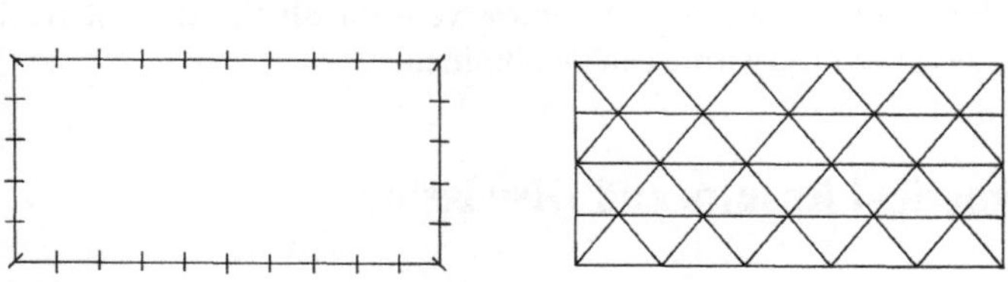

a b

Fig. 4a, b. Discretization for whole rectangular domain. **a** Boundary elements. **b** Internal cells

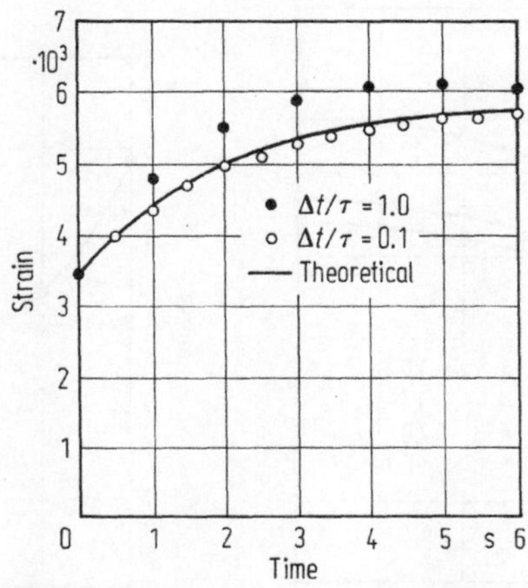

Fig. 5. Time variation of strain in loading direction

constant boundary elements are used, and the internal cell subdivision for evaluation of the volume integral in Eq. (4.12) by use of fortyfour internal cells.

Figure 5 shows the time variation of tensile strain ε_{11} in the loading direction. When a smaller time step interval is used, the results obtained approach closer to the theoretical ones.

Next, the creep problem of a thick-walled circular cylinder subjected to the internal pressure $p = p_0 H(t)$. The outside boundary is assumed stress free. Because of symmetry a quarter of the tube cross-section is analyzed, and this portion is discretized by thirty constant boundary elements and ninety internal cells as shown in Figs. 6a and b.

It is assumed that $R_i = 50$ mm, $R_0 = 100$ mm, $p_0 = 1$ MPa, $\mu_0 = 945$ MPa, $a = 0.25$, $b = 0.75$ and $c = -4$.

The results obtained for the time step interval $\Delta t/\tau_0 = 1/20$ are shown in Fig. 7 for time variations of radial displacements. Fairly good agreement can be recognized between the numerical and theoretical results.

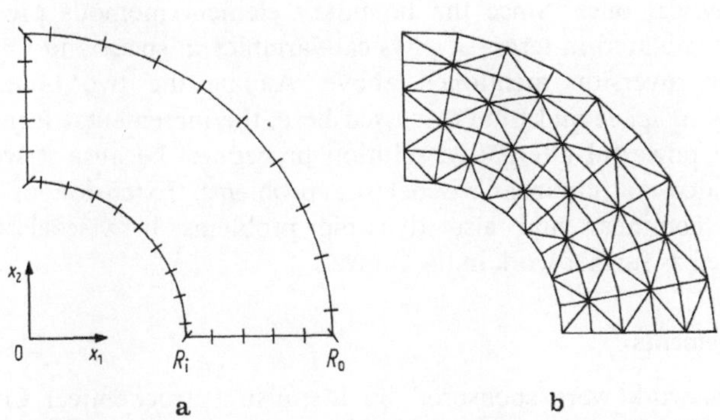

Fig. 6 a, b. Discretization for quarter of tube cross-section. **a** Boundary Elements. **b** Internal cells

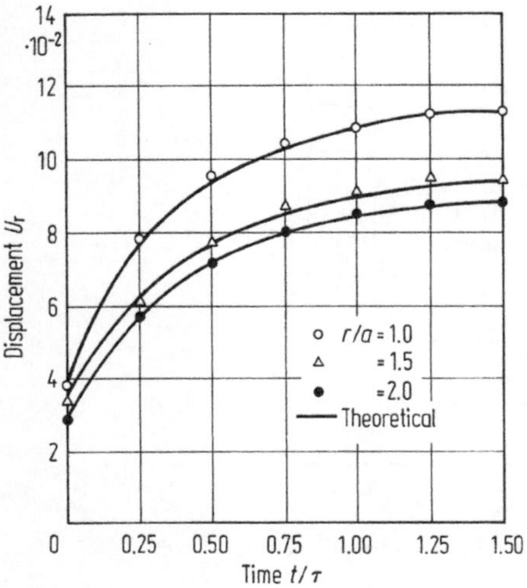

Fig. 7. Time variations of radial displacements

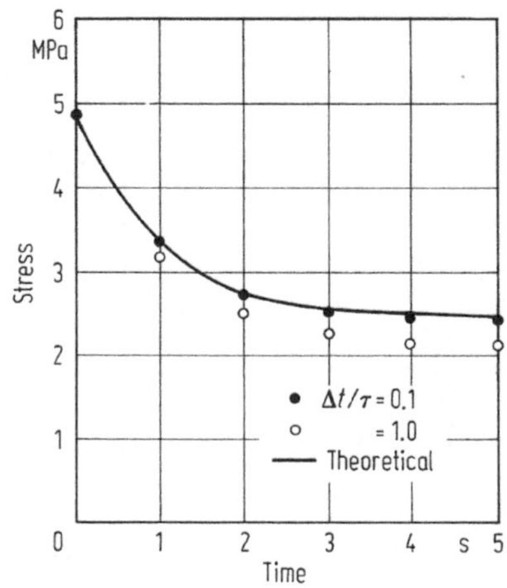

Fig. 8. Time variation of stress in loading direction

Finally, the method of solution is applied to a stress relaxation problem, which can be computed with equal ease by means of the same computer program as above. Figure 8 shows the results obtained for the stress relaxation problem of the same rectangular plate as in the first example illustrated in Fig. 3. It can be seen that excellent numerical results are obtained using smaller time step intervals.

2.6 Concluding Remarks

New integral equation formulations with no use of the Laplace transform have been discussed in this paper. The boundary element method so far proposed is based on the Laplace transform and the correspondence principle. Therefore, the method inevitably encounters difficulties in inversion of the transformed variables into the physical ones. Since the boundary element methods presented in this paper are formulated in terms of physical variables in space and time, there is no need of the inversion mentioned above. Among the two boundary element formulations in space and time discussed here, the incremental formulation could be the most powerful alternative solution procedure, because it would be more easily extended to nonlinear viscoelastic problems. Extension of the proposed method to nonlinear and also dynamic problems in viscoelasticity can be recommended as further work in the future.

Acknowledgements

Parts of this work were sponsored by Idemitsu Petrochemical Co. The author wishes to express his cordial thanks to Messrs. M. Kimura and T. Shiojima of this company for their helpful support in numerical computation.

References

1 Tanaka, M., Recent advances in boundary element methods. Appl. Mech. Revs. **36,** 627–634

2 Brebbia, C.A., *The Boundary Element Method for Engineers.* London, Pentech Press, 1978

3 Brebbia, C.A. (ed.), *Recent Advances in Boundary Element Methods.* London, Pentech Press, 1978

4 Brebbia, C.A. (ed.), *New Developments in Boundary Element Methods.* Southampton, CML Publications, 1980

5 Brebbia, C.A. (ed.), *Boundary Element Methods.* Berlin London New York Tokyo, Springer-Verlag, 1981

6 Brebbia, C.A. (ed.), *Boundary Element Methods in Engineering.* ditto, 1982

7 Brebbia, C.A., Futagami, T., and Tanaka, M. (eds.), *Boundary Elements.* ditto, 1983

8 Brebbia, C.A. (ed.), *Boundary Elements VI.* ditto, 1984

9 Rizzo, F.J. and Shippy, D.J., An application of the corresponding principle of linear viscoelasticity theory. SIAM J. Appl. Math. **21,** 321–330, 1971

10 Kusama, T. and Mitsui, Y., Boundary element method applied to linear viscoelastic analysis. Appl. Math. Modelling **6,** 285–290, 1982

11 Christensen, R.M., *Theory of Viscoelasticity.* 2nd ed., 1982

12 Ziegler, H., *An Introduction to Thermodynamics.* 2nd. ed., 1983, pp. 193–215

13 Brebbia, C.A., *Basic principles.* pp. 3–28 of Ref. [7]

14 Banerjee, P.K. and Butterfield, R., *Boundary Element Methods in Engineering Science.* London, McGraw-Hill (UK), 1981, pp. 243–248

15 Kaneko, N., Shinokawa, T., Yoshida, N., and Kawahara, M., *Numerical Analysis of Viscoelasticity Using Boundary Element Method.* Proc. 4th Int. Conf. Appl. Numerical Modeling, Dec. 27–29, 1984, Tainan/Taiwan, pp. 437–477

Chapter 3

Numerical Integration

by H. Pina

3.1 Introduction

The Boundary Element Method (BEM) requires the evaluation of integrals of different kinds. It is therefore important to have a sound knowledge of the tools available to perform this task as efficiently as possible. By this we mean that numerical integration should provide sufficiently accurate values without incurring in excessive computing time. In this expository chapter we review the fundamentals of numerical integration and describe how to apply this technique to BEM problems.

Before proceeding towards this objective, it is useful to recognize the origins and the types of integrals arising in BEM. We take as an illustrative example the two dimensional Poisson's equation in a bounded domain Ω with a smooth boundary Γ,

$$\text{lap}\, u + f = 0 \quad \text{in} \quad \Omega \subset R^2 \tag{1}$$

with the following boundary conditions,

$$u = \bar{u} \quad \text{on} \quad \Gamma_u \tag{2a}$$

$$\frac{\partial u}{\partial n} = \bar{q} \quad \text{on} \quad \Gamma_q \tag{2b}$$

where f is a function over Ω representing *a* source, $\bar{u}$ is a prescribed function over Γ_u specifying the potential and $\bar{q}$ is a prescribed function over Γ_q specifying the flux. Γ_u and Γ_q are disjoint but complementary parts of the boundary Γ and $\boldsymbol{n}$ is the exterior unit normal.

The above problem can be reformulated, as it is now well known [Brebbia, Brebbia et al., Jaswon and Symm], as an integral equation. In fact, if we put

$$\boldsymbol{q} = \text{grad}\, u \quad \text{and} \quad q = \boldsymbol{q} \cdot \boldsymbol{n} \tag{3a}$$

$$\boldsymbol{q}^* = \text{grad}\, u \quad \text{and} \quad q^* = \boldsymbol{q}^* \cdot \boldsymbol{n} \tag{3b}$$

where $\boldsymbol{u}^*$ is the fundamental solution of Laplace's equation in the plane

$$u^* = \frac{1}{2\pi} \ln(1/r) \quad \text{with} \quad \boldsymbol{r} = \boldsymbol{x} - \boldsymbol{x}', \quad r = \| \boldsymbol{r} \| \tag{4a}$$

$$\boldsymbol{q}^* = \text{grad}\, u^* = -\boldsymbol{r}/(2\pi r^2) \tag{4b}$$

and $\| \boldsymbol{r} \|$ represents the euclidean norm of $\boldsymbol{r}$, the integral representation associated with Poisson's equation is

$$u(\boldsymbol{x}) = \int_{\Omega} f(\boldsymbol{x}')\, u^*(\boldsymbol{x}, \boldsymbol{x}')\, d(\boldsymbol{x}') + \int_{\Gamma} [u^*(\boldsymbol{x}, \boldsymbol{x}')\, q(\boldsymbol{x}') - q^*(\boldsymbol{x}, \boldsymbol{x}')\, u(\boldsymbol{x}')]\, d\Gamma(\boldsymbol{x}'). \tag{5}$$

By taking the point $\boldsymbol{x}$ to the boundary the integral representation (5) can be used to set an integral equation over the boundary Γ, which can be written as

$$c\, u = \int_{\Omega} f\, u^*\, d\Omega + \int_{\Gamma} (u^*\, q - q^*\, u)\, d\Gamma. \tag{6}$$

In the above expression, the notation of which has been somewhat abbreviated, the variable c equals 1 for interior points and is equal to the interior angle (measured in fractions of 2π) subtended by Γ at boundary points ($c = \frac{1}{2}$ for smooth boundary points).

The discretization of such equation entails the need to evaluate integrals over domain finite elements Ω_e and boundary finite elements Γ_e, the subscript e denoting the current element. More specifically, we end up with integrals of the following types.

a)
$$\int_{\Omega} u^*\, f\, d\Omega. \tag{7a}$$

This is a domain integral with an integrand involving a $\ln(1/r)$ factor. If the collocation point belongs to the domain element Ω_e, this integral becomes singular.

b)
$$\int_{\Gamma} u^*\, q\, d\Gamma. \tag{7b}$$

This is a boundary integral with an integrand exhibiting also a $\ln(1/r)$ factor. When the collocation point belongs to the boundary element Γ_e, this integral becomes singular.

c)
$$\int_{\Gamma} q^*\, u\, d\Gamma. \tag{7c}$$

This is a boundary integral, its integrand involving the function q^*. At first sight this integrand looks singular when $r \to 0$, but a simple analysis based in elementary geometry and the application of L'Hopital rule shows that it is indeed bounded.

There are cases where the above integrals can be evaluated analytically. However if one wishes to explore the full capabilities of second and higher order isoparametric finite elements this approach is no longer feasible.

3.2 Integration over One-Dimensional Domains

In this section we review briefly the methods available for the numerical evaluation of

$$I(f) = \int_a^b f(x)\, dx \tag{8}$$

where $[a, b]$ is a finite interval on the real line and $f(x)$ is a real function defined over this interval. A full account of numerical integration theory can be found in [Davis and Rabinowitz, Engels, Krylov].

In all the methods for the numerical computation of $I(f)$ that shall be presented here, this value is replaced by a formula of the type

$$\tilde{I}(f) = \sum_{i=0}^{n} A_i f(x_i) \tag{9}$$

where $\tilde{I}(f)$ is intended as an approximation to $I(f)$. The x_i's are the integration points and the A_i's the integration weights and expressions like this are known as integration formulas or quadrature rules. As we can see, the exact value of the integral is expected to be approximated by an weighted sum of values of the integrand at some points, the objective of numerical integration theory being to determine the location of the integration points and the values of the integration weights in order to make the integration formula as accurate as possible. We mention in passing that it is possible to generalize expression (9) to incorporate also values of derivatives of the integrand, but since the rules thus obtained seem to have no special interest in BEM applications, we shall not develop this point any further.

The error of the integration formula is defined by

$$E(f) = I(f) - \tilde{I}(f). \tag{10}$$

It is important to have upper bounds for $|E(f)|$ and to know the convergence properties of formula (9) as the number n of integration points increases.

3.2.1 Basic Rules

One important class of quadrature rules is based on the idea of using the polynomial $p_n(x)$ of degree less or equal to n that interpolates the integrand $f(x)$ at some preassigned points x_i, $i = 0, 1, \ldots, n$, to replace $f(x)$. Expressing the interpolating polynomial in terms of the Lagrange polynomials L_i's, we have

$$p_n(x) = \sum_{i=0}^{n} f(x_i) L_i(x) \tag{11}$$

and therefore

$$I(p_n) = \sum_{i=0}^{n} A_i f(x_i) \tag{12a}$$

with

$$A_i = \int_a^b L_i(x)\, dx. \tag{12b}$$

Thus once the integration points are chosen, the calculation of the integration weights follows easily by expression (12b). Now we declare that

$$\tilde{I}(f) = I(p_n). \tag{13}$$

We say that an integration formula has *degree n* if it is exact for all polynomials of degree less or equal to n and there exists at least one polynomial of degree $n+1$

for which the formula is not exact. The way formula (9) was constructed guarantees that the degree of $\tilde{I}(f)$ is at least n, but as we shall see below, it can sometimes be greater than n depending on the location of the integration points.

If we take the interpolating polynomial to have degree zero, that is

$$p_0(x) = p(x_0) \tag{14a}$$

for some x_0, we obtain

$$\tilde{I}(f) = I(p_0) = (b-a)\, f(x_0). \tag{14b}$$

The usual choices of x_0 are

$$x_0 = a \quad \text{or} \quad x_0 = b \quad \text{or} \quad x_0 = (a+b)/2. \tag{14c}$$

In the first case we have the so called *left rectangular* rule

$$\tilde{I}(f) = (b-a)\, f(a) \tag{15a}$$

for which

$$E(f) = (b-a)^2 f'(\eta)/2 \tag{15b}$$

for some η in the interval [a, b]. Similarly, the second case yields the *right rectangular* rule

$$\tilde{I}(f) = (b-a)\, f(b) \tag{16a}$$

for which

$$E(f) = -(b-a)^2 f'(\eta)/2. \tag{16b}$$

For the third case we obtain the *midpoint* rule

$$\tilde{I}(f) = (b-a)\, f\frac{(a+b)}{2} \tag{17a}$$

for which

$$E(f) = (b-a)^3 f''(\eta)/24. \tag{17b}$$

As it is apparent, both rectangular rules have degree zero but the midpoint rule has degree one, an unexpected bonus. Also its error looks the most favourable of the three.

Increasing the degree of the interpolating polynomial to one and selecting the interpolating points as

$$x_0 = a \quad \text{and} \quad x_1 = b \tag{18a}$$

we arrive at the *trapezoidal* rule

$$\tilde{I}(f) = \frac{(b-a)}{2}[f(a) + f(b)] \tag{18b}$$

for which

$$E(f) = -(b-a)^3 f''(\eta)/12. \tag{18c}$$

This formula has an error that looks about twice the error of the midpoint rule (note that the value of η need not be the same in both formulas). Since the trapezoidal rule requires two function evaluations instead of one as for the midpoint rule, we may be led to the conclusion that the former is inferior. This is certainly true if the trapezoidal rule is used as above, but if used as a compound rule the conclusion is not so clear-cut.

The next formula is derived by choosing a polynomial of degree two interpolating at

$$x_0 = a, \quad x_1 = (a+b)/2 \quad \text{and} \quad x_2 = b. \tag{19a}$$

The resulting formula is the celebrated *Simpson's* rule

$$\tilde{I}(f) = \tfrac{1}{6}(b-a)[f(a) + 4f(a+b) + f(b)] \tag{19b}$$

for which

$$E(f) = -(b-a)^5 f^{(4)}(\eta)/2880. \tag{19c}$$

This formula possesses degree three, one more than we should expect.

Proceeding in this fashion, formulas of increasing degree can be derived. When the integration points are chosen to be equidistant, as is usual, we obtain the so called Newton-Cotes integration formulas whose errors are given by the expressions

$$E(f) = c_k (b-a)^{(2k+1)} f^{(2k)}(\eta) \tag{20a}$$

for $2k-1$ integration points and

$$E(f) = d_k (b-a)^{(2k+1)} f^{(2k)}(\eta) \tag{20b}$$

for $2k$ integration points, for some η in $[a, b]$ and where c_k and d_k are some numerical constants. Hence it can be confirmed that formulas with an odd number of integration points yield a bonus in their degree.

3.2.2 Gauss Formulas

As we have said in the previous section, integration rules derived via the interpolating polynomial of degree n which interpolates at $n+1$ pre-selected points have degree at least n. Some formulas like the midpoint or Simpson's for instance show that the degree can be greater than n. Thus it is legitimate to raise the question of what is the highest degree that can be achieved with a formula employing $n+1$ integration points $x_0, x_1, \ldots, x_n$, assuming that we are free to locate them in the interval $[a, b]$. The answer is provided by the following .

Theorem 1. The degree of an integration formula of type (9) employing $n+1$ distinct integration points cannot exceed $2n+1$.

Proof. See [Davis and Rabinowitz, Engels].

In the so called *Gauss* type integration formulas the integration points and weights are chosen in such a way that this upper bound is attained. The construction of these formulas is intimately related to the theory of orthogonal polynomials and is outside the scope of this chapter. We limit ourselve to present the more relevant properties of Gauss formulas from the applications standpoint.

Theorem 2. Let $w(x) > 0$ be a weight function defined over the interval $[a, b]$. Then the Gauss integration formula

$$\tilde{I}(f, w) = \sum_{i=0}^{n} A_i f(x_i) \qquad \text{21a)}$$

associated with the integral

$$I(f, w) = \int_a^b w(x) f(x) \, dx \tag{21 b}$$

has positive weights and the integration points are the $n+1$ distinct real zeros of the orthogonal polynomials (with respect to weight function $w(x)$). This zeros are located in the interval $[a, b]$. The error of this formula is given by

$$E_n(f, w) = \frac{f^{(2n+2)}(\eta)}{(2n+2)! \, K_{n+1}^2} \tag{21 c}$$

where K_n is the leading coefficient of the n-th orthogonal polynomial.
Proof. See [Davis and Rabinowitz, Engels].

References [Abramowitz and Stegun, Krylov, Stroud and Secrest] provide extensive tables for the abscissas x_i and weights A_i for the most common Gauss type formulas. We mention that the case $w(x) = 1$, which yields the *Gauss-Legendre* rules, is the most frequent in BEM. Also it must be pointed out that the Gauss type formulas are derived with respect to a standard interval, usually $[-1, 1]$ or $[0, 1]$ and thus a change of coordinates has to be performed to map $[a, b]$ onto this interval.

3.2.3 Singular Integrals

We have remarked in the introduction that BEM makes use of integrands that might be unbounded. It is then necessary to reformulate the usual definition of the integral to accomodate bounded as well as unbounded functions. It is convenient to have the singular integrals written in the form

$$I(f, w) = \int_a^b w(x) f(x) \, dx \tag{22}$$

where $f(x)$ is smooth and $w(x)$ carries the singularity of the integrand.

Definition 1. Let $f(x)$ and $w(x)$ be functions over $[a, b]$, where $f(x)$ is smooth and $w(x)$ is unbounded for $x \to a$. We define

$$I(f, w) = \int_a^b w(x) f(x) \, dx = \lim_{\varepsilon \to 0} \int_{a+\varepsilon}^b w(x) f(x) \, dx \tag{23}$$

whenever the above limit exists. A similar definition applies when $w(x)$ becomes unbounded for $x \to b$.

There are instances where the integrand is unbounded at a point c in the interior of $[a, b]$. In this case we have

Definition 2. Let $f(x)$ and $w(x)$ be functions over $[a, b]$, where $f(x)$ is smooth and $w(x)$ is unbounded for $x \to c$, where c is an interior point of $[a, b]$. We define the Cauchy principal value of the integral as

$$I(f, w) = \int_a^b w(x) f(x)\, dx = \lim_{\varepsilon \to 0} \left[\int_a^{c-\varepsilon} w(x) f(x)\, dx + \int_{c+\varepsilon}^b w(x) f(x)\, dx \right] \tag{24}$$

whenever the above limit exists.

The accurate calculation of singular integrals is critical for the success of BEM so we shall give a brief description of the techniques available to deal with such integrals.

a) Integrable Singularities

It may happen that that both limits in the right hand side of expression (24) exist by themselves, in which case we say that the singularity is integrable and Definition 1 applies. Whenever this is the case the best course is to look for Gauss type rules that have $w(x)$ as weight function thereby taking automatically care of the singularity. Such formulas exist for some of the integrands appearing in BEM. [Mineur, Stroud and Secrest] present tables for $w(x) = \ln(1/x)$ which have application in two dimensional potential and elasticity problems.

When no integration rules exist for the type of singularity at hand, a technique that can be sometimes be used is to "subtract out" the singularity [Davis and Rabinowitz]. Suppose we wanted to compute the integral in expression (23). We can write it as

$$I(f, w) = \int_a^b w(x) [f(x) - f(a)]\, dx + f(a) \int_a^b w(x)\, dx. \tag{25}$$

The integrand in the first integral of the right hand side of this expression is continuous and thus we are in the presence of a regular integral that can be computed numerically by using some suitable integration formula. If the second integral can be evaluated analytically then the problem of calculating the value of expression (23) is solved.

Example 1. Let

$$I(f, w) = \int_0^1 \ln(1/x) f(x)\, dx. \tag{26a}$$

Then employing the above method we can write

$$I(f, w) = \int_0^1 \ln(1/x) [f(x) - f(0)]\, dx + f(0) \int_0^1 \ln(1/x)\, dx. \tag{26b}$$

But from elementary calculus we know that

$$\int_0^1 \ln(1/x)\, dx = [x - x \ln x]_0^1 = 1 \tag{26c}$$

and

$$\lim_{x \to 0} \ln(1/x) [f(x) - f(0)] = 0 \tag{26d}$$

this later relation showing that for smooth $f(x)$ the first integral in expression (26 b) is indeed regular.

b) Cauchy Principal Value Integrals

For integrals existing only in the Cauchy principal value sense the method of "subtracting out the singularity" can still be applied. We illustrate with an example.

Example 2. Let

$$I(f, w) = \int_a^b f(x)/x \, dx \tag{27a}$$

where the interval $[a, b]$ contains the origin. According to Definition 2 and elementary calculus we can put

$$I(f, w) = \lim_{\varepsilon \to 0-} \left[\int_a^{-\varepsilon} f(x)/x \, dx + \int_\varepsilon^b f(x)/x \, dx \right] \tag{27b}$$

$$= \lim_{\varepsilon \to 0+} \left[\int_a^{-\varepsilon} (f(x) - f(0))/x \, dx + \int_\varepsilon^b (f(x) - f(0))/x \, dx \right.$$

$$\left. + f(0) \int_a^{-\varepsilon} 1/x \, dx + f(0) \int_\varepsilon^b 1/x \, dx \right]. \tag{27c}$$

However for smooth functions $f(x)$ we have that

$$\lim_{x \to 0} [f(x) - f(0)]/x = f'(0). \tag{27d}$$

Moreover

$$\int_a^{-\varepsilon} dx/x = [\ln |x|]_a^\varepsilon = \ln \varepsilon/|a| \tag{27e}$$

$$\int_\varepsilon^b dx/x = [\ln |x|]_a^\varepsilon = \ln |b|/\varepsilon. \tag{27f}$$

Introducing this results in expression (27c) we conclude that

$$I(f, w) = \int_a^b (f(x) - f(0))/x \, dx + f(0) \ln |b/a|. \tag{27g}$$

c) Special Integration Rules

If the interval $[a, b]$ is symmetric relative to the location c of the singularity, then as [Piessens] has shown, Gauss-Legendre integration rules with an even number of integration points can be used without any further loss of accuracy. Should the interval not be symmetric we can split the integral in the following way

$$I(f, w) = \int_a^{c-s} (\cdot) + \int_{c-s}^{c+s} (\cdot) + \int_{c+s}^b (\cdot) \tag{28}$$

for some suitable positive s. We are thus left with two regular integrals and one Cauchy principal value integral over a symmetric interval. The implementation of this technique implies however a nontrivial departure from the usual methods of programming the BEM.

3.3 Integration for Two-Dimensional Problems

As we have mentioned in the introduction, the use of isoparametric elements of order greater than one requires that the integrals be evaluated numerically. In this section we present the details for the two dimensional case.

3.3.1 Regular Boundary Integrals

Let us then concern ourselves here with integrals of the following types

$$I(q, u^*, \boldsymbol{x}) = \int_{\Gamma} u^*(\boldsymbol{x}, \boldsymbol{x}')\, q(\boldsymbol{x}')\, d\Gamma(\boldsymbol{x}') \tag{29a}$$

$$I(u, q^*, \boldsymbol{x}) = \int_{\Gamma} q^*(\boldsymbol{x}, \boldsymbol{x}')\, u(\boldsymbol{x}')\, d\Gamma(\boldsymbol{x}') \tag{29b}$$

In the BEM the boundary curve is discretized in finite elements. This discretization can lead to an approximate boundary different from Γ. However, to keep the notation as simple as possible, we shall continue to refer to the discretized boundary as Γ. Over Γ the functions u or q are approximated by finite element functions and likewise we shall continue to denote these approximations by the same symbols. Figure 1 gives a schematic view of this process. According to the isoparametric technique, the details of which can be seen in [Zienckiewicz, Ciarlet], each finite element Γ_e is mapped onto a reference finite element $\hat{\Gamma}$. We represent this one-to-one mapping by

$$\boldsymbol{x} = \Phi_e(\hat{x}), \quad \Phi_e\colon \hat{\Gamma} \mapsto \Gamma_e. \tag{30}$$

Given such map, the restriction of any function $u(x)$ over Γ to Γ_e can be transformed into a function over $\hat{\Gamma}$. We shall write this by

$$[u(\boldsymbol{x})]_e = u(\Phi_e(\hat{x})). \tag{31}$$

As we have already noted, the integral (29a) above is a singular integral with an integrable singularity and the integral (29b) is a regular one. Thus we can put

$$I(q, u^*, \boldsymbol{x}) = \sum_e I_e(q, u^*, \boldsymbol{x}) \tag{32a}$$

$$I(u, q^*, \boldsymbol{x}) = \sum_e I_e(u, q^*, \boldsymbol{x}) \tag{32b}$$

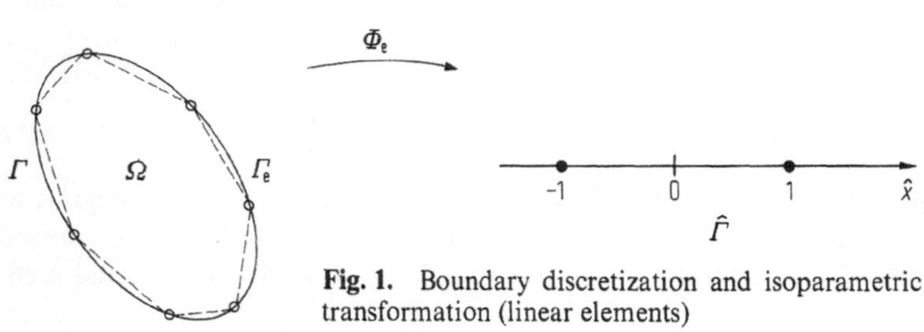

Fig. 1. Boundary discretization and isoparametric transformation (linear elements)

where

$$I_e(q, u^*, x) = \int_{\Gamma_e} u^*(x, x')\, q(x')\, d\Gamma_e(x') \tag{33 a}$$

$$I_e(u, q^*, x) = \int_{\Gamma_e} q^*(x, x')\, u(x')\, d\Gamma_e(x'). \tag{33 b}$$

Making the isoparametric change of coordinates we transform in the following way these integrals into integrals over the reference element,

$$I_e(u, q^*, x) = \int_{\hat{\Gamma}} q^*(x, x')\, u(x')\, J_e(\hat{x}')\, d\hat{\Gamma}(\hat{x}') \tag{34 a}$$

$$I_e(q, u^*, x) = \int_{\hat{\Gamma}} u^*(x, x')\, q(x')\, J_e(\hat{x}')\, d\hat{\Gamma}(\hat{x}') \tag{34 b}$$

where $J_e(\hat{x}')$ is the jacobian of the transformation which for the present case is

$$J_e(\hat{x}) = \left[\left(\frac{\partial \Phi_{e1}}{\partial \hat{x}}\right)^2 + \left(\frac{\partial \Phi_{e2}}{\partial \hat{x}}\right)^2\right]^{1/2}. \tag{35}$$

These integrals can be computed by any of the integration rules referred to in section 2, namely the Gauss-Legendre formulas.

3.3.2 Singular Boundary Integrals

When the collocation point belongs to the current element, the integral (29 a) develops a $\ln(1/r)$ type singularity while the integral (29 b) remains regular. To be able to use the formulas developed for logarithmic weight functions (see [Abramowitz and Stegun, Mineur]) we need to make some changes in integral (29 a). First we note that the integration formulas for this case apply to the interval [0, 1] which may not coincide with the reference element and therefore may require a special reference element and associate coordinate transformation. We denote this transformation by $\boldsymbol{\Phi}_{es}$. Also the singularities $\ln(1/r)$ and $\ln(1/\hat{x})$ are not the same. The following modification solves this difficulty. Putting

$$\ln(1/r) = \ln(\hat{x}/r) + \ln(1/\hat{x}) \tag{36}$$

we can write then

$$I_e(q, u^*, x) = I_{er}(q, u^*, x) + I_{es}(q, u^*, x) \tag{37}$$

with

$$I_{er}(q, u^*, x) = \int_0^1 \frac{1}{2\pi} \ln(\hat{x}/r)\, q(x')\, J_e(\hat{x}')\, d\hat{x}'. \tag{38 a}$$

$$I_{es}(q, u^*, x) = \int_0^1 \frac{1}{2\pi} \ln(1/\hat{x})\, q(x')\, J_e(\hat{x}')\, d\hat{x}'. \tag{38 b}$$

Since $\hat{x}/r(\hat{x})$ remains bounded when $x' \to x$, the integral (38 a) is regular. The singular integral (38 b) has now the proper form.

3.3.3 Domain Integrals

In this section we examine the computation of the domain integrals arising from the source term in expression (7a). The integrand includes singular functions and therefore it is necessary to assess the meaning of such integrals. The following definition given in [Mikhlin] generalizes the concept of Cauchy principal value to multi-dimensional integrals.

Definition 3. Let Ω be a manifold in R^n and $f(x)$ a smooth function over Ω and $w(\boldsymbol{x}, \boldsymbol{x}')$ a function over Ω which becomes unbounded as $\boldsymbol{x}' \to \boldsymbol{x}$, and $B_\varepsilon(x)$ a ball of radius ε centered at x. We put

$$
\begin{aligned}
I(f, w) &= \int_\Omega w(\boldsymbol{x}, \boldsymbol{x}')\, f(\boldsymbol{x}')\, d\Omega(\boldsymbol{x}') \\
&= \lim_{\varepsilon \to 0} \int_{\Omega_\varepsilon} w(\boldsymbol{x}, \boldsymbol{x}')\, f(\boldsymbol{x}')\, d\Omega(\boldsymbol{x}')
\end{aligned} \tag{39}
$$

with $\Omega_\varepsilon = \Omega - B_\varepsilon(\boldsymbol{x})$, whenever the above limit exists. $I(f, w)$ is called the singular integral of f over Ω with weight w.

In [Mikhlin] conditions for the existence of singular integrals are also given which show that singularities of the $\ln(1/r)$ type are integrable in the above sense for any bounded domain so that we can write

$$
\begin{aligned}
I(f, w) &= \int_\Omega w(\boldsymbol{x}, \boldsymbol{x}')\, f(\boldsymbol{x}')\, d\Omega(\boldsymbol{x}') \\
&= \sum_e \int_{\Omega_e} w(\boldsymbol{x}, \boldsymbol{x}')\, f(\boldsymbol{x}')\, d\Omega(\boldsymbol{x}') \\
&= \sum_e I_e(f, w)
\end{aligned} \tag{40}
$$

where the summation is taken over all domain finite elements. This means that we can take one element at a time as usual.

a) Regular Integrals

When the collocation point $\boldsymbol{x}$ does not belong to the finite element Ω_e we are in the presence of a regular integral, for which we can extend without difficulty the technique described above for one dimensional isoparametric elements. Denoting by $\hat{\Omega}$ the reference element and by J_e the jacobian of the coordinate transformation we have that

$$
\begin{aligned}
I_e(f, w) &= \int_{\Omega_e} w(\boldsymbol{x}, \boldsymbol{x}')\, f(\boldsymbol{x}')\, d\Omega(\boldsymbol{x}') \\
&= \int_{\hat{\Omega}} w(\boldsymbol{x}, \boldsymbol{x}')\, f(\boldsymbol{x}')\, J_e(\hat{\boldsymbol{x}}')\, d\hat{\Omega}(\hat{\boldsymbol{x}}').
\end{aligned} \tag{41}
$$

In [Stroud] there are various integration formulas for triangles and squares that can be used to calculate the right hand side of this expression.

b) Singular Integrals

When the collocation point $\boldsymbol{x}$ belongs to the finite element Ω_e the integral $I_e(f, w)$ becomes singular. To evaluate it we can proceed in the following manner. Making the usual isoparametric transformation we can write that

$$\int_{\Omega_e} \frac{1}{2\pi} \ln(1/r)\, f(\boldsymbol{x}')\, d\Omega(\boldsymbol{x}')$$
$$= \int_{\hat{\Omega}} \frac{1}{2\pi} \ln(1/r)\, f(\boldsymbol{x}')\, J_e(\hat{x}')\, d\hat{\Omega}(\hat{\boldsymbol{x}}') = I_{es} + I_{er}$$

where

$$I_{es} = \int_{\hat{\Omega}} \frac{1}{2\pi} \ln(1/r)\, g(\boldsymbol{x}')\, d\hat{\Omega}(\hat{\boldsymbol{x}}')$$
$$I_{er} = \int_{\hat{\Omega}} \frac{1}{2\pi} \ln(\hat{r}/r)\, g(\boldsymbol{x}')\, d\hat{\Omega}(\hat{\boldsymbol{x}}') \tag{42a}$$
$$g(\boldsymbol{x}') = f(\boldsymbol{x}')\, J_e(\hat{\boldsymbol{x}}').$$

The second integral in the right hand side is regular and therefore we can employ some adequate two dimensional integration formula, as for instance, those given in [Stroud]. The first integral in expression above is however singular. To exemplify how to evaluate it we consider specifically the case of linear triangles. Changing to polar coordinates r and θ in the isoparametric plane (see Fig. 2), we transform this integral accordingly

$$I_{es} = \int_0^{\alpha} \int_0^{R(\theta)} r \ln(1/r)\, g(r, \theta)\, d\theta\, dr$$
$$= \int_0^{\alpha} d\theta \int_0^{R(\theta)} r \ln(1/r)\, g(r, \theta)\, dr. \tag{42b}$$

Since the function $r \ln(1/r)$ is bounded as $r \to 0$, the integral in r is regular and we can apply one dimensional integration rules to it. Similarly the integration in the θ coordinate can be performed by one dimensional rules. Thus this integral can be evaluated by a product integration rule obtained as a cartesian product of two rules, one in the r coordinate and the other in the θ coordinate.

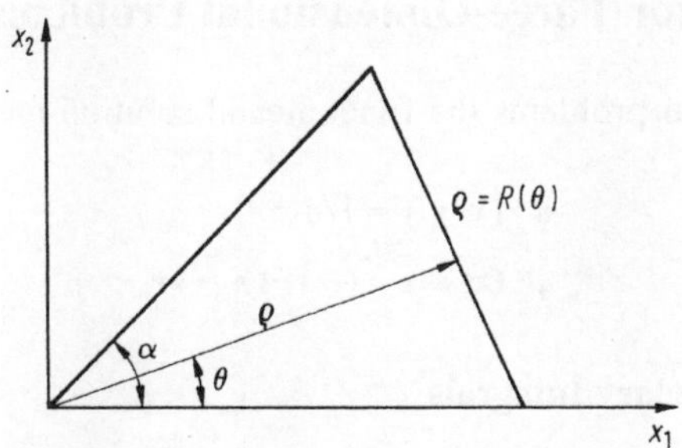

Fig. 2. Evaluation of singular integrals using polar coordinates (linear triangles). The singularity is at the origin

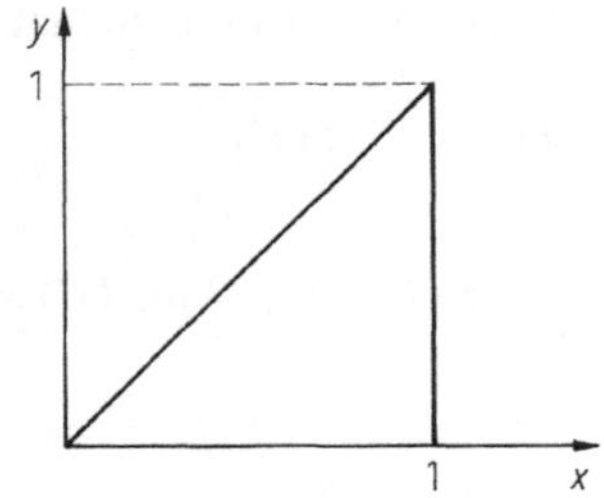

Fig. 3. Special element for singular integration. The singularity is at the origin

To avoid the awkward upper limit of integration in expression (42b) the following idea can be used. Instead of the standard reference element we resort to a special element as shown in Fig. 3 and make the change of coordinates introduced in [Duffy]

$$\hat{x} = \xi \quad \text{and} \quad \hat{y} = \xi \, \eta \tag{43a}$$

Then as can be easily verified,

$$I_{es} = \int_0^1 d\hat{x} \int_0^1 \ln(1/\hat{r})\, g(\hat{x}')\, d\hat{\Omega}$$

$$= \int_0^1 d\xi \int_0^1 \ln(1/\hat{r})\, g(\xi, \xi\,\eta)\, d\eta$$

$$= \int_0^1 \int_0^1 \xi \ln(1/\hat{r})\, g(\xi, \xi\,\eta)\, d\xi\, d\eta \tag{43b}$$

with

$$\hat{r} = (\hat{x}^2 + \hat{y}^2)^{1/2} = \xi\,(1 + \eta^2)^{1/2}. \tag{43c}$$

Not only has the singularity disappeared but we have now an integration over a square which can be dealt with by cartesian product rules or any adequate two dimensional rules as those given in [Stroud] already referred to.

3.4 Integration for Three-Dimensional Problems

For three dimensional problems the fundamental solution and associated flux are respectively

$$u^*(\boldsymbol{x}, \boldsymbol{x}') = 1/4\,\pi\, r \tag{44a}$$

$$\boldsymbol{q}^*(\boldsymbol{x}, \boldsymbol{x}') = (-1/4\,\pi\, r^3)\, \boldsymbol{r} \tag{44b}$$

3.4.1 Regular Boundary Integrals

All that was said above for the two dimensional case applies here with only minor adaptations.

3.4.2 Singular Boundary Integrals

The singularity $1/r$ is integrable so we can proceed as before, calculating the integrals in each element by itself. We can write then

$$I_{es}(q, u^*) = \int_{\Gamma_e} \frac{1}{4\pi\hat{r}} (\hat{r}/r)\, q(\boldsymbol{x}, \boldsymbol{x}')\, J_e(\hat{\boldsymbol{x}}')\, d\Gamma(\hat{\boldsymbol{x}}') \tag{45}$$

or more simply by putting

$$g(\boldsymbol{x}') = (\hat{r}/4\pi r)\, q(\boldsymbol{x}, \boldsymbol{x}')\, J_e(\hat{\boldsymbol{x}}') \tag{46 a}$$

$$I_{es}(q, u^*) = \int_{\Gamma_e} (1/\hat{r})\, g(\boldsymbol{x}')\, d\Gamma(\hat{\boldsymbol{x}}') \tag{46 b}$$

Special integration formulas for the common isoparametric elements with a $1/r$ singularity have been developed by [Critescu and Loubignac, Pina et al.] which can be applied to this case.

3.4.3 Domain Integrals

a) Regular Integrals

Everything that we have said for the two dimensional case applies here with straightforward generalizations.

b) Singular Integrals

The techniques explained for the two dimensional domain integrals can also be used in principle for three dimensional problems, its pratical implementation is however less straightforward. [Duffy] however explains how to apply a transformation of coordinates similar to (43a) to a pyramid with a singularity at the vertex and also to a cube with a singularity at a corner. Combining this with an isoparametric transformation over a suitable reference element we can evaluate numerically the singular integrals arising in three dimensional BEM.

Acknowledgements

The author thanks J. L. M. Fernandes for several helpful discussions concerning the subject of this chapter. This work was partially supported by Instituto Nacional de Investigação Cientifica, Lisbon.

References

1 Abramowitz, M. and Stegun, I.A., Handbook of Mathematical Functions with Formulas, Graphs and Mathematical Tables, Dover 1968
2 Brebbia, C.A., Telles, J., and Wrobel, L., Boundary Element Techniques, Theory and Applications in Engineering. Springer Verlag, Berlin, NY, 1984
3 Brebbia, C.A., The Boundary Element Method for Engineers, Pentech Press, London 1978
4 Ciarlet, P.G., The Finite Element Method, North-Holland 1979

5 Critescu, M. and Loubignac, G., Gaussian Quadrature Formulas for Functions with Singularities in 1/R over Triangles and Quadrangles, Recent Advances in Boundary Element Methods, C.A. Brebbia (ed.), Pentech Press 1978

6 Davis, P.J. and Rabinowitz, P., Methods of Numerical Integration, Academic Press 1975

7 Duffy, M.G., Quadrature Over a Pyramid or Cube of Integrals with a Singularity at a Vertex, SIAM J. Numer. Anal., **19,** 6, 1260–1262, 1982

8 Engels, H., Numerical Quadrature and Cubature, Academic Press 1980

9 Jawson, M.A. and Symm, G. T., Integral Equations Methods in Potential Theory and Elastostatics, Academic Press 1977

10 Krylov, V.I., Approximate Calculation of Integrals, MacMillan, 1962

11 Mineur, H., Techniques de Calcul Numerique, Dunod, 1966

12 Mikhlin, S.G., Multidimensional Singular Integrals and Integral Equations, Pergamon, 1965

13 Piessens, R., Numerical Evaluation of Cauchy Principal Values, BIT, **10,** 476–480, 1970

14 Pina, H.L.G., Fernandes, J.L.M., and Brebbia, C.A., Some Numerical Integration Formulae over Triangles and Squares with a $1/R$ Singularity, Appl. Math. Modeling, **5,** 209–211, 1981

15 Stroud, A.H. and Secrest, D., Gaussian Quadrature Formulas, Prentice-Hall, 1966

16 Stroud, A.H., Approximate Calculation of Multiple Integrals, Prentice-Hall, 1971

17 Zienkiewicz, O.C., The Finite Element Method in Engineering Science, McGraw-Hill, 1971

Chapter 4

Computational Aspects of the Boundary Element Method

by M. Doblaré

4.1 Introduction

In this chapter, some specific aspects of the Boundary Element Method (B.E.M.) will be examined from a computational perspective. The most common techniques, used in the implementation of a computer B.E. program will also be briefly discussed.

The paucity of available information on this topic is surprising, specially in view of the extensive body of information, concerning the B.E.M. in general. We think that this is mainly due to the fact that in the short period of time, elapsed since the publication of the first papers on the method, most research efforts have been directed towards expanding its applications to an increasing number of different situations in Engineering and Mathematical Physics.

For computational problems, on the other hand, the tendency has been to employ already existent general routines, rather than develop particular techniques (e.g. the computational apparatus for the Finite Element Method – F.E.M. – has been widely used).

The theoretical characteristics of the B.E.M., however, are completely different from those of that method, so, although some specific techniques have been already developed (numerical integration schemes, treatment of boundary conditions, etc.), much work remains to be done in these fields, as well as in other topics which have not been so well studied (e.g. the solution of the system of linear equations). For these reasons, it seems certain that the succesive development of new specific techniques for the B.E.M. will remarkably improve its applications.

The domain methods are well established, and have been profusely applied to many different problems in Physics and Engineering, however there are certain areas where the disadvantages of these methods vis-a-vis the B.E.M. are obvious (e.g. unbounded domains, stress concentrations, singularities, etc.). Although attempts have been made in order to improve the shortcomings of these methods in such situations, no completely satisfactory solution have been found [1]. For these and also for other cases, the B.E.M. has some important advantages, mainly the necessity of discretizing only the boundary of the domain, while for domain methods the entire volume must be discretized. This means a smaller amount of input data and consequently a valuable reduction in the time required for its preparation, which is a fundamental aspect in many problems (3-D). On the other hand, the followers of the domain methods claim that the ease of developing a F.E.M. computer program, and the greater speed in solving the same problem (which is not usually the case) argue for the use of domain methods instead of boundary ones. However, the use of progressively faster computers, and the reduction in the cost of C.P.U. time in micro and minicomputers makes the first of the advantages mentioned above more attractive, compared with the previous so-called disadvantages. It is possible then to foresee a substantial increase in the fields of application for the B.E.M., displacing the F.E.M. in some areas and complementing it in others. With this in mind, it is easy now to understand the phrase of Wendland [2]: 'it is the combination of both methods which seems to promise a new generation of extremely efficient numerical methods'.

To study the computational topics which arise in any numerical method, different approaches can be followed. For example, a specific computer program can be written for a particular application or, the different available techniques to solve each case can be analyzed. Both extremes have advantages and disadvantages; if the first approach is easier for the nonexpert reader, the second option is more convenient for the specialist. In this chapter, we have opted to follow an intermediate alternative. We will describe different techniques for solving the computational problems of the B.E.M., but we will also study, in greater detail, those which have been used in a particular computer program.

We have to decide now what type of program to describe. In this case, we have chosen a somewhat sophisticated program which shows the reader, regardless his background, some interesting features of the method. The program selected solves the problem on an elastic axisymmetric medium subjected to axisymmetric loads acting on the plane r, z, by using the so-called Direct B.E.M. Several reasons prompted us to chose this problem. First of all, it includes not only all the main characteristics which appear in a classical B.E.M. case, but also others that the 2-D and 3-D cases do not (i.e. semianalytical integration). Secondly, the implementation of such a program would allow the reader to solve not only the axisymmetric case, but also the plane strain problem by simply placing the axis of symmetry at an adequate distance from the section to be analyzed.

In Sect. 2 the basic elasticity equations for the Direct B.E.M. will be briefly examined, including several references to other important applications. Section 3 will study the organization of a typical B.E.M. computer program, and the special characteristics of the one enclosed. Each of the following sections is related to one of the main parts of any B.E. program: storage and management of the data, input of these data, computation of the matrices of the problem, solution of the system of linear equations. Moreover, each section explains some computational particularities concerning other applications, with the reader sometimes being directed to appropriate additional references. They also include some parts of the above-mentioned program, which allow a better understanding of the explanations.

Finally, it should also be pointed out that for the purposes of this analysis, it will be assumed that the reader is sufficiently familiarized with the theoretical aspects of the B.E.M. in the fields under discussion. Some basic references for the axisymmetric case may be found in [3–7].

4.2 Formulation and Numerical Treatment

The initial expression for the solution of the elastostatic case using the Direct B.E.M. is the well-known Somogliana's identity, which expresses the displacement vector $u_i(\boldsymbol{x})$ at any point $\boldsymbol{x}$ of a domain Ω as a function of the displacements $u_i(\boldsymbol{y})$ and tractions $t_i(\boldsymbol{y})$ at points $\boldsymbol{y}$ of the boundary $\delta\Omega$ of this domain, and the body forces X_i.

$$(\delta_{ik} - C_{ik})\, u_i(\boldsymbol{x}) = \int_{\delta\Omega} U_{ik}(\boldsymbol{x},\boldsymbol{y})\, t_i(\boldsymbol{y})\, dS_y - \int_{\delta\Omega} T_{ik}(\boldsymbol{x},\boldsymbol{y})\, u_i(\boldsymbol{y})\, dS_y + \int_{\Omega} U_{ik}(\boldsymbol{x},\boldsymbol{y})\, X_i(\boldsymbol{y})\, d\Omega_y. \tag{1}$$

The expressions for U_{ik}, T_{ik}, C_{ik} may be found in Hartmann [6]. The reader should bear in mind that this system of equations is a singular integral equations one, due to the singularity of the kernel T_{ik}, and the corresponding integral must be understood in terms of its Cauchy Principal Value (see Mikhlin [8]).

By applying the Navier's operator to (1) it is possible to compute the stresses at any internal point of Ω.

$$\sigma_{ij}(\boldsymbol{x}) = \int_{\delta\Omega} D_{ijk}(\boldsymbol{x},\boldsymbol{y})\, t_k(\boldsymbol{y})\, dS_y - \int_{\delta\Omega} S_{ijk}(\boldsymbol{x},\boldsymbol{y})\, u_k(\boldsymbol{y})\, dS_y + \int_{\Omega} D_{ijk}(\boldsymbol{x},\boldsymbol{y})\, X_i(\boldsymbol{y})\, d\Omega_y; \quad \boldsymbol{x} \in \Omega. \tag{2}$$

The expressions for D_{ijk}, S_{ijk} are given in [6] for the two-dimensional and three-dimensional cases, and in [7] for the axisymmetric one. It should be also pointed out that it is not possible to apply (2) when $\boldsymbol{x} \in \delta\Omega$ due to the strong singularity of the tensor S_{ijk}, which does not allow the use of an efficient numerical approach. An interesting way to compute these stresses, when necessary is shown in Hartmann [6].

Another important topic concerns the domain integrals which appear in (1) and (2), related to the body forces X_i. For the general case it would be necessary to discretize the whole domain, in order to perform the numerical computation of these integrals. Fortunately, in most cases, these body forces satisfy the following restrictions:

$$\operatorname{div} X = X_{i,i} = K_0 = \text{const.}$$
$$\operatorname{curl} X = \nabla \wedge X = 0. \tag{3}$$

It can then be shown [3] that the volume integral in (1) may be expressed as the sum of the two boundary integrals:

$$\int_{\Omega} U_{ik}(\boldsymbol{x},\boldsymbol{y})\, X_i(\boldsymbol{y})\, d\Omega_y = \int_{\delta\Omega} E_{ik}(\boldsymbol{x},\boldsymbol{y})\, X_i(\boldsymbol{y})\, dS_y + K_0 \int_{\delta\Omega} F_k(\boldsymbol{x},\boldsymbol{y})\, dS_y \tag{4}$$

where E_{ik} and F_k may be found, for the three cases in Elasticity (bi, tri-dimensional and axisymmetric) in [3]. In the same way (2) is modified under conditions (3) as:

$$\int_{\Omega} D_{ijk}(\boldsymbol{x},\boldsymbol{y})\, X_k(\boldsymbol{y})\, d\Omega_y = \int_{\delta\Omega} G_{ijk}(\boldsymbol{x},\boldsymbol{y})\, X_k(\boldsymbol{y})\, dS_y + K_0 \int_{\delta\Omega} H_{ij}(\boldsymbol{x},\boldsymbol{y})\, dS_y \tag{5}$$

where G_{ijk} and H_{ij} may be found again in Doblaré [3].

Unlike the Theory of Plasticity and some approaches to dynamical problems, in which a domain integral, which cannot be reduced to the boundary, appears, the other applications do not include these integrals, provided that the body forces satisfy (3), or its equivalent, which happens in most cases.

Analogous formulations to (1)–(5) can be developed for other applications. We will mention only those sources which summarizes the main aspects and provide additional references: Plate Theory [6], [9], Theory of Shallow Shells [9], Elastodynamics [10], Potential Theory [11], Time-Dependent Potential Theory [12], Electromagnetism [13], Theory of Plasticity [14].

In nonlinear problems, it is usually necessary to employ an incremental approach, the equations are therefore written in an incremental form. For the Theory of Plasticity, for example, the equivalent equations to (1) and (2) are given by:

$$(\delta_{ik} - C_{ik})\, \dot{u}_i(\boldsymbol{x}) = \int_{\delta\Omega} U_{ik}\, \dot{t}_i\, dS_y - \int_{\delta\Omega} T_{ik}\, \dot{u}_i\, dS_y + \int_{\Omega} U_{ik}\, \dot{X}_i\, d\Omega_y + \int_{\Omega} \Sigma_{ijk}\, \dot{\varepsilon}^P_{ij}\, d\Omega_y \tag{6}$$

$$\dot{\sigma}_{ij}(\boldsymbol{x}) = \int_{\delta\Omega} D_{ijk}\, \dot{t}_k\, dS_y - \int_{\delta\Omega} S_{ijk}\, \dot{u}_k\, dS_y + \int_{\Omega} D_{ijk}\, \dot{X}_k\, d\Omega_y + \int_{\Omega} \Sigma_{ijkl}\, \dot{\varepsilon}^P_{kl}\, d\Omega_y + I\, \dot{\varepsilon}^P_{ij}(\boldsymbol{x}) \tag{7}$$

where the dots denote the increment of the corresponding variable. The expressions for the tensors Σ_{ijk}, Σ_{ijkl} so as for the constant I, as well as the way for arriving at them are shown in [14] for the 2-D and 3-D and in [7] for the axisymmetric one. In (6) and (7) $\dot{\varepsilon}^P_{ij}(\boldsymbol{x})$ denotes the increment of plastic strains at the point $\boldsymbol{x} \in \Omega$. Alternative formulations to that, proposed in (6), (7) may be found in [14], [15].

The next step is then to solve the system of equations (1) or its equivalents. In general, it is not possible to obtain the solution of (1) in a closed form, except for simple geometries and boundary conditions. Therefore an approximate method must be used.

The B.E.M. assumes that the unknown variables (u_i and t_i at the boundary, in Elasticity) are approximated in the form:

$$\Phi_i = \sum_{k=1}^{K} N^k_{\Phi_i}\, \Phi^k_i \tag{8}$$

where $N^k_{\Phi_i}$ are explicit known functions (usually polynomial splines) defined over the boundary (or domain), the so-called 'approximation functions', and Φ^k_i are the new unknowns of the problem. Of course if K tends to infinity and Φ_i belongs to a complete space of function for which $N^k_{\Phi_i}$ $(k=1, 2, \ldots)$ form a basis, equation (8) would not be an approximation but an infinite expansion of Φ_i. An important additional property of these approximation functions is that their support is closely linked to the definition of the geometry of the boundary, as will be shown.

In most real cases it is not possible to define the global geometry of the domain surface by a simple expression. Rather, a set of 'elements' with a simple geometry is considered a preferable form of definition. Therefore it is easy to establish an application between each of these elements and a closed subset of R^{n-1} (n is the dimension of the space Ω). Even so, the geometry of the boundary is sometimes so

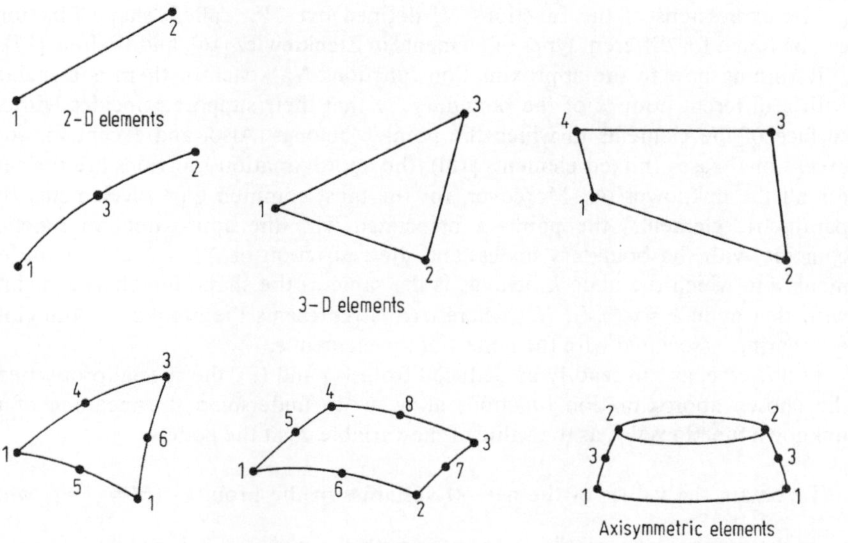

Fig. 1

complex that too many elements would require to define it. The process is thus rendered uneconomical, and even sometimes impossible. In these cases, an approximate geometry is used; the boundary is defined in the manner described above, with a relatively small number of elements, provided that the geometric approximation is good enough to obtain accurate results.

The geometry is usually defined by the coordinates of a number of points inside it ('nodes') as is shown in Fig. 1.

As was said before the most important property of the elements is the possibility of establishing a simple application h: $V_e \to \delta\Omega_e$, where V_e is a closed subset of R^{n-1}. The expression of this application for the elements used in the B.E.M. is given in Fig. 2.

$$V_e \to \delta\Omega_e;\; x_i^e(\xi_j) = \sum_{l=1}^{L_e} x_i\,[d(e,l)]\,N_x^l(\xi_j)\,. \tag{9}$$

$i = 1, 2, \ldots n$

$j = 1, 2, \ldots n-1$

L_e = number of nodes of the element e

$d(e, l)$ = global number of the node l of the element e.

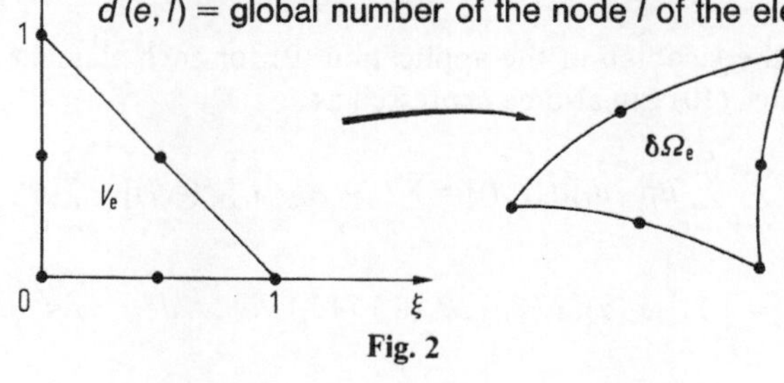

Fig. 2

The expressions of the functions N_x^l defined over V_e, called 'shape functions', can be found for different types of elements in Zienkiewicz [16], and Brebbia [17].

Returning now to the approximation functions $N_{\Phi_i}^k$, each of them is associated with a different point k of the boundary, so that their support coincides with the surface of the elements to which the point k belongs. Also, and except for some exceptional cases (mixed elements [18]), the approximation functions are the same for all the unknowns Φ_i. Moreover, for the most common type of elements (isoparametric elements) the points k associated with the approximation functions coincide with the boundary nodes, and the restriction of $N_{\Phi_i}^k$ to each of the elements e to which the node k belongs is the same as the shape functions associated with that node $k = d(e, l)$, N_x^l, where $d(e, l)$ represents the number, in the global numbering, associated with the node 1 of the element e.

In this case, as can readily be deduced from (8) and (9), the special properties of the chosen approximation functions allow us to understand the meaning of the unknown functions Φ_i^k, as the value of the variable Φ_i at the node k.

To obtain the values of the new M variables of the problem ($M = \sum_{k=1}^{K} v_k$, where v_k is the number of variables associated with the node k), it is necessary to use a numerical algorithm. For the B.E.M. the most frequently used have been: Galerkin [2], [19], Collocation-Galerkin [2], [20, 21], and because of its simplicity and lesser number of integrations required, Collocation. From now on, we will be concerned only with this last method.

By establishing the fulfilment of the Eq. (1) at the K nodes of the boundary, $n \times K$ equations are obtained which, together with the given boundary conditions, allow us to solve the problem. Assuming that the discretization of the boundary includes p elements, the Eq. (1) can be expressed for a certain node α as

$$[\delta_{ik} - C_{ik}(\alpha)]\, u_i(\alpha) + \sum_{e=1}^{p} \sum_{l=1}^{L_e} u_i[d(e,l)] \int_{V_e} T_{ik}[\boldsymbol{x}(\alpha), \boldsymbol{y}(\xi_j)]\, N_\Phi^l(\xi_j)\, J(\xi_j)\, d\xi_j \ldots d\xi_{n-1}$$

$$= \sum_{e=1}^{p} \sum_{l=1}^{L_e} t_i[d(e,l)] \int_{V_e} U_{ik}[\boldsymbol{x}(\alpha), \boldsymbol{y}(\xi_j)]\, N_\Phi^l(\xi_j)\, J(\xi_j)\, d\xi_1 \ldots d\xi_{n-1}$$

$$+ \sum_{e=1}^{p} \sum_{l=1}^{L_e} \left[\int_{V_e} E_{ik}[\boldsymbol{x}(\alpha), \boldsymbol{y}(\xi_j)]\, X_i(\xi_j)\, J(\xi_j)\, d\xi_i \ldots d\xi_{n-1} \right.$$

$$\left. + K_0 \int_{V_e} F_k[\boldsymbol{x}(\alpha), \boldsymbol{y}(\xi_j)]\, J(\xi_j)\, d\xi_1 \ldots d\xi_{n-1} \right] \tag{10}$$

where $J(\xi_j)$ is the jacobian of the application (9) for each element e. By grouping some of the terms, (10) can also be expressed as

$$\sum_{e=1}^{p} \sum_{l=1}^{L_e} B_{ik_{el}}^{\alpha}\, u_i[d(e,l)] = \sum_{e=1}^{p} \sum_{l=1}^{L_e} A_{ik_{el}}^{\alpha}\, t_i[d(e,l)] + \sum_{e=1}^{p} P_{k_e}^{\alpha} \tag{11a}$$

$$B_{ik_{el}}^{\alpha} = \int_{V_e} T_{ik}[\boldsymbol{x}(\alpha), \boldsymbol{y}(\xi_j)]\, N_\Phi^l(\xi_j)\, J(\xi_j)\, d\xi_1 \ldots d\xi_{n-1} \quad \alpha \neq d(e,l)$$

$$B^{\alpha}_{ik_{el}} = \int_{V_e} T_{ik}[\boldsymbol{x}(\alpha), \boldsymbol{y}(\xi_j)]\, N^l_{\Phi}(\xi_j)\, J(\xi_j)\, d\xi_1 \dots d\xi_{n-1} + [\delta_{ik} - C_{ik}(\alpha)]\ \alpha = d(e, l)$$

$$A^{\alpha}_{ik_{el}} = \int_{V_e} U_{ik}[\boldsymbol{x}(\alpha), \boldsymbol{y}(\xi_j)]\, N^l_{\Phi}(\xi_j)\, J(\xi_j)\, d\xi_1 \dots d\xi_{n-1} \tag{11b}$$

$$P^{\alpha}_{k_e} = \int_{V_e} \{E_{ik}[\boldsymbol{x}(\alpha), \boldsymbol{y}(\xi_j)]\, X_i(\xi_j) + K_0\, F_k[\boldsymbol{x}(\alpha), \boldsymbol{y}(\xi_j)]\}\, J(\xi_j)\, d\xi_1 \dots d\xi_{n-1}$$

while by writing (11) for $\alpha = 1, 2, \dots K$ and collecting the different terms, the system of linear Eq. (12) is obtained.

$$\boldsymbol{B\,u} = \boldsymbol{A\,t} + \boldsymbol{P} \tag{12}$$

with $n \times K$ equations and M variables. Now $v_k - n$ boundary conditions for each node k are required in order to solve the problem. Alternative Eq. to (10)–(12) can be obtained for other applications (see references).

Once the system (12) has been solved, the Eq. (1), (2), can be applied for computing the displacement and stresses at internal points.

In the theory of plasticity, it is necessary to extend the approximation (8) to the volumen unknowns $\dot{\varepsilon}^P_{ij}$, in order to compute the volume integrals which appear in (6) and (7). The approximation functions $N^k_{\dot{\varepsilon}^P}(\xi_i;\, i)\ 1, 2, \dots n)$ are now associated to an internal node h, $h \in \bar{\Omega}$, and its support coincides with the internal cells containing h. The geometry of these cells is also defined by the application (9) over a closed subset $U_c \subset R^n$.

By following an identical procedure to that used for the elastic case, the expression below is obtained.

$$\boldsymbol{B\,\dot{u}} = \boldsymbol{A\,\dot{t}} + \boldsymbol{\dot{P}} + \boldsymbol{C\,\dot{\varepsilon}}^P \tag{13}$$

where the matrix C is defined as

$$C^{\alpha}_{ijk_{cb}} = \int_{V_c} \Sigma_{ijk}[\boldsymbol{x}(\alpha), \boldsymbol{y}(\xi_i)]\, N^b_{\dot{\varepsilon}^P}(\xi_i)\, J'(\xi_i)\, d\xi_1 \dots d\xi_n \tag{14}$$

with $J'(\xi_i)$ the jacobian of the mapping of U_c into Ω_c.

The system of nonlinear Eq. (13) cannot be solved using only the boundary conditions, but some other equations must be used. The increment of plastic strains $\dot{\varepsilon}^P_{ij}$ can be expressed in function of $\dot{\sigma}_{ij}$ by means of the chosen plastic flow rule. If this expression is substituted in (7) the increment of plastic strains at the internal nodes can be expressed in function of $\boldsymbol{u}, \boldsymbol{t}$ at the boundary nodes, being possible now to solve (13). A detailed exposition of this procedure may be found in [14, 15].

4.3 A Boundary Element Program Organization

The general problem of developing a computer program, capable of solving a specific problem can be best understood when subdivided into the following basic topics:

(1) Theoretical approach.
(2) Theoretical and computational numerical analysis.

(3) Development of the organization of the program.
(4) Programming techniques and development of the final programs.
(5) Checking and examples.

The first topic, and part of the second, have been analyzed in previous sections. Some other aspects of the second point will be treated in subsequent epigraphs. The third subject will be studied in these lines, along with some particular problems of the programming techniques. Finally, the development and checking of the final programs, and the collection of all the information needed by the user to manage the program, require a specific treatment for each program, and will therefore not be included in this chapter. The interested reader can find some of these aspects for particular programs in [3, 22].

The final objective is then to obtain an effective, easy-to-use and easy-to-modify computer program. For this, one must develop an exhaustive study of the three fundamental parts which can be distinguished in any program:

- Input and management of the data.
- Computation.
- Treatment of the results.

The strategy adopted varies, depending on the complexity of these parts, but the final goal is always arrive at a fast-running, generally-adaptable and easy-modifiable computer program. The limitations of the computer being used must also be taken into consideration.

There are three main possibilities:

(1) An in-core program. This is, no doubt, the simplest and most effective in terms of time, solution, although it is limited by the amount of data and the capacity of the computer.
(2) An integrated system of different programs. This is the best solution for very general, large-dimensioned programs. The complexity and high cost of this scheme (similar solutions in F.E. analysis required about 10 – 100 man. years) must, however, be taken into account.
(3) Intermediate structures, which can be closer or farther away from one of the two previous solutions, depending on the particular characteristics of the problem.

The final decision will depend on the dimensions and generality of the required program.

A B.E. computer program entails long and complex computations and usually a large amount of data. (The dimensions of the data structure will be proportional to the number of elements and nodes of the discretization, so that, for a medium size-computer with 64 available K-words, the maximum number of elements with constant approximation functions in axisymmetric elasticity is about 200. This can be insufficient for some applications.) Due to this and also in order to show some useful computation techniques, a type 3-approach has been chosen. Type 2 has been discarded because of its complexity (for such a solution see Danson and Brebbia [23]), while most of the B.E. programs already developed have used a type 1-scheme.

The main topics to be considered here are:

(1) Establishment of the structure of the overall program, its division in modules, the interfaces among them and the connections with the data structure.

(2) Establishment of the data structure and the access procedures from each of the programming modules.

The first point will now be taken up, while the second will be studied in next sections.

Without any doubt, one of the most important recommendations of the 'Structured Programming' is the idea of 'modularity'. This means that the program must be composed of a set of independent modules managed by another one, which controls the running of the whole program. Some of the advantages of this scheme are:

– Flexibility during the programming, checking and running phases due to the independence of the modules.

– Possibility of new enlargements and modifications with a minimum of time and difficulty.

– For computers which allow a segmentation procedure, it is possible to obtain a substantial reduction of the needed memory by overlapping the different modules during the running of the program.

The general scheme for the program is then composed by a Main Module and several Procedure Modules, which can be run independently. The Main Module is usually composed of a Main Program and a Control Routine. The first defines the in-core areas size, while the second initializes the control areas for the storage arrays (see Sect. 4), and links the rest of the modules during the running of the program. This module can also include some other general routines used in different Procedure Modules. It also must be the only one accessible to the user, so that he can modify the dimensions needed for the storage areas, in light of the problem being solved. For some computers, which allow the user to modify the dimensions of the arrays during the running of the program, this module should also be inaccessible with all the needed parameters being defined in the input data.

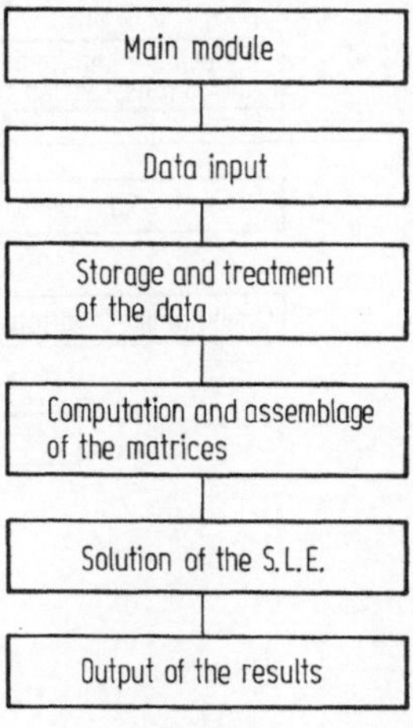

Fig. 3

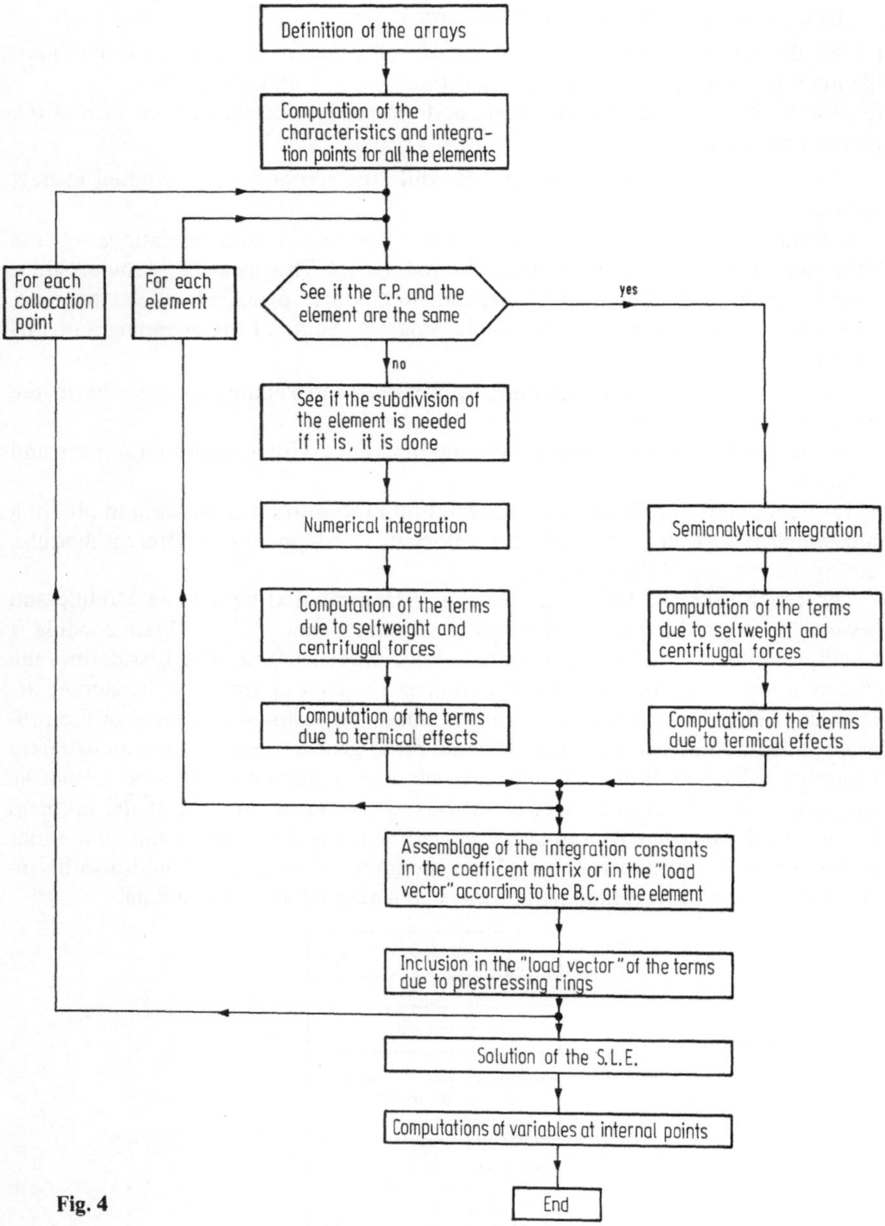
Definition of the arrays
Computation of the characteristics and integration points for all the elements
For each collocation point
For each element
See if the C.P. and the element are the same
yes
no
See if the subdivision of the element is needed if it is, it is done
Numerical integration
Computation of the terms due to selfweight and centrifugal forces
Computation of the terms due to termical effects
Semianalytical integration
Computation of the terms due to selfweight and centrifugal forces
Computation of the terms due to termical effects
Assemblage of the integration constants in the coefficent matrix or in the "load vector" according to the B.C. of the element
Inclusion in the "load vector" of the terms due to prestressing rings
Solution of the S.L.E.
Computations of variables at internal points
End

Fig. 4

The Procedure Modules are usually composed of a general routine with no arguments, which controls the work of a set of other subroutines, either specific for this module or shared with others. A typical scheme for such a B.E. program is shown in Fig. 3.

Each of the proposed modules will be studied in detail below.

The enclosed program has been written in accordance with the previous scheme, and its general flow diagram has been shown in Figure 4. It is written in FORTRAN V, has been implemented in a UNIVAC 1108, and is composed of about 100 routines and 3000 lines.

```
C
C                             *************************
C                             *   PROGRAM   AXELAST   *
C                             *************************
C
C      ANALYSIS OF AXISYMMETRIC ELASTIC STRUCTURES UNDER AXISYMMETRIC LOADS
C      BY USING THE BOUNDARY ELEMENT METHOD
C
C
C
C
C
C************************** VARIABLES IN COMMON /CONST/ *******************
C
C      LEC----- READING LOGIC UNIT IN THE PROGRAM
C      IMP----- PRINTING LOGIC UNIT IN THE PROGRAM
C      MEMO---- NUMBER OF INTEGER WORDS FOR ARRAYS IN COMMON /CIA/
C      MEMOB--- NUMBER OF INTEGER WORDS FOR ARRAYS IN COMMON /CIAB/
C      NBP----- NUMBER OF BITS PER INTEGER WORD
C      NPER---- NUMBER OF INTEGER WORDS PER REAL VARIABLE
C      NPEDP--- NUMBER OF INTEGER WORDS PER DOUBLE PRECISION VARIABLE
C      NCON---- NUMBER OF DISJOINT BOUNDARIES OF THE PROBLEM
C      NUDOSC-- TOTAL NUMBER OF BOUNDARY NODES
C      NPI----- NUMBER OF INTERNAL POINTS FOR WHICH THE SOLUTION IS COMPUTED
C      NUDOS--- TOTAL NUMBER OF NODES: NUDOSC+NPI
C      NELEM--- NUMBER OF BOUNDARY ELEMENTS
C      NAPOYO-- NUMBER OF DEGREE OF FREEDOM WITH RESTRICTIONS
C      IPARAG-- THIS PARAMETER SAYS IF THERE IS SYMMETRY WITH RESPECT TO
C               THE Z AXIS (IPARAG=1) OR NOT (IPARAG=0)
C      YOUNG--- YOUNG'S MODULUS
C      POIS---- POISSON RATIO
C      PORANA-- PERCENTAGE OF THE LENGTH OF EACH ELEMENT USED FOR ANALYTICAL
C               INTEGRATION
C
C**************************************************************************
C
C
C
C
C**************************** COMMON /CIA/ VARIABLES **********************
C
C     NELE(NCON)---------------- NUMBER OF ELEMENTS FOR EACH BOUNDARY
C     COOR(2,NUDOS)------------- NODE COORDINATES
C     CONTOR(3,NELEM)----------- BOUNDARY CONDITIONS FOR EACH ELEMENT
C     NANU(NAPOYO)-------------- DEGREES OF FREEDOM WITH RESTRICTIONS
C     CAREL(15,NELEM)----------- CHARACTERISTICS OF INTEGRATION FOR EACH
C                                BOUNDARY ELEMENT
C     VEC(4*NELEM)-------------- RESULTS ALONG THE BOUNDARY
C     MOVIN(2*NPI)-------------- DISPLACEMENTS FOR EACH INTERNAL POINT
C     TENIN(4*NPI)-------------- STRESSES FOR EACH INTERNAL POINT
C
C**************************************************************************
C
C
C
C
C
C************************** COMMON /CIAB/ VARIABLES ***********************
C
C     MATRIX A(2*NELEM,2*NELEM)-- MATRIX OF THE SYSTEM OF EQUATIONS
C
C**************************************************************************
```

```
C************************ DEFINITION OF MEMO AND MEMOB *******************
C
C      MEMO=3*NUDOS+NCON+22*NELEM+NAPOYO+6*NPI+40
C      MEMOB=35000-MEMO
       PARAMETER MEMO=2000,MEMOB=7000
C
C*************************************************************************
C
C
C
C          DIMENSIONING AND DEFINITION OF ARRAYS
C
       COMMON /CIA/ IA(MEMO)
       COMMON /CIAB/ IAB(MEMOB)
       COMMON /CONST/ IAC(14),AC(3)
       COMMON /WORK/ IAD(1)
       COMMON /CPUT/ TCPU,CPUT,CPU1,CPU2,CPU3,CPU4,CPU5,CPU6,CPU7,CPU8
C
C          READING AND PRINTING LOGIC UNITS
C
       LEC=5
       IMP=6
C
C          CHARACTERISTICS OF THE COMPUTER
C
       NBP=36
       NPER=1
       NPEDP=2
C
C          OPENING OF THE FILES
C
       CALL ERTRAN(6,'@ASG,T AXPLAS*1')
C
C          IDENTIFICATION OF THE VARIABLES OF THE COMMON /CONST/
C
       IAC(1)=LEC
       IAC(2)=IMP
       IAC(3)=MEMO
       IAC(4)=MEMOB
       IAC(5)=NBP
       IAC(6)=NPER
       IAC(7)=NPEDP
C
C          CALL TO THE 'MASTER' ROUTINE
C
       CALL MASTER
C
C          CLOSING THE FILES
C
       CALL ERTRAN(6,'@FREE AXPLAS*1')
C
C          END OF THE PROCESS
C
       STOP
       END

       SUBROUTINE MASTER
C
C      'MASTER' ROUTINE. THIS SUBROUTINE CALLS THE DIFFERENT MODULES OF
C      THE PROCESS
C
C
C          DIMENSIONING AND DEFINITION OF THE ARRAYS
C
       COMMON /CIA/ IA(1)
       COMMON /CIAB/ IAB(1)
       COMMON /CONST/ LEC,IMP,MEMO,MEMOB,NBP,NPER,NPEDP,NCON,NUDOSC,
      *               NPI,NUDOS,NELEM,NAPOYO,IPARA6,YOUNG,POIS,PORANA
       COMMON /WORK/ NFIL1,NCOL1,NR1,NTR1
       COMMON /CPUT/ TCPU,CPUT,CPU1,CPU2,CPU3,CPU4,CPU5,CPU6,CPU7,CPU8
C
C          INITIALIZATION OF THE CONTROL AREA OF COMMON /CIA/
C
       CALL IAINIC(MEMO)
C
C          GENERAL DATA OF THE PROGRAM
```

```
C
      CALL DATOS
C
C           DEFINITION OF THE PARAMETERS OF THE FILES
C
      NFIL1=2*NELEM
      NCOL1=MEMOB/NFIL1
      NTR1=NCOL1*NFIL1
      NR1=NFIL1/NCOL1+1
C
C           INITIALIZATION OF THE CONTROL AREA OF COMMON /CIAB/
C
      CALL IABIN(NR1,NTR1)
C
C           ASSEMBLAGE OF THE MATRICES
C
      CALL MATELA
C
C           SOLUTION PROCESS
C
      CALL SOLVE
C
C           PRINTING THE RESULTS ON THE BOUNDARY
C
      CALL ELACON
C
C           DISPLACEMENTS AND STRESS AT INTERNAL POINTS
C
      CALL ELAINT
C
C           END OF THE PROCESS
C
      RETURN
      END
```

4.4 Storage and Management of Data

In Sect. 3, the importance of the data structure was made clear, as was the relationship between this structure and the architecture of the whole program. Therefore, once the overall scheme for the program has been defined, the largest arrays to be managed must be studied. The possible interrelationships and overlappings among these data must likewise be examined with a view towards determining the most adequate data structure for the problem.

4.4.1 Variables in a Boundary Element Program

Although there are many arrays and variables to be used in a B.E. program, this section will review only those that, in light of their dimensions, merit careful consideration. It will begin with the arrays used in every B.E. program regardless of its application or complexity, and later will examine some other arrays and variables which are only used in specific ones.

Among the first group are the following:

- Co-ordinates of the Boundary Nodes.
- Connectivity between Boundary Nodes and Boundary Elements (in 2-D with correlative numbering this array is not needed).
- Boundary Conditions.
- Matrix of the System of Linear Equations (S.L.E.).
- Vector of the Right Hand Side in the S.L.E.

All of them are large arrays which are used in different parts of the program. However, the coefficient matrix of the S.L.E. is undoubtedly the largest, differing from the rest in one additional important characteristic: the two times it appears (configuration and solution of the S.L.E.) it is used sequentially, unlike the other arrays (except for the vector of independent terms) which are used randomly throughout the program. From this, if it is necessary to employ auxiliary memory to store all these arrays, it is generally more convenient to use an in-core storage for all the arrays, unlike the coefficient matrix that will need in most occasions a mass memory device for its storage.

There are many other arrays that can appear even in the simplest of B.E. programs, but they are generally small ones with fixed dimensions used in specific parts of each module. For them, any of the well-known in-core storage techniques can be used.

For more complex applications (Plasticity, Piecewise heterogeneous Bodies, etc.), other large arrays are needed. Each of them must also be analyzed, in light of its dimensions and use along the program, in order to decide how to manage and store it. For instance, in the case of subregionalization the following are needed: the characteristics of the materials which compose each of the homogeneous regions; an array of pointers which indicate the initial and last elements of each of these regions (assuming correlative numbering), an other one for the interfaces between subregions; an array to store the diagonal terms of the submatrices (see 6.1.4), etc.

In Plasticity the following additional information is needed: the coordinates of the interior nodes: the connectivity of the interior cells; the matrices D, S, Σ in (7); and the vectors of plastic strains and stresses at the internal nodes. (It must not be forgotten that for each array of results in Plasticity, two identical storage arrays are needed, to account for the incremental and global results.)

Finally, for very sophisticated programs, it is possible that some other auxiliar arrays may be needed (e.g. to help during the imposition of boundary conditions, or to include some other properties of the elements). Here it is necessary to decide how to store them, according to their specific features. Situations of this kind can be found in [3, 7, 22, 24] and some case will be commented on Sect. 6.

4.4.2 Data Structure

Once the overall scheme for the data structure and the specific treatment that each array will receive have been commented. We can define the former's characteristics more precisely. For this, it is absolutely necessary to consider the kind of computer the program is going to be run on (if it has virtual memory or if it does not. Whether it has buffers with fixed or variable dimensions, what are the possibilities for reassigning the array's dimensions during the running of the program, etc.) Each programmer then, must design the best data structure for his specific situation.

As it is not possible to study here all of the data structures which can be used, the reader is referred to the monographies on this subject by Felippa [25, 26], Bettes [27]. Date [28], Kamel et al. [29], Katzan [30], McCornick and Wright [31] with additional references.

We will consider here for illustrative purpose that data structure adopted in the attached program. Its main characteristics are:

(1) Use of auxiliary memory with fixed size registers and fixed number of them for each direct access file.
(2) Restrictions on the dimensions of the in-core storage.
(3) Easy-checking and access to each array.

In general, it can be said that these characteristics are common to many programs designed for both mini and microcomputers. The following detailed scheme can therefore be useful for many of these situations.

To solve these three problems we are going to introduce the following concepts:

– Use of the auxiliary storage area by means of a variable number of buffer areas, which allows an access to the data independent of the type or size of the files and computer.
– Dynamic assignation of the in-core storage is available during the running of the program.
– Use of a set of routines which allow a simple checking process of any array at any time.

There are two main approaches to manage large arrays. The first one uses an in-core storage area in which part of the whole array is stored, and which is refilled when it is necessary to gain access to a different part of the array not stored in that area. This approach is specially effective when the access to the array is made sequentially, so that the number of times that the buffer is changed is minimal. Moreover the scheme is very simple.

For arrays which need random access, the former system becomes inefficient. In these cases a 'pagination scheme' is more effective. Basically it involves subdividing the in-core storage area into several equally dimensioned parts, the so-called 'pages', so that each of them contains a part of an array or even several arrays at any time. In order to gain access to a specific element, it is necessary to replace one of the pages with another, that includes the sought element. This process, as one would imagine, is more complex than the first, but it is also more general and allows a greater freedom for managing large arrays. It is usually used by Operative Systems which have a Virtual Memory Scheme. With it, the external user does not need to worry about the size of the computer. He must bear in mind however that some problems may arise from an indiscriminate use of this system, specially the possibility that the whole system might get jammed. To avoid this happening, a careful design of the data structure is necessary.

The method proposed here, which we consider well adapted to the characteristics of the B.E.M. and the restrictions initially assumed for the computer being used, is a combination of the two set forth above; it is possible then to get the simplicity of the buffer area method together with some of the efficiency of the pagination scheme for nonsequential processes. Basically it consists of assigning several buffer areas to each array managed by the system, while the actual arrays are stored in one direct access file with registers of the same size as the buffer areas. The information that is needed to run the system is controlled by an overall control area for the whole system and other particular ones (one for each array to be treated by this process). All of these control and buffer areas are included in a common in-core 'working area'.

A slightly different scheme involves defining several direct access files, one for each array (they can be sequential files if the arrays to store are managed in that way), so that each buffer area can have a different size, although all of them are kept in the same working area. Where only one buffer area is defined for each array, this approach is very similar to the pagination scheme, with the pages varying in size. On the other hand, if there is just one array with only one buffer area, the method is exactly the same as the first one mentioned above. A scheme for the general and particular control areas for the second of the two cases explained is shown below.

General Control Area

(1) Dimension of the general control area.
(2) Dimension of the particular control area.
(3) Size of the registers of the first file.
(4) Size of the registers of the second file.
⋮
(2 + N) Size of the registers of the Nth file.
(3 + N) Number of registers of the first file.
⋮
(2 + 2 N) Number of registers of the Nth file.
(3 + 2 N) Number of filled files at the current time.

Particular Control Areas

(1) Name of the array.
(2) – (4) Dimensions N 1 × N 2 × N 3 of the array.
(5) Number of integer words hold by each element of the array.
(6) Number of the file in which the array is stored.

The use of one or another system depends on the number, size and treatment of each array as well as on the optimum size of the register which is required by each of them. When there are many relatively small arrays the first scheme is preferred, while for a few large ones with different optimum register size the second approach is better.

The number of buffer areas used by each array, depends on the way this array is used along the program. Arrays with almost sequential access will require just one buffer area (matrix of coefficients). Arrays with simultaneous double sequential access will need two buffer areas (coordinates of the boundary nodes) and arrays with almost random access will precise a greater number of smaller areas. The scheme chosen for the proposed program is that where, just one buffer area is allotted to each array (in this case only one file has been scheduled).

The management of this system is controlled by a set of routines with the following tasks:

– Initialization of the global control area (subroutine IABIN).
– Creation of a specific array (subroutine CREARF). This fills the particular control area of the array and reserves the corresponding size in the working array.
– Location of a specific array in the working area (Function LOCAF).
– Location of a specific element in the working area, and if necessary, replacing

the buffer area with the register in which this element is located (Subroutines IOB, IOBF).

With this system we have an identical management for each array regardless of its characteristics. The format is also very similar to that used for the in-core arrays. Furthermore, the system is very general, so that the same programming scheme can be used for any computer, with but a small number of changes in the interface routines, in places such as the number of buffer areas for each array, the number of files, the size of the registers, etc. (This scheme was first proposed for the B.E.M. by A. Martin [32], and the author has used it profusely [3, 7], proving its efficiency.)

With respect to the system that should be developed for an effective management of the available in-core area, there exist many different possibilities. We will limit this discussion to that which has been chosen for the proposed program. It consists of using all the available area as a "working array", in which all the arrays needed by the program are included. Under this option, it is possible to create, use and erase at any time any of the arrays in this area by means of a set of routines (IAINIC, CREAR, LOCA, BORRA). This area is also managed by a global control area and several particular ones – one for each array – as is shown below.

General Control Area
- (1) Dimension of the general control area.
- (2) Dimension of the particular control areas.
- (3) Dimension of the working area already filled.
- (4) Dimension of the whole working area.
- (5) Number of arrays, presently stored.

Particular Control Areas
- (1) Name of the array.
- (2) Dimension of the array.

This system provides an optimum use of the available in-core memory at any moment, which is singularly important for programs with great capabilities and large dimensions, so that in these programs the memory required while running is very variable.

Both the set of buffer areas and the in-core working area are located in a COMMON area defined in the main program, so that it is possible to adjust the memory by changing the dimensions of these COMMON areas in the main program. The definition of the size of each buffer area and consequently of the characteristics of each of the files is performed during the running of the program, according to the input data (see Sect. 3).

Finally, the last requirement of those proposed above, corresponds to the development of a set of routines which allow for an easy checking of each array at any time. This is accomplished using some printing routines, without parameters, so that they can be included at any point of the program during the checking phase. Some of these routines are included in this chapter (see Sect. 7).

As was previously mentioned, this data structure may be modified for other particular programs and computers; it is however flexible and general enough to be easily adapted to almost any situation. For this reason, we feel it to be very useful for any kind of program we are concerned with here.

```
      SUBROUTINE IAINIC(MEMO)
C
C     INITIALIZES THE GLOBAL CONTROL AREA OF THE WORKING COMMON /CIA/
C
C******************** DICTIONARY
C
C     MEMO  - TOTAL SIZE OF THE AREA /CIA/
C     IA(1) - SIZE OF THE GLOBAL CONTROL AREA
C     IA(2) - SIZE OF THE PARTICULAR CONTROL AREA FOR EACH ARRAY
C     IA(3) - CURRENT NUMBER OF INTEGER WORDS FILLED IN /CIA/
C     IA(4) - MEMO
C     IA(5) - CURRENT NUMBER OF ARRAYS IN /CIA/
C
C*******************************
C
C              DEFINITION OF THE PARAMETERS OF THE AREA /CIA/
C
C
      COMMON /CIA/ IA(5)
      COMMON /CONST/ LEC,IMP,MEMO,MEMOB,NBP,NPER,NPEDP
      DO 10 I=1,MEMO
   10 IA(I)=0
      ITAMA=6/NPEDP
      ITAMA=(ITAMA+1)*NPEDP
      IA(1)=ITAMA
      IA(2)=NPEDP
      IA(3)=IA(1)
      IA(4)=MEMO
      IA(5)=0
C     IA(6) TO IA(ITAMA) - NOT USED
      RETURN
      END
```

```
      SUBROUTINE CREAR(NAME,N1,N2,N3,NENTE,IER)
C
C     FILLS THE PARTICULAR CONTROL AREA FOR THE ARRAY "NAME" IN THE
C     WORKING AREA /CIA/ AND RESERVES IN THIS AREA THE WORDS FOR THAT ARRAY
C
C******************** DICTIONARY
C
C     NAME     = NAME OF THE ARRAY TO BE CREATED
C     N1,N2,N3 = DIMENSIONS OF THE ARRAY
C     NENTE    = NUMBER OF INTEGER WORDS PER ELEMENT OF THE ARRAY
C     IER      = ERROR CODE
C                IER > 0 NORMAL EXIT. THE ARRAY HAS BEEN CREATED, OR THIS
C                        ARRAY ALREADY EXISTS. IER IS THE FIRST POSITION OF
C                        THE ARRAY IN /CIA/
C                IER < 0 THERE IS NOT ENOUGH SPACE IN /CIA/ TO INCLUDE THE
C                        ARRAY. IER WORDS ARE NEEDED
C
C*******************************
C
C
      COMMON /CIA/ IA(5)
      COMMON /CONST/ LEC,IMP,MEMO,MEMOB,NBP,NPER,NPEDP
      IF(IA(5).EQ.0) GO TO 11
      ILOC=LOCA(NAME)
      IF(ILOC.GT.0) GO TO 21
C
C              THE ARRAY IS NOT IN /CIA/.,IT CAN BE CREATED
C              THE TOTAL NUMER OF WORDS OF THE ARRAY MUST BE MULTIPLE OF
C              NPEDP IN ORDER TO PORTABILITY
C
   11 ITAMA=N1*N2*N3*NENTE
      ITOTD2=(ITAMA+IA(2))/NPEDP
      ITOTD2=(ITOTD2+1)*NPEDP
      ITAMA=ITOTD2-IA(2)
      ITOT=ITAMA+IA(2)+IA(3)
      IF(ITOT.GT.IA(4)) GOT TO 31
C
C              THERE IS ENOUGH SPACE IN /CIA/. THE ARRAY IS CREATED
C
      I=IA(3)
      IA(I+1)=NAME
      IA(I+2)=ITAMA
```

```
      IA(5)=IA(5)+1
      IA(3)=ITOT
      IER=(I+IA(2))/NENTE+1
      RETURN
C
C                 THE ARRAY IS ALREADY CREATED
C
   21 IER=ILOC
      RETURN
C
C                 THERE IS NOT ENOUGH SPACE IN /CIA/. THE ARRAY CAN NOT BE
C                 CREATED
C
   31 IER=IA(4)-ITOT
      WRITE(IMP,20000)NAME
      RETURN
20000 FORMAT(10X,10(1H*),'THE ARRAY ',A6,' CAN NOT BE CREATED.',
     *       ' THERE IS NOT ENOUGH SPACE IN THE WORKING AREA /CIA/')
      END
```

```
      SUBROUTINE BORRA(NAME)
C
C     ERASE THE ARRAY 'NAME' FROM THE WORKING AREA /CIA/
C
C******************** DICTIONARY
C
C     NAME = NAME OF THE ARRAY TO BE ERASED
C
C*******************************
C
C
      COMMON /CIA/ IA(5)
C
C                IF IA(5)=0 THERE IS NO ARRAY IN /CIA/
C
      IF(IA(5).EQ.0) RETURN
C
C                IF ILOC.LE.0 THE ARRAY IS NOT IN /CIA/
C
      ILOC=LOCA(NAME)
      IF(ILOC.LE.0) RETURN
C
C                SEE IF THE LAST ELEMENT OF THE ARRAY 'NAME' IS ALSO THE LAST
C                OF THE ARRAY /CIA/
C
      NI=ILOC+IA(ILOC-1)-1-IA(3)
      IF(NI.GT.0) GO TO 11
C
C                THE ARRAY IS NOT THE LAST OF THOSE STORED IN /CIA/. IT IS ERASED
C
      NI=ILOC+IA(ILOC-1)
      NF=IA(3)
      J=ILOC-IA(2)
      DO 10 I=NI,NF
         IA(J)=IA(I)
   10 J=J+1
      IA(3)=J-1
      GO TO 21
C
C                THE ARRAY IS THE LAST ONE STORED. IT IS ERASED
C
   11 IA(3)=ILOC-IA(2)-1
   21 IA(5)=IA(5)-1
      RETURN
      END
```

```
      FUNCTION LOCA(NAME)
C
C     LOCATES AN ARRAY IN /CIA/
C
C******************** DICTIONARY
C
C     NAME = NAME OF THE ARRAY TO BE FOUND
C
C*******************************
```

```
C
C
      COMMON /CIA/ IA(5)
      IF(IA(5).EQ.0) GO TO 11
      NC=IA(5)
C
C               SEARCHES THE NAME OF THE ARRAY IN /CIA/
C
      LOCA=IA(1)
      DO 10 K=1,NC
          IF(NAME.EQ.IA(LOCA+1)) GO TO 21
   10 LOCA=LOCA+IA(2)+IA(LOCA+2)
C
C               THE ARRAY HAS NOT BEEN FOUND
C
      LOCA=-1
      RETURN
C
C               THERE IS NO ARRAY IN /CIA/
C
   11 LOCA=0
      RETURN
C
C               THE FIRST POSITION OF THE ARRAY 'NAME' IS OBTAINED
C
   21 LOCA=LOCA+IA(2)+1
      RETURN
      END
```

```
      SUBROUTINE IABIN(NR1,NTR1)
C
C     INITIALIZATION OF THE CONTROL AREA FOR COMMON /CIAB/
C     AND DEFINITION OF THE FILE 1.
C
C*******************DICTIONARY
C
C     NR1           =NUMBER OF REGISTERS OF THE FILE 1.
C     NTR1          =SIZE OF EACH REGISTER OF FILE 1.
C******************************
C
C                   DIMENSIONING AND DEFINITION OF ARRAYS
C
      COMMON /CIAB/ IAB(10)
      COMMON /CONST/ LEC,IMP,MEMO,MEMOB,NBP,NPER,NPEDP,NCON,NUDOSC,
     *              NPI,NUDOS,NELEM,NAPOYO,IPARA6,YOUNG,POIS,PURANA
C
C                   DEFINITION OF FILES
C
      DEFINE FILE 1(NR1,NTR1,U,NR10)
C
C                  INITIAL PARAMETERS
C
      IAB(1)=10
      IAB(2)=10
      IAB(3)=NTR1
C     IAB(4)=NOT USED
C     IAB(5)=NOT USED
      IAB(6)=NR1
C     IAB(7)=NOT USED
C     IAB(8)=NOT USED
C     IAB(9)=NOT USED
      IAB(10)=0
      RETURN
      END
```

```
      SUBROUTINE CREARF(NOMBRE,N1,N2,N3,NENTE,NFILE,IER)
C
C     GENERATES THE CONTROL AREA OF AN ARRAY IN COMMUN /CIAB/
C
C*******************DICTIONARY
C
C     NOMBRE  =NAME OF THE ARRAY TO CREATE
C     N1,N2,N3=DIMENSIONS OF THE ARRAY TO CREATE
C     NENTE   =NUMBER OF INTEGER WORDS PER EACH ELEMENT OF THE ARRAY
C     NFILE   =NUMBER OF THE FILE WHERE THE ARRAY IS STORED
C     IER     =FIRST POSITION OF THE ARRAY IN COMMON /CIAB/
C******************************
```

```
C
C                  DIMENSIONING AND DEFINITION OF ARRAYS
C
      COMMON /CIAB/ IAB(10)
      COMMON /CONST/ LEC,IMP
      COMMON /WORK/ ILOC,ITAMA,NBL,IW,NFILM1
C
C                    LOCATION OF THE ARRAY IN COMMON /CIAB/
C
      IF(IAB(10).EQ.0) GO TO 1
      ILOC=LOCAF(NOMBRE)
      IF(ILOC.GT.0) GO TO 12
C
C                    THE ARRAY HAS NOT BEEN CREATED
C
    1 ITAMA=N1*N2*N3*NENTE
      NBL=ITAMA/IAB(2+NFILE)+1
      IF(NBL.GT.IAB(5+NFILE)) GO TO 31
C
C                  THERE IS ENOUGH STORING AREA. THE ARRAY IS CREATED
C
      IW=IAB(1)+(NFILE-1)*IAB(2)
      NFILM1=NFILE-1
      IF(NFILM1.EQ.0) GO TO 11
      DO 10 I=1,NFILM1
         IW=IW+IAB(2+I)
   10 CONTINUE
   11 IAB(IW+1)=NOMBRE
      IAB(IW+2)=N1
      IAB(IW+3)=N2
      IAB(IW+4)=N3
      IAB(IW+5)=NENTE
      IAB(IW+6)=NFILE
      IAB(IW+7)=0
C     IAB(IW+8)=NOT USED
C     IAB(IW+9)=NOT USED
C     IAB(IW+10)=NOT USED
      IAB(10)=IAB(10)+1
      IER=IW+1
      GO TO 101
C
C                  THE ARRAY IS ALREADY CREATED
C
   12 IER=ILOC
      GO TO 101
C
C               THERE IS NOT ENOUGH STORING AREA FOR THE ARRAY
C
   31 WRITE(IMP,20001)NOMBRE,NFILE,NBL
      IER=-NBL
C
C               END OF THE PROCESS
C
  101 RETURN
20001 FORMAT(10X,'THERE IS NOT ENOUGH STORING AREA FOR THE ARRAY',A6,' T
     *HE NUMBER OF NEEDED REGISTERS FOR FILE',A6,' IS',I5)
      END
```

```
      FUNCTION LOCAF(NOMBRE)
C
C     LOCATES THE FIRST POSITION OF AN ARRAY IN THE CONTROL AREA
C     OF COMMON /CIAB/
C
C****************** DICTIONARY
C
C     NOMBRE = NAME OF THE ARRAY TO LOCATE
C
C******************************
C
C
C               DIMENSIONING AND DEFINITION OF THE ARRAYS
C
      COMMON /CIAB/ IAB(10)
C
C               LOCATION OF THE ARRAY
```

```
C     IF(IAB(10).EQ.0) GO TO 11
      NC=IAB(10)
      LOCAF=IAB(1)+1
      DO 10 K=1,NC
            IF(NOMBRE.EQ.IAB(LOCAF)) GO TO 21
            LOCAF=LOCAF+IAB(2)+IAB(2+K)
   10 CONTINUE
C
C                 THE ARRAY HAS NOT BEEN CREATED
C
      LOCAF=-1
      RETURN
C
C                 THERE IS NO ARRAY IN THE CONTROL AREA
C
   11 LOCAF=0
   21 RETURN
      END
```

```
      FUNCTION IOB(NOMBRE,I,J,K)
C
C     LOCATION OF AN ELEMENT IN COMMON /CIAB/
C
C*******************DICTIONARY
C
C     NOMBRE=NAME OF THE ARRAY TO WHOM THE ELEMENT BELONGS
C     I,J,K =INDICES OF THE ELEMENT TO LOCATE
C*****************************
C
C
C                    DIMENSIONING AND DEFINITION OF ARRAYS
C
      COMMON /CIAB/ IAB(10)
C
C                    PARAMETERS OF THE BUFFER TO CONSIDER
C
      ILOC=LOCAF(NOMBRE)
      IF(ILOC.LE.0) GO TO 101
      NABUF=IAB(ILOC+5)
      IJK=((K-1)*IAB(ILOC+1)*IAB(ILOC+2)+(J-1)*IAB(ILOC+2)+I)*
     *IAB(ILOC+4)
      IBL=IJK/IAB(2+NABUF)+1
      IBU=IJK-(IBL-1)*IAB(2+NABUF)
      IF(IBU.NE.0) GO TO 11
      IBL=IBL-1
      IBU=IAB(2+NABUF)
C
C                    SEARCHING OF THE BUFFER
C
   11 IF(IBL.EQ.IAB(ILOC+6)) GO TO 21
C
C                    THE BUFFER IS NOT IN THE CONTROL AREA
C                    SEARCHING OF THE BUFFER TO CHANGE
C
      NR=IAB(ILOC+6)
      IAB(ILOC+6)=IBL
C
C                    INTERCHANGE OF THE BUFFERS
C
      KI=ILOC+IAB(2)
      KF=KI+IAB(2+NABUF)-1
      CALL IOBF (NR,IBL,KI,KF,1,NABUF)
      GO TO 31
C
C                    THE BUFFER IS IN THE CONTROL AREA
C
   21 KI=ILOC+IAB(2)
C
C                    POSITION OF THE ELEMENT UNDER CONSIDERATION
C
   31 IOB=(KI-1+IBU)/IAB(ILOC+4)
      RETURN
C
C                    THE ARRAY HAS NOT BEEN CREATED
C
  101 IOB=-1
      RETURN
      END
```

```
      SUBROUTINE IOBF(IRA,IRN,KI,KF,NCOD,NFILE)
C
C     MULTIPLE BUFFER AREAS METHOD
C     INPUT/OUPUT PROCESS FOR FILE NFILE.
C
C********************DICTIONARY
C
C     IRA  =NUMBER OF THE OLD BUFFER
C     IRN  =NUMBER OF THE NEW BUFFER
C     KI   =FIRST POSITION OF THE BUFFER IN THE CONTROL AREA
C     KF   =LAST POSITION
C     NCOD =PARAMETER THAT DEFINE THE TYPE OF OPERATION
C           <0 ONLY READING OF THE NEW BUFFER
C           >0 WRITTING OF THE OLD BUFFER AND
C              READING OF THE NEW ONE
C     NFILE=NUMBER OF THE STORING FILE
C
C********************
C
C     CONTROL AREA
C
      COMMON /CIAB/ IAB(10)
C
      IF(NCOD.LT.0)    GO TO 110
C
C     WRITTING THE OLD BUFFER
C
      IF(IRA.EQ.0)     GO TO 100
      WRITE(NFILE'IRA)(IAB(K),K=KI,KF)
      GO TO 110
C
C     READING THE NEW BUFFER
C
 100  WRITE(NFILE'IRN)(IAB(K),K-KI,KF)
 110  READ(NFILE'IRN,ERR=120)(IAB(K),K=KI,KF)
 120  RETURN
      END
```

4.5 Data Input

Once the different needed data for developing a B.E. program have been analyzed, we will be concerned with obtaining an efficient data input, mainly for those cases which need a great amount of initial data for their definition, specially the geometry and boundary conditions for 3-D bodies.

In fact, although the discretization with B.E. is much simpler than that with F.E. it is, nevertheless, cumbersome and requires tedious checking, when three-dimensional bodies have to be managed. Even for easier cases, such that 2-D and axisymmetric ones, it is possible to simplify this process, with a minimum extra-work, for those cases (which are the majority) which have a regular discretization and simple boundary conditions. This is made effective by an additional routine that acts as an automatic mesh generator. An example of such routine for plane (and axisymmetric) straight elements is shown along these lines.

A more complex problem occurs when we have curve planar elements or three-dimensional ones. In these cases however, some of the experience gained by previous F.E. researches in this area can be taken into account. We will describe along the next paragraphs one automatic mesh generation processor that we have developed for B.E. [33]. Specifically, this program generates 3-D parabolic

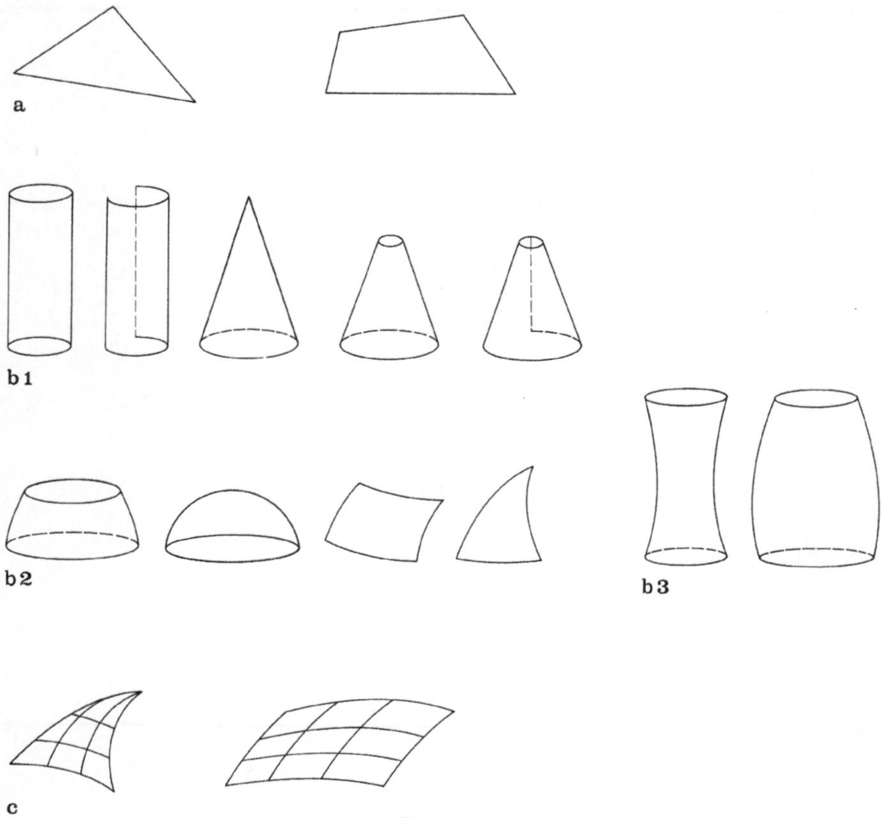

Fig. 5 a, b

boundary meshes, although many of the ideas can become useful for other problems in two dimensions or for simpler 3-D elements.

The basis of the program is the concept of isoparametric macro-element that, after subdivision according to certain rules, are adjusted to the actual geometry through the use of transfinite ("blended") interpolation as defined by Birkhoff et al. [34], Gordon [35], or Gordon et al. [36, 37].

The boundary then, is subdivided into several macro-elements S_i such that:

$$U S_i = S$$
$$S_i \cap S_j = L_{ij} \tag{15}$$

where S is the whole boundary and L_{ij} is the border between macro-elements i and j. Although it is not strictly necessary we will consider for simplicity a compatible division for S_i and S_j along the line L_{ij} (same number of divisions for both macro-elements along the line).

The choice of the macro-elements is generally dictated by the body shape and boundary conditions (we will consider macro-elements with the same kind of boundary condition for all its points). On the other hand the processor must con-

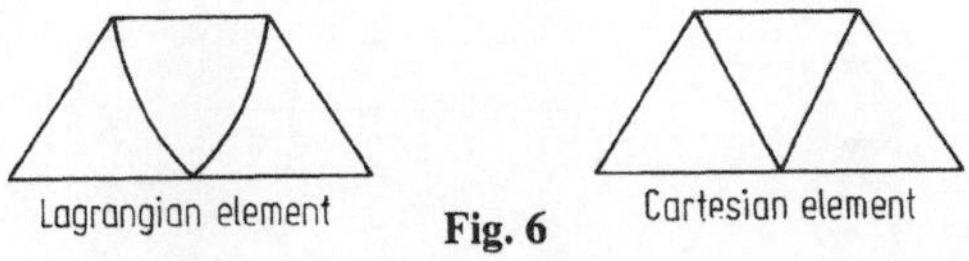

Fig. 6

tain the most common shapes and also the possibility of a general macro-elements which can approximate any complex shape with the required accuracy. The catalogue of the macro-elements developed in this program is shown in Fig. 5, and can be grouped in the following way:

I. Plane elements: Interpolation is made in Cartesian coordinates.
  I a. Triangular elements.
  I b. Quadrilateral elements.

II. Nonplane elements with particular coordinates: Interpolation is made in the corresponding intrinsic coordinates.
  II a. Conical and cylindrical (triangles, quadrilaterals, troncoconical surfaces).
  II b. Spherical (triangles, quadrilaterals, and any other spherical surfaces).
  II c. Quadric Surfaces.

III. Isoparametric Lagrangian elements (triangles and quadrilaterals): Interpolation is made in natural coordinates.

The first group is motivated for simplicity and also in order to get a more natural division in several cases than provided by the third group in the same case (see Fig. 6).

The second class attempts a better approximation of the geometry in cases which are quite usual in practical applications, while the third group provide the generality needed for many applications.

With respect to the lines L_{ij} between macro-elements, Fig. 7 shows the different types used in this processor, as well as the data needed for their definition. In addition to those data it is necessary to specify the number of sub-division in every line as well as the kind of variation of this division along the line. The relation given by

$$\xi(J, R) = \bar{\xi} + \frac{R}{2}(\bar{\xi}^2 - 1) \tag{16}$$

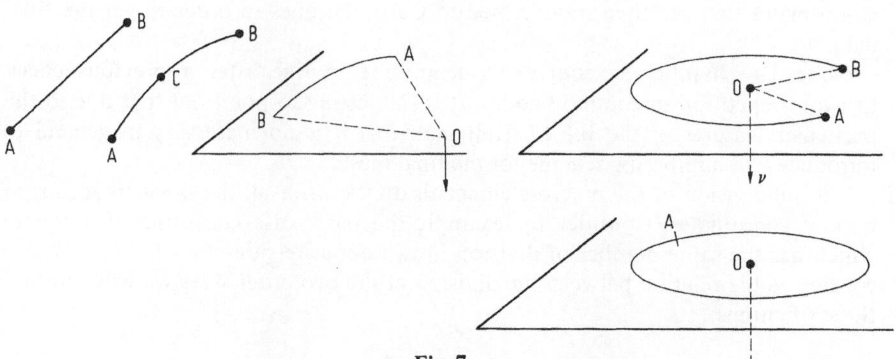

Fig. 7

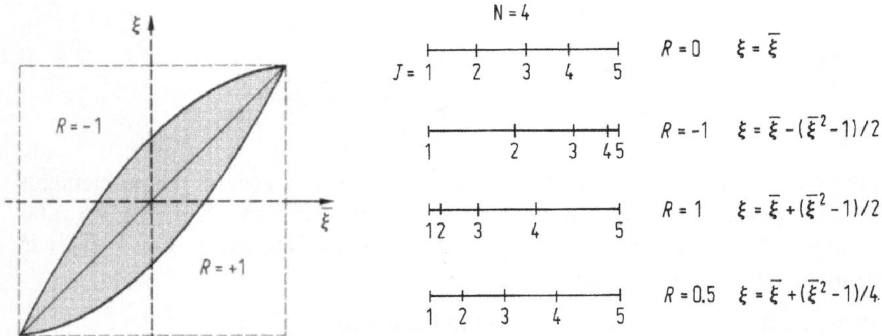

Fig. 8

has proved to be useful for defining the variation of the length of the divisions along the line, so that with only the parameter R it is possible to get a rich family of densification possibilities (see Fig. 8). In (16) ξ is the required natural coordinate and

$$\bar{\xi} = \frac{2(J-1)}{N} - 1, \qquad J = 1, 2, \ldots, N+1,$$

N being the number of desired subdivisions.

This subdivision is intended to model only the extreme points of every element. The third node, necessary for the parabolic interpolation is subsequently introduced midway between the previous ones.

After subdivision, a reverse transformation to cartesian coordinates produces for straight lines

$$\boldsymbol{x} = \frac{1-\xi}{2}\,\boldsymbol{x}_A + \frac{1+\xi}{2}\,\boldsymbol{x}_B$$

for parabolic lines (18)

$$\boldsymbol{x} = \tfrac{1}{2}\,\xi\,(1-\xi)\,\boldsymbol{x}_A + (1-\xi^2)\,\boldsymbol{x}_B + \tfrac{1}{2}\,\xi\,(1+\xi)\,\boldsymbol{x}_C.$$

Circular arcs and circumferences are interpolated as in (18) but using circular co-ordinates that are then transformed to Cartesian ones in order to get the final data.

Nodes are numbered sequentially from line to line after a previous check to avoid repetition in common nodes. It is interesting to point out that due to the particular features of the B.E.M. (full matrices) it is not necessary in general to introduce a re-numbering scheme for the final mesh.

The subdivision of the macro-elements is attacked for all the types by means of natural coordinates. Consider for example the basic case consisting of a square which has the same number of division in two opposite sides $L_2 = L_4$ (Fig. 9). According to the relation between the division of the two order sides we will consider three situations:

(1) $L_3 < 2\,L_1$

(2) $$L_3 > 2\,L_1;\quad \frac{L_3}{L_1} \leqq 2^{L_2-1} \tag{19}$$

(3) $$\frac{L_3}{L_1} > 2^{L_2-1}$$

where L_1, L_3 denotes the number of division of both sides.

In order to solve these three possibilities, three different schemes have been developed:

(a) Each segment in line j is related to another one in line $j+1$ (Fig. 9a).
(b) One segment in line j is related to two in $j+1$ (Fig. 9b).
(c) One segment in j is related to n ones in $j+1$ (Fig. 9c).

For example using schemes (a) and (b) it is possible to solve the case (1). Starting from line 3 with L_3 segments, one progresses with procedure (b) until the row where the number L_1 is reached and then follows with rule (a) (Fig. 9d). Case (2) is as before but now there will be several rules (b) inside each row. The first additional one is located at side 4, while the following are distributed as

$$p_k = \left(\frac{N_i}{n}\right) k, \qquad k = 1, 2, \ldots, n-1, \tag{20}$$

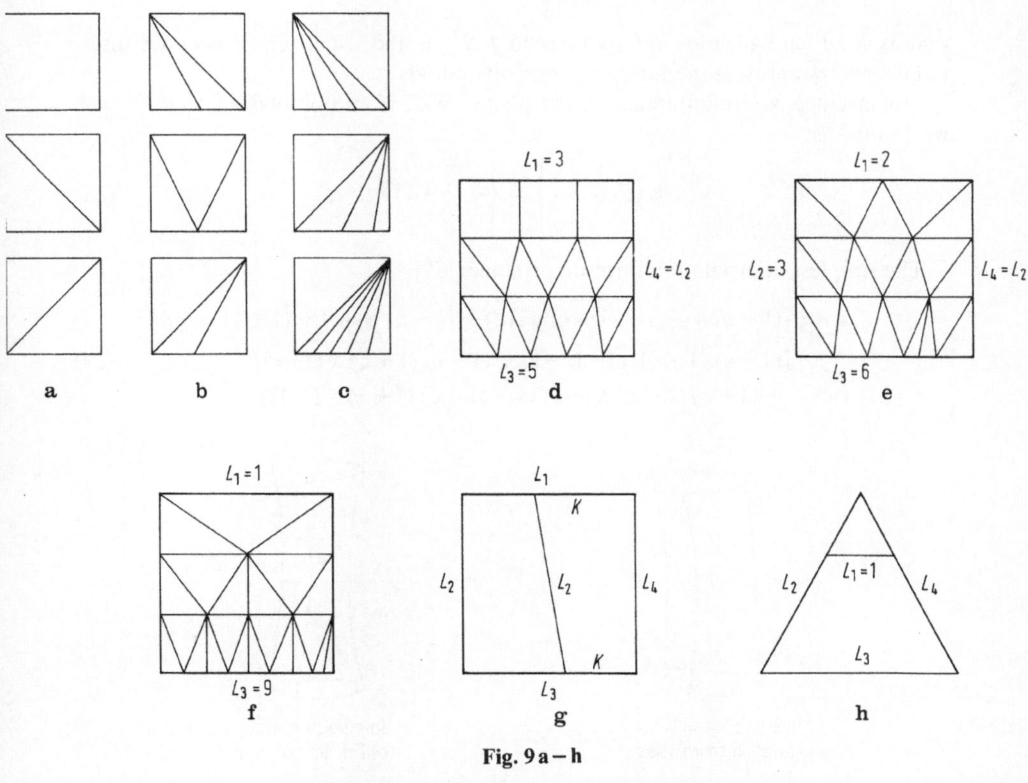

Fig. 9a–h

where n is the number of rules type (b) needed at row i; N_i is the total number of segments in row i and p_k is the position of the rule (b) k (see Fig. 9e).

Finally, case (3) is solved with rules (c) adopted when the available rules type (b) have been exhausted (Fig. 9f). When the basic square has a different number of segments on every side it is easy to reduce this situation to the commented above by using an initial division of the macro-element into two parts (see Fig. 9g). The same is true for triangular macro-elements, by dividing of one of the tips of the figure into a single element (Fig. 9h).

Once the number of subdivisions in every row has been established the computation of the natural co-ordinates is straight-forward. After this they are transformed to cylindrical, spherical or Cartesian ones by means of the transfinite method. This use the linear combination:

$$B f = I_\xi f + I_\eta f - I_{\xi\eta} f \tag{21}$$

called the blended spline approximation to f, where:

$$\begin{aligned} I_\xi f(\xi, \eta) &= \sum_{i=1}^{n} f(\xi_i, \eta)\, p_i(\xi) \\ I_\eta f(\xi, \eta) &= \sum_{j=1}^{n} f(\xi, \eta_j)\, p_j(\eta) \\ I_{\xi\eta} f &= \sum_{\substack{0<i<n \\ 0<j<n}} f(\xi_i, \eta_j)\, p_i(\xi)\, p_j(\eta) \end{aligned} \tag{22}$$

that is I_ξ, I_η are blended interpolates to f: $I_{\xi\eta}$ is the usual tensor product interpolate, and p represent the interpolating polynomials.

For instance, a circumference in the plane $z = C$, from points $\theta = 0$ to $\theta = \frac{\pi}{2}$ can be defined as:

$$\psi(\xi) = \begin{Bmatrix} r \\ \theta \\ z \end{Bmatrix} = \begin{Bmatrix} R \\ \frac{\pi}{4}(\xi + 1) \\ C \end{Bmatrix}. \tag{23}$$

The simplest interpolation formula produces [35].

$$\begin{aligned} x = \tfrac{1}{2}[&(1-\eta)\,\psi_1(\xi) + (1+\eta)\,\psi_3(\xi) + (1-\xi)\,\psi_2(\eta) + (1+\xi)\,\psi_4(\eta)] \\ -\tfrac{1}{4}[&(1-\eta)(1-\xi)\,x(-1,-1) + (1-\eta)(1+\xi)\,x(1,-1) \\ &+ (1+\eta)(1-\xi)\,x(-1,1) + (1+\xi)(1+\eta)\,x(1,1)] \end{aligned} \tag{24}$$

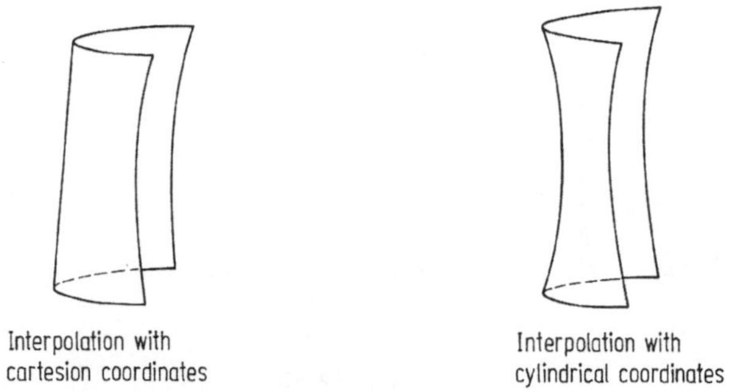

Fig. 10

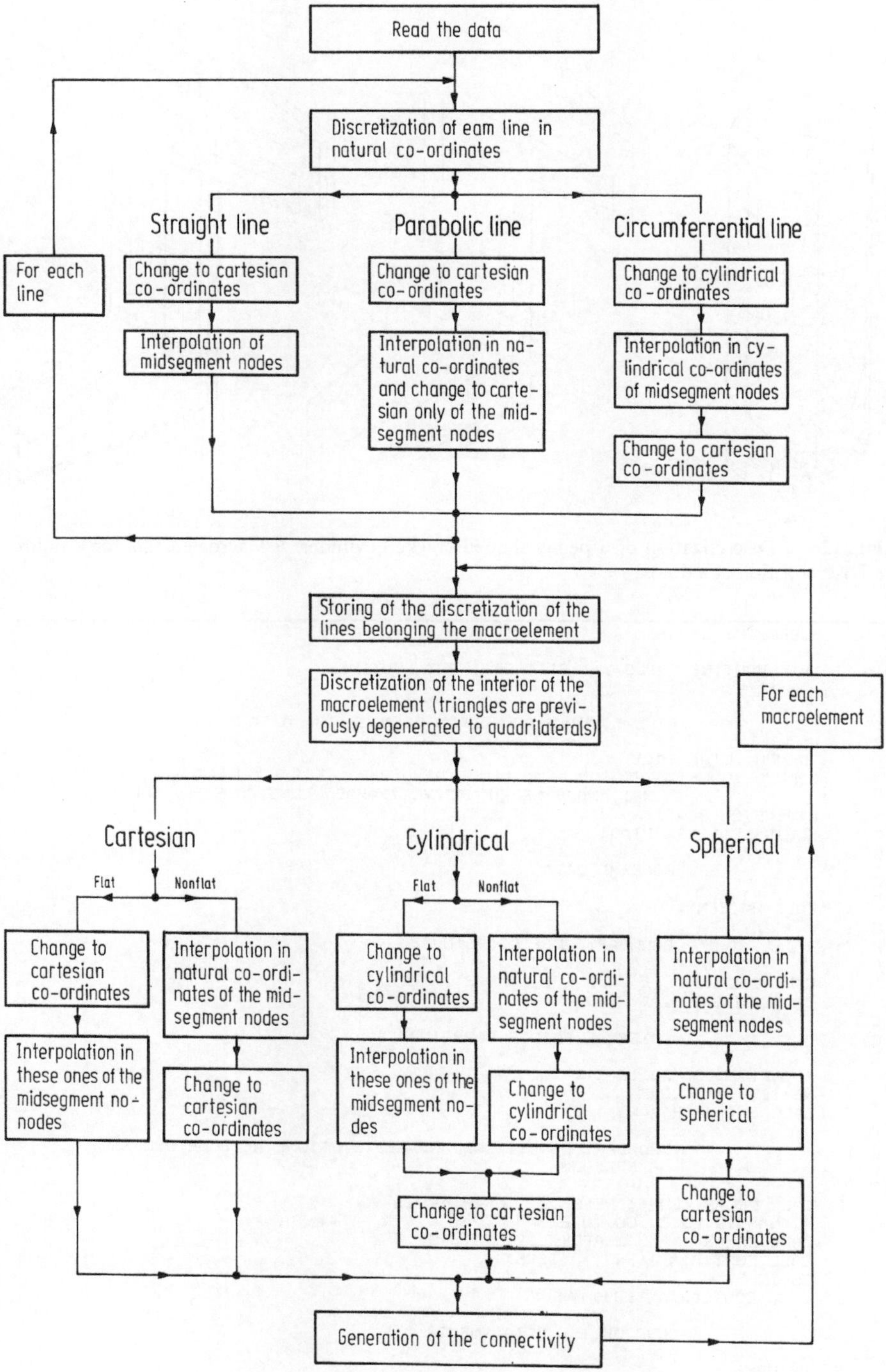
Read the data
Discretization of eam line in natural co-ordinates
Straight line
Parabolic line
Circumferrential line
For each line
Change to cartesian co-ordinates
Interpolation of midsegment nodes
Change to cartesian co-ordinates
Interpolation in natural co-ordinates and change to cartesian only of the midsegment nodes
Change to cylindrical co-ordinates
Interpolation in cylindrical co-ordinates of midsegment nodes
Change to cartesian co-ordinates
Storing of the discretization of the lines belonging the macroelement
Discretization of the interior of the macroelement (triangles are previously degenerated to quadrilaterals)
For each macroelement
Cartesian
Cylindrical
Spherical
Flat
Nonflat
Flat
Nonflat
Change to cartesian co-ordinates
Interpolation in natural co-ordinates of the midsegment nodes
Interpolation in these ones of the midsegment no-nodes
Change to cartesian co-ordinates
Change to cylindrical co-ordinates
Interpolation in natural co-ordinates of the midsegment nodes
Interpolation in these ones of the midsegment nodes
Change to cylindrical co-ordinates
Interpolation in natural co-ordinates of the midsegment nodes
Change to spherical
Change to cartesian co-ordinates
Change to cartesian co-ordinates
Generation of the connectivity

Fig. 11

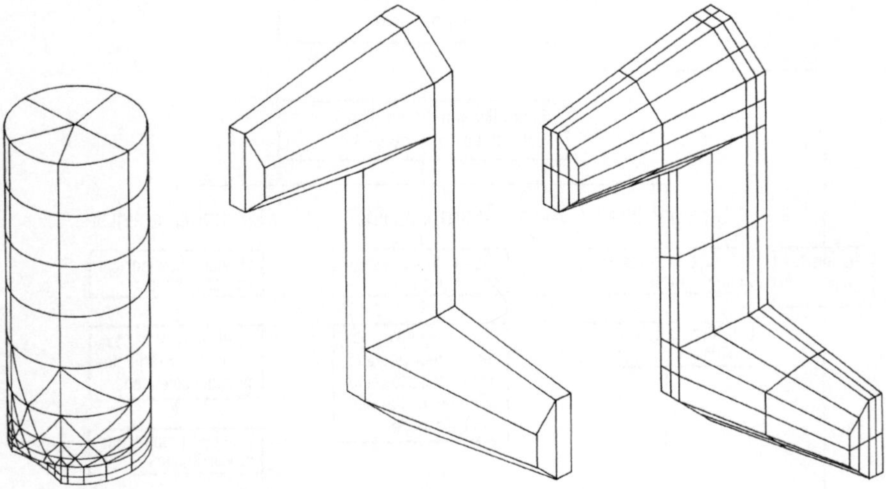

Fig. 12. **a** Discretization of a penny-shaped cracked cylinder. **b** Macroelements for a dolos. **c** Discretization of a dolos.

```
      SUBROUTNE DATOS
C
C     THIS ROUTINE CALLS THE REST OF INPUT ROUTINES
C
C
C                  DIMENSIONING AND DEFINITION OF THE ARRAYS
C
      COMMON /CIA/ IA(1)
      COMMON /CONST/ LEC,IMP,MEMC,MEMOB,NBP,NPER,NPEDP,NCON,NUDOSC,
     *               NPI,NUDOS,NELEM,NAPOYO,IPARA6,YOUNG,POIS,PORANA
      DIMENSION RIA(1)
      EQUIVALENCE(IA,RIA)
C
C                  GENERAL DATA
C
      CALL DATGEN
C
C                  BOUNDARY NODE COORDINATES
C
      CALL CREAR('COOR',NUDOS,2,1,NPER,IER2)
      CALL COOCON(RIA(IER2))
C
C                  INTERNAL POINTS COORDINATES
C
      IF(NPI.EQ.0) GO TO 11
      N3=IER2+2*NUDOSC
      CALL DATINT(RIA(N3))
C
C                  BOUNDARY CONDITIONS AND RESTRICTIONS OF SOME DEGREE
C                  OF FREEDOM
C
   11 CALL CREAR('CONT',NELEM,3,1,NPER,IER4)
      IF(NAPOYO.EQ.0) GO TO 21
      CALL CREAR('NANU',NAPOYO,1,1,1,IER5)
      CALL CONT(RIA(IER4),IA(IER5))
      GO TO 31
   21 CALL CONTO(RIA(IER4))
C
C                  END OF THE DATA INPUT
C
   31 RETURN
      END
```

```
      SUBROUTINE DATGEN
C
C     READS AND PRINTS THE GENERAL DATA OF THE PROGRAM
C
      COMMON /CIA/ IA(1)
      COMMON /CONST/ LEC,IMP,MEMO,MEMOB,NBP,NPER,NPEDP,NUDOSC,
     *                NPI,NUDOS,NELEM,NAPOYO,IPARA6,YOUNG,POIS,PORANA
      COMMON /WORK/ TITL(20)
C
C                    READS AND PRINTS THE TITLE OF THE PROBLEM
C
      READ(LEC,10000)TITL
      WRITE(IMP,20000)
      WRITE(IMP,20001)TITL
C
C                    READING OF THE GENERAL VARIABLES
C
      READ(LEC,10001)NCON,NUDOSC,NPI,NAPOYO,IPARA6
      READ(LEC,10001)YOUNG,POIS,PORANA
C
C                    COMPUTATION OF THE TOTAL NUMBER OF NODES AND ELEMENTS
C
      NUDOS=NUDOSC+NPI
      NELEM=NUDOSC-NCON
C
C                    PRINTING OF THE GENERAL VARIABLES
C
      WRITE(IMP,20002)NCON,NUDOSC,NPI,NUDOS,NELEM,NAPOYO
      WRITE(IMP,20003)YOUNG,POIS,PORANA
      WRITE(IMP,20004)IPARA6
C
C                    READING AND PRINTING OF THE NUMBER OF ELEMENTS OF
C                    EACH BOUNDARY
C
      CALL CREAR('NELE',NCON,1,1,1,IER)
      IER1=IER+NCON-1
      READ(LEC,10001)(IA(J),J=IER,IER1)
      WRITE(IMP,20005)
      DO 10 J=IER,IER1
           JJ=J-IER+1
           WRITE(IMP,20006)JJ,IA(J)
   10 CONTINUE
      RETURN
10000 FORMAT(20A4)
10001 FORMAT()
20000 FORMAT(1H1,2X,113('*')/3X,'PROGRAM AXELAST FOR THE ANALYSIS OF ELA
     *STIC AXISYMMETRIC STRUCTURES BY USING THE BOUNDARY ELEMENT METHOD'
     */3X,104('*')//)
20001 FORMAT(20X,20A4///)
20002 FORMAT(7X,46('*')/7X,'* * *  GENERAL VARIABLES OF THE PROGRAM  * *
     * *'/14X,32('*')///6X,'NUMBER OF BOUNDARIES',43('-'),I5/
     *6X,'NUMBER OF BOUNDARY NODES',36('-'),I5/
     *6X,'NUMBER OF INTERNAL POINTS',35('-'),I5/
     *6X,'TOTAL NUMBER OF NODES',39('-'),I5/
     *6X,'NUMBER OF BOUNDARY ELEMENTS',33('-'),I5/
     *6X,'NUMBER OF RESTRICTED DEGREES OF FREEDOM ON THE BOUNDARY',
     *5('-'),I5/)
20003 FORMAT(7X,'YOUNG"S MODULUS OF THE MATERIAL',29('-'),E10.4/
     *6X,'POISSON RATIO OF THE MATERIAL',31('-'),E10.4/
     *6X,'PERCENTAGE OF THE LENGTH FOR ANALYTICAL INTEGRATION',
     *9('-'),E10.4/)
20004 FORMAT(7X,'SYMMETRY PARAMETER',44('-'),I5/
     *29X,' * THERE"S NO SYMMETRY WITH RESPECT TO Z ( 0 )'/
     *29X,' * THERE"S SYMMETRY WITH RESPECT TO Z ( 1 )'/)
20005 FORMAT(1H1,/8X,'* * *  NUMBER OF ELEMENTS OF EACH BOUNDARY  * * *'
     */16X,35('*')////10X,'BOUNDARY',5X,'NUMBER OF ELEMENTS'/10X,8('-'),
     *5X,19('-')/)
20006 FORMAT(12X,I3,13X,I3/)
      END
```

```
      SUBROUTINE COOCON(COOR)
C
C     READS AND PRINTS THE BOUNDARY NODES COORDINATES
C
C******************** DICTIONARY
C
C     COOR(2,NUDOSC) = BOUNDARY NODES COORDINATES
C
C*******************************
C
C              DIMENSIONING AND DEFINITION OF THE ARRAYS
C
      COMMON /CONST/ LEC,IMP,MEMO,MEMOB,NBP,NPER,NPEDP,NCON,NUDOSC
      COMMON /WORK/ I,L,M,K,INCRE,IN,AR,AZ
      DIMENSION COOR(2,NUDOSC)
C
C              READING AND GENERATION OF THE COORDINATES
C
      K=0
      DO 100 I=1,NUDOSC
           IF(K.EQ.NUDOSC) GO TO 101
           READ(LEC,10000)L,COOR(1,L),COOR(2,L)
           IF(L.NE.K+1) GO TO 11
           K=K+1
           GO TO 100
   11      INCRE=L-K
           AR=(COOR(1,L)-COOR(1,K))/INCRE
           AZ=(COOR(2,L)-COOR(2,K))/INCRE
           IN=INCRE-1
           DO 20 M=1,IN
                COOR(1,K+M)=COOR(1,K)+AR*M
                COOR(2,K+M)=COOR(2,K)+AZ*M
   20      CONTINUE
           K=K+INCRE
  100 CONTINUE
C
C                   PRINTING OF THE COORDINATES
C
  101 CALL IMCOCO
      RETURN
10000 FORMAT()
      END
```

```
      SUBROUTINE CONT(CONTOR,NANU)
C
C     READS AND PRINTS THE BOUNDARY CONDITIONS
C
C******************** DICTIONARY
C
C     NCON                = NUMBER OF DISJOINT BOUNDARIES
C     NELEM               = TOTAL NUMER OF BOUNDARY ELEMENTS
C     CONTOR(1,NELEM)     = CODE OF THE BOUNDARY CONDITIOS OF EACH ELEMENT
C     CONTOR(2,NELEM)
C                         = BOUNDARY DISPLACEMENT OR TRACTION DATA
C     CONTOR(3,NELEM)
C     NAPOYO              = NUMBER OF FREEDOM WITH RESTRICTIONS
C     NANU(NAPOYO)        = RESTRICTED DEGREE OF FREEDOM
C                           2*NUDO-1 .... RESTRICTED THE RADIAL DISPLACEMENT
C                           2*NUDO....... RESTRICTED THE AXIAL DISPLACEMENT
C
C*******************************
C
C
C              DIMENSIONING AND DEFINITION OF THE ARRAYS
C
      COMMON /CONST/ LEC,IMP,MEMO,MEMOB,NBP,NPER,NPEDP,NCON,NUDOSC,
     *               NPI,NUDOS,NELEM,NAPOYO
      COMMON /WORK/ I,II,L,J,M,K,INCRE,AR,AZ
      DIMENSION CONTOR(3,NELEM),NANU(NAPOYO)
C
C              READING AND GENERATION OF THE CODE AND DATA CORRESPONDING
C              TO THE BOUNDARY CONDITIONS
C
      K=0
      DO 100 I=1,NELEM
           IF(K.EQ.NELEM) GO TO 101
```

```
         READ(LEC,10000)L,CONTOR(1,L),CONTOR(2,L),CONTOR(3,L)
         IF(L.NE.K+1) GO TO 11
         K=K+1
         GO TO 100
  11     INCRE=L-K
         AR=(CONTOR(2,L)-CONTOR(2,K))/INCRE
         AZ=(CONTOR(3,L)-CONTOR(3,K))/INCRE
         DO 20 M=1,INCRE
            CONTOR(1,K+M)=CONTOR(1,K)
            CONTOR(2,K+M)=CONTOR(2,K)+AR*M
            CONTOR(3,K+M)=CONTOR(3,K)+AZ*M
  20     CONTINUE
         K=K+INCRE
 100 CONTINUE
C
C                 READING OF THE RESTRICTIONS OF SOME DEGREE OF FREEDOM
C
 101 READ(LEC,10000)(NANU(II),II=1,NAPOYO)
C
C                 PRINTING OF THE BOUNDARY CONDITIONS
C
     CALL IMCCON
     RETURN
10000 FORMAT()
     END
```

```
     SUBROUTINE CONTO(CONTOR)
C
C    READS AND PRINTS THE BOUNDARY CONDITIONS
C
C******************** DICTIONARY
C
C    NCON              = NUMBER OF DISJOINT BOUNDARIES
C    NELEM             = TOTAL NUMER OF BOUNDARY ELEMENTS
C    CONTOR(1,NELEM) = CODE OF THE BOUNDARY CONDITIOS OF EACH ELEMENT
C    CONTOR(2,NELEM)
C                      = BOUNDARY DISPLACEMENT OR TRACTION DATA
C    CONTOR(3,NELEM)
C
C********************************
C
C
C                DIMENSIONING AND DEFINITION OF THE ARRAYS
C
     COMMON /CONST/ LEC,IMP,MEMO,MEMOB,NBP,NPER,NPEDP,NCON,NUDOSC,
    *               NPI,NUDOS,NELEM
     COMMON /WORK/ I,II,L,J,M,K,INCRE,AR,AZ
     DIMENSION CONTOR(3,NELEM)
C
C                READING AND GENERATION OF THE CODE AND DATA CORRESPONDING
C                TO THE BOUNDARY CONDITIONS
C
     K=0
     DO 100 I=1,NELEM
         IF(K.EQ.NELEM) GO TO 101
         READ(LEC,10000)L,CONTOR(1,L),CONTOR(2,L),CONTOR(3,L)
         IF(L.NE.K+1) GO TO 11
         K=K+1
         GO TO 100
  11     INCRE=L-K
         AR=(CONTOR(2,L)-CONTOR(2,K))/INCRE
         AZ=(CONTOR(3,L)-CONTOR(3,K))/INCRE
         DO 20 M=1,INCRE
            CONTOR(1,K+M)=CONTOR(1,K)
            CONTOR(2,K+M)=CONTOR(2,K)+AR*M
            CONTOR(3,K+M)=CONTOR(3,K)+AZ*M
  20     CONTINUE
         K=K+INCRE
 100 CONTINUE
C
C                PRINTING OF THE BOUNDARY CONDITIONS
C
     CALL IMCCO
     RETURN
10000 FORMAT()
     END
```

where the vector x can be represented in either of the types of coordinates mentioned above. This kind of interpolation allows a very accurate approximation of the geometry and boundary conditions. The reason for using a different system of coordinates for each macro-element now become apparent: the interpolation among lines can defigure basic features of the actual surfaces (Fig. 10).

Figure 11 shows the flow chart of this processor and several results can be seen in Fig. 12.

4.6 Computation of the Matrices

Once the characteristics of the program and the initial data, have been defined, one can then proceed to compute the matrices A, B, P in (12), and when necessary C in (13). The introduction of the boundary conditions, together with an appropriate assemblage algorithm, allow the obtention of a system of linear algebraic equations.

$$Mx = F. \tag{25}$$

Here x is a vector in which all the unknowns of the problem are included. The solution of this system will be studied in Sect. 7.

In almost every B.E. program already developed, the procedure for obtaining the matrices and the assemblage are simultaneous, meaning that, once the coefficients corresponding to the integration over an element from a collocation point (CP) have been computed, they are included at once in their place in the system (25), according to the particular boundary conditions of the element on which the integration has been carried out.

In views of this, this section is divided into two clearly differentiated parts. The first one will analyze the different approaches for obtaining the elements of the matrices A, B, P, while the second will study the application of boundary conditions and the assemblage process. Also this section will attempt to analyze the different ways in which the computations can be sequenced in order to get a simple and effective assemblage.

Two particular aspects which decisively influence the determination of the system (25) are also included: the symmetry of the domain with respect to any of the coordinate planes and the solution of piecewise heterogenous bodies by means of a subregionalization approach.

4.6.1 Computation of the Integration Coefficients

As was seen in Sect. 2, in order to get the system (25) it is necessary to compute a set of coefficients whose expressions are given by (11); this entails integrating some functions which depend on the coordinates of the collocation and integration points.

The first possibility for computing these integrals is the use of a general numerical algorithm, which can be a standard one, provided that the sets V_e U_c, defined in (11) and (14) respectively, are the standard sets on which the integration is per-

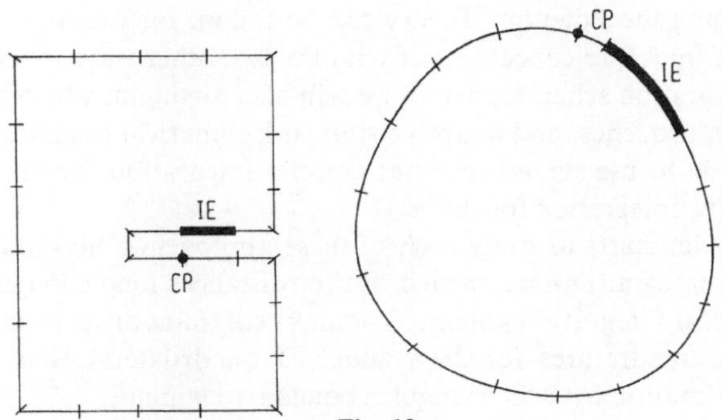

Fig. 13

formed for the chosen quadrature. In this case, however, there is an important additional problem: the singularity of the functions of the integrand, as it was studied in Sect. 2. Although this singularity was saved theoretically by considering the integrals in its Cauchy Principal Value sense, it is obvious that a standard numerical integration process would provide inaccurate results when the distance between the collocation point $\mathbf{x}(\xi)$ and the integration point $\mathbf{y}(\xi)$ on the element e, over which the integration is carried out, tends to zero, and moreover when $\mathbf{x}$ belongs to e, so the functions to integrate are not bounded. The same considerations can be maintained for the volume integrals in (14). Of course, when the distance between $\mathbf{x}$ and $\mathbf{y}$ is long enough, a standard numerical integration scheme can be used.

The first problem arises then when trying to define the minimum distance beyond which this standard scheme can be used. This question does not need an answer, when an analytical integration scheme is employed for all the elements. However, this approach is only possible for very simple elements (usually only 2-D problems).

A second answer would be to forget this distance, considering that, maximum influence appears only in those elements to which the collocations point belongs. A special scheme would thus be devised for these elements and a standard one for the rest. This method is more general than the first, because an analytic, semi-analytic or numerical integration scheme can be developed for these "singular elements". However it has also some problems when we perform the integration over elements, which are very close to the singular point, as in the case of the discretizations shown in Fig. 13.

The third approach is to use special numerical integration schemes for the singular element and also for the closest of the rest, according to some specific criterion (eg. ratio between the distance collocation point-element of integration, and the size of this element, limitation of the truncation error of the integration, etc.).

A fourth solution is to use a standard numerical integration scheme, but increasing the number of points of integration over the closest elements to the collocation point, according to one of the previous criteria (adaptive numerical integration).

We will not consider here the global analytical integration scheme, since it does not include computational but analytic problems. Some results of analytical integrations concerning the Potential Theory can be found, for example, in Paris [38]. This section will therefore concern itself with the two other ways to solve the problem: special integration schemes, where we will also distinguish between analytical and numerical approaches, and adaptive standard numerical schemes. Naturally it would be possible to use mixed schemes (special integration for the singular element and adaptive integration for the rest).

Before the reader starts to study each of these approaches, he should remember that most of the integrations are carried out for distances long enough to involve a standard numerical integration scheme. The most common of such schemes are the Gauss-Legendre quadratures for 2-D, and 3-D quadrilateral elements, and the Hammer approximation for 3-D triangular boundary elements.

$$\int_{-1}^{1} f(x)\,dx = \sum_{k=1}^{n} w_k f(x_k); \quad \int_{-1}^{1}\int_{-1}^{1} f(x,y)\,dx\,dy = \sum_{j=1}^{n_1}\sum_{k=1}^{n_2} w_j w_k f(x_j y_k)$$

Gauss-Legendre

$$\int_{0}^{1} f(x)\,dx = \sum_{k=1}^{n} w'_k f(x'_k); \quad \int_{0}^{1} dx \int_{0}^{1-x} f(x,y)\,dy = \sum_{j=1}^{n} w'_j f(x_j, y_j) \tag{26}$$

Hammer

where w_k and x_k are the well-known values of the weights and abscissas of the integration points in these quadratures, which depend only on the order, n, and type of quadrature. With a Gauss-Legendre approximation over 2-D boundary elements, a five points quadrature usually provides enough accuracy for most applications. When the distance is long enough, a second or third order quadrature is usually enough. For 3-D quadrilateral elements, these numbers change to 3×3 and 2×2. For Hammer quadratures (triangles) a seven-point quadrature is accurate enough, while the four-point one is useful for most cases.

Tables for values of w_k and x_k for the most common quadratures can be found throughout the literature; we will however, only cite Davis and Rabinowitz [39], [40], Abramowitz and Stegun [41], and Zienkiewicz [16].

When the distance between the collocation point and the element of integration is very short, as in Fig. 13, these quadratures do not provide accurate results. In this case it is possible to use adaptive numerical integration schemes (see 6.1.2), or a special numerical technique.

4.6.1.1 Analytical and Semianalytical Integration over an Element Containing the Collocation Point

As was mentioned before, despite the fact that this type of integrations has a wider range of applications than does global analytical integration, carrying out this process is little more than a somewhat complicated exercise in analysis, depending on the type of elements used, and the function to be integrated. Results in Potential theory can be found: 1) with 2-D boundary elements and constant approximation functions in Brebbia [17]; 2) with linear elements in Paris [38]; 3) with parabolic ones in García-Benítez [42]. The results for 3-D elements with constant interpolation in Anza [43] and Paris [38]. The Theory of Elasticity with 2-D constant elements is discussed in Dominguez [44] and Brebbia [17]. Linear interpolation is taken up in Paris [38]. Telles [45] and Cruse [46]. Other results for different applications may be found in [9–14].

Computationally more interesting than this is the semianalytical integration scheme. This approach is specially attractive when the complete analytical integration is very complex, but can be performed with respect to the singularity-inducing variable, following a numerical scheme for the rest of the variables. It is also of interest when the function to be integrated can be simplified in the neighborhood of the collocation point using an asymptotic approximation.

The first case occurs, for instance, in the Theory for Plasticity [14, 15] for the 2-D cells and for the boundary and volume integrals in 3-D Plasticity [15], although it certainly can be applied to many other kinds of problems involving the integration of functions in two or more dimensions.

This approach is based on the fact that the singularity depends only on the variable r (distance from the collocation point), so that an analytical integration process can be carried out with respect to this variable, while for the rest (θ in Fig. 14a and θ, ϕ in Fig. 14b) a numerical scheme can be used.

$$\int_{V_e} f(r,\theta)\, dV_e = \int_{\theta_1}^{\theta_2} d\theta \left[\lim_{\varepsilon \to 0} \int_{\varepsilon}^{r(S'_\theta)} f(r,\theta)\, dr \right] = \int_{\theta_1}^{\theta_2} I\, d\theta$$

$$I = \lim_{\varepsilon \to 0} \int_{\varepsilon}^{r(S'_\theta)} f(r,\theta)\, dr \tag{27}$$

$$\int_{\Omega_c} f(r,\theta,\phi)\, dV_c = \int_{S'_{\theta\phi}} d\theta\, d\phi \left[\lim_{\varepsilon \to 0} \int_{\varepsilon}^{r(S'_{\theta\phi})} f(r,\theta,\phi)\, dr \right] = \int_{S'_{\theta\phi}} I\, d\theta\, d\phi$$

$$I = \lim_{\varepsilon \to 0} \int_{\varepsilon}^{r(S'_{\theta\phi})} f(r,\theta,\phi)\, dr. \tag{28}$$

The computation of the integral I in (27), (28) is that which has to be done analytically. Some results for the Kernels of the Theory of Elasticity and Plasticity can be found in [14, 15]. Once the integral I has been evaluated, the integration with respect to the other variables can be performed using a standard numerical procedures as in (26).

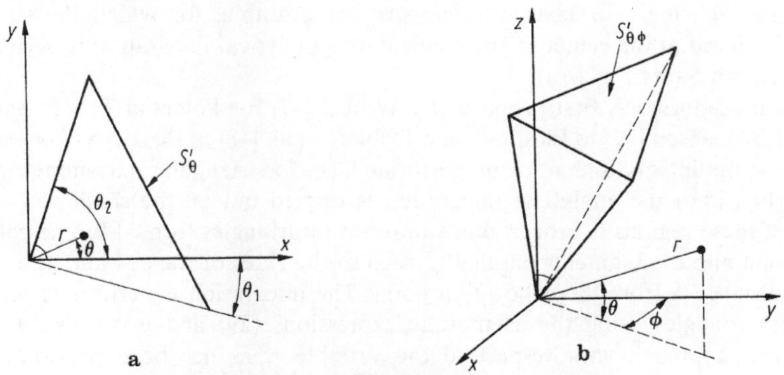

Fig. 14 a, b

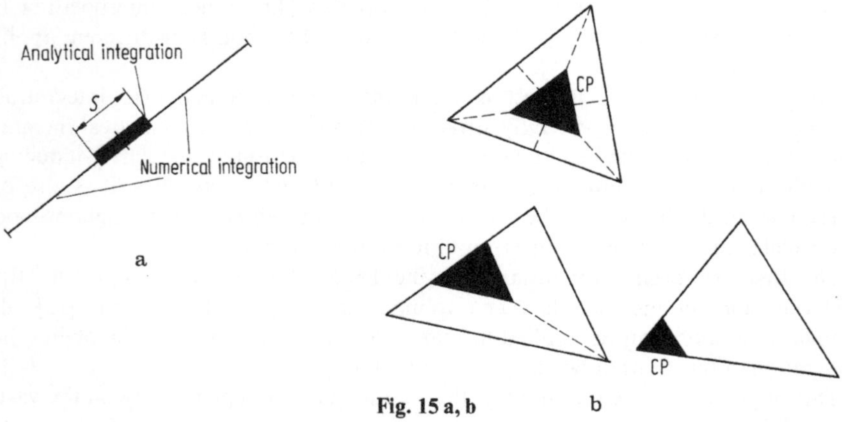

Fig. 15 a, b

The other type of semianalytical integration is resorted when the integrand is a very complex function, whose asymptotic approximation when r tends to zero is known. This is true, for instance, with the axisymmetric case, for which the fundamental solution is a function of the Legendre functions of the second kind (in Elasticity $Q_{1/2}, Q_{-1/2}$ [6]) whose asymptotic approximation is given by [41].

$$Q_{-1/2}(\gamma) \to -\tfrac{1}{2}\ln\frac{\gamma-1}{32}, \quad Q_{1/2}(\gamma) \to -\tfrac{1}{2}\ln\frac{\gamma-1}{32} - 2 \qquad \gamma \to 1$$

$$\gamma = 1 + \frac{(R-r)^2 + (Z-z)^2}{2Rr} \tag{29}$$

where (R, Z) are the coordinates of the collocation point (collocation ring) and (r, z) those of the field point. Clearly these expressions allow a much easier analytical integration than do the Legendre functions.

Using these results it is possible to get an approximation of the integral over an element by analytically integrating over part of it (the neighborhood of the singularity) where the Legendre functions can be approximated by (29) with minimum error, and adding the result of evaluating the integral over the rest of the element. The computation of this last integral can be carried out using a standard numerical scheme. For constant elements, for example, for which the singular point is placed at the center of the element, the analytical integration is performed over a length S (see Fig. 15 a).

This procedure was first proposed by Wrobel [47] for Potential Theory and extended by Danson [23] to Elasticity and Doblaré et al. [48] to Plasticity. For the last case, volume integrations must be performed, so, for triangular axisymmetric sections (Fig. 15 b) the analytical integration is carried out on the shadowed zones. Each of these regions is broken down into several triangles formed by the collocation point and the segments parallel to each of the sides of the cell and at a maximum distance S from the collocation point. The integration is performed on each of these triangles using the asymptotic expressions (29) and using also a semianalytical approach with respect to the variable r, as has been previously explained. A numerical procedure is followed for the rest of the cell so that all of these results are added to get the final integral over the whole cell. The complete expressions for Elastoplasticity may be found in Gomez-Lera [7].

The value S is usually chosen as a percentage of the total length of the element (contiguous side in Fig. 16 b). The best results are obtained with values between 4 and 10%. This integration procedure for Elasticity has been followed in the discussed program.

4.6.1.2 Special Numerical Integration Techniques over an Element Containing the Collocation Point

As has been mentioned before, in order to obtain a sufficiently accurate numerical solution of a B.E. problem, it is necessary to compute with a great accuracy the values of the integrals, for the elements that contain the collocation point. When an analytical or semianalytical approach cannot be used, because of its difficulty, or even its impossibility, a numerical one must be employed. For this the weights and abscisses of the quadrature formulae to be used are needed, so it is the evaluation of the error of these formulae.

An additional difficulty arises in this case: the unboundedness of the integrands. Therefore a standard quadrature procedure cannot be used, at least with the confidence of obtaining accurate results. Despite of this fact, some authors have used this approach, ignoring the singularity of the function to integrate. A sophisticated method of this type is studied in the next section.

A different approach consists in using a special numerical approximation for the integral, taking into account the singularity. Some of these methods will be briefly studied along the next lines, directing the reader to appropriate references for more details.

We must distinguish again between the two kinds of integrals which appear in the B.E.M.: a weakly singular integral and an actual singular one with Cauchy Kernel. The numerical approximation of the former case has been better studied than the latter. It is usually based on standard quadrature techniques, using the singular term as a weighting function (i.e. computation of the roots of the orthogonal polynomials to the weighting function over the integration interval, and computation of the weight coefficients for these abscissas).

For one dimension (2-D boundary elements) the weighting function corresponds to the logarithmic function. A table of the weights and abscissas for this quadrature, with 1 to 10 points of integration, is given by Andersen [49] and by Krylow and Pal'tsen [50]. Andersen includes also the evaluation of the error of this quadrature, which can be expressed as:

$$\int_{-1}^{1} \ln \frac{1}{r} f(\xi)\, d\xi = \sum_{k=1}^{k} w_k f(\xi_k) + R_k. \tag{30}$$

Formulae for the solution of 3-D bodies (2-D quadratures) are not so extended, but some work has been done during the last years, getting formulae in the form:

$$\begin{aligned} &\int_0^1 d\xi_1 \int_0^{1-\xi_1} \frac{1}{r} f(\xi_1, \xi_2)\, d\xi_1\, d\xi_2 = \sum_{k=1}^{K} w_k f(\xi_{1k}, \xi_{2k}) + R_K \\ &\int_{-1}^{1}\int_{-1}^{1} \frac{1}{r} f(\xi, \eta)\, d\xi\, d\eta = \sum_{i=1}^{I} \sum_{j=1}^{J} w_i w_j f(\xi_i \eta_j) + R_{IJ} \end{aligned} \tag{31}$$

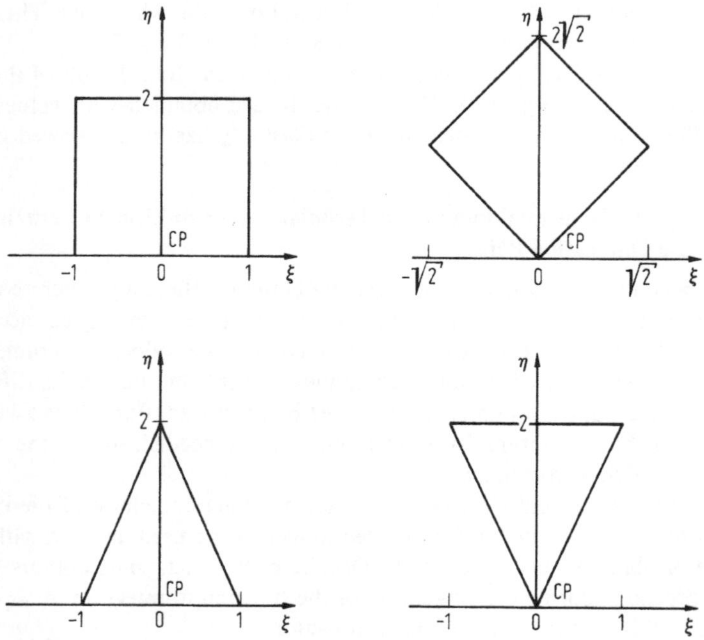

Fig. 16. Singular numerical integration over triangles and quadrilaterals

Tables of weights and abscissas for quadratures of different orders, over quadrilaterals and triangles may be found in Critescu and Loubignac [51] and Pina et al. [52] (see Fig. 16).

Another type of weakly singular integral which appears in the B.E.M. corresponds to the axisymmetric case due to the singularity that the Legendre functions introduce, as can be seen in (29). The integration of these functions can be done using a simple trick: the multiplication and division of the integrand by $\ln(\gamma - 1)$ as in (32):

$$\int_{-1}^{1} Q_{-1/2}(\gamma)\, d\xi = \int_{-1}^{1} Q_{-1/2} \frac{\ln(\gamma-1)}{\ln(\gamma-1)}\, d\xi = \int_{-1}^{1} \ln(\gamma-1)\, f(\gamma)\, d\xi . \tag{32}$$

Now $f(\gamma)$ is a bounded continuous function, and the same quadratures as in (30) can therefore be used, in order to integrate $f(\gamma)$ with $\ln(\gamma - 1)$ as weighting function.

Concerning the evaluation of Cauchy Principal Value integrals, various methods have been devised. For the two-dimensional case, a simple change of coordinates (bilinear transformation) allows us to write the integral

$$\int_{a}^{b} \frac{\gamma(x)}{x - x_0}\, dx \qquad a < x_0 < b \tag{33 a}$$

as

$$\int_{-1}^{1} \frac{f(x)}{x}\, dx . \tag{33 b}$$

A number of methods for evaluating integrals of this type have been suggested (a brief survey can be found in Davis and Rabinowitz [53] and in Van der Sluis and Zweerus [54]). Some of them use a preliminar analytic manipulation of the integral in order to remove the singularity [55]. Others use a direct numerical approach, and although some of them are based on non-Gaussian quadratures [56], the most interesting for us are those which use a special type of Gaussian quadratures. The simplest of this kind was proposed by Piessens [57]. It uses the following break-down of the integral.

$$\int_{-1}^{1} \frac{f(x)}{x}\,dx = \lim_{\varepsilon\to 0}\left(\left[\int_{-1}^{\varepsilon} + \int_{\varepsilon}^{1}\right]\frac{f(x)}{x}\,dx\right) = \lim_{\varepsilon\to 0}\int_{\varepsilon}^{1}\frac{f(x)-f(-x)}{x}\,dx$$

$$= \int_{0} \frac{f(x)-f(-x)}{x}\,dx. \tag{34}$$

If the density function $f(x)$ is $(4n+1)$ times differentiable in the segment $(-1, 1)$, the function $\phi(x) = 1/x\,[f(x) - f(-x)]$ is an even, continuous, $4n$ times differentiable function in $(-1, 1)$ (see Lebedev [58]). For an even function $\phi(x)$ the quadrature formula of the highest algebraic degree of accuracy will be the formula.

$$\int_{0}^{1} \phi(x)\,dx = \sum_{i=1}^{n} A_i\,\phi(x_i) + R_n \tag{35}$$

where A_i, x_i $(i = \pm 1, \ldots, \pm n;\ x_i = -x_i;\ A_i = A_{-i})$ are the weights and abscissas of the Gauss-Legendre quadrature of order $2n$ for the segment $(-1, 1)$. Consequently, for the integral (34) the formula of highest degree of accuracy will be

$$\int_{-1}^{1} \frac{f(x)}{x}\,dx = \sum_{i=-n}^{n} \frac{A_i}{x_i}\,f(x_i) + R_n \tag{36}$$

where the remainder R_n has the form:

$$|R_n| \leqq \frac{2^{4n+1}|\phi^{(4n)}(\xi)|}{(4n+1)!}\left[\frac{(2n)!}{(2n+1)\ldots 4n}\right]^2, \quad -1<\xi<1. \tag{37}$$

This term is equal to zero for even polynomials of degree lower than $4n-2$.

A generalization of these formulae has been developed by Hunter [59], for an odd number of points of integration:

$$\int_{-1}^{1} \frac{f(x)}{x}\,dx = A_1 f'(0) + \sum_{r=2}^{n} \frac{A_r}{x_r} f(x_r) + R_n, \qquad n\,(\text{odd}), \tag{38}$$

where x_r and A_r have the same meaning that those in (36). The convergence of this quadrature was proved by Elliot [60]. Other interesting extension was suggested by Paget and Elliot [61], including the case of several singularities along the interval of integration, and another one by Chawla and Ramakrishnan [62] which solves the same case that Piessens and Hunter but for the case of a Gauss-Jacobi quadrature.

$$I = \int_{-1}^{1} (1-x)^{\alpha}(1+x)^{\beta}\frac{f(x)}{x-t}\,dx, \qquad -1<t<1, \quad \alpha, \beta > -1. \tag{39}$$

This integral includes those of Hunter and Piessens for $\alpha=\beta=0$. The quadrature is expressed by:

$$I=\sum_{j=1}^{n}\frac{A_j f(x_j)}{x_j-t}-\frac{2\,\psi_n^{(\alpha,\beta)}(t)\,f(t)}{P_n^{(\alpha,\beta)}(t)}+R_n, \tag{40}$$

where the nodes $\{x_j\}_{j=1}^{n}$ are the zeroes of the Jacobi polynomials, the coefficients A_j coincide with the weights of the standard Gauss-Jacobi rule and $\psi_n^{(\alpha,\beta)}$ are expressed as

$$\psi_n^{(\alpha,\beta)}(t)=\tfrac{1}{2}\int_{-1}^{1}\frac{(1-x)^{\alpha}(1+x)^{\beta}P_n^{(\alpha,\beta)}(x)}{t-x}\,dx,\ t\in[-1,1]. \tag{41}$$

The expression for the remainder R_n can be found in [63].

The numerical evaluation of two-dimensional Cauchy Principal Value integrals is not so well studied, and it has not been until recently that some methods have arise for its computation [63–67]. We will analyze here the approach proposed by Theocaris and Kazantzakis [63–65].

This method considers the 2-D integral as an iterated one by decomposing the second order pole $(1/r^2)$ into two complex first poles. These iterated integrals can then be solved using some of the numerical methods, studied above for one-dimensional singular integrals.

If we consider, as an example, the integration over a quadrilateral, the integral to solve can be expressed, after a suitable change of coordinates as:

$$I=\int_{-1}^{1}\int_{-1}^{1}K(\boldsymbol{x},\xi,\eta)\,\phi(\xi,\eta)\,J(\xi,\eta)\,d\xi\,d\eta \tag{42}$$

where $\boldsymbol{x}$ is the vector of coordinates of the singular point, $K(\boldsymbol{x},\xi,\eta)$ the singular Kernel of the integral, $\phi(\xi,\eta)$ the density function and $J(\xi,\eta)$ the jacobian of the mapping between cartesian (x,y,z) and normalized coordinates (ξ,η).

Following [63] the integral (42) can be written as:

$$I=\int_{-1}^{1}\int_{-1}^{1}\left[K(\boldsymbol{x},\xi,\eta)-\frac{f(\boldsymbol{x},\theta)}{r^2}\right]\phi(\xi,\eta)\,J(\xi,\eta)\,d\xi\,d\eta-I_1(\boldsymbol{x})$$

$$I_1(\boldsymbol{x})=\int_{-1}^{1}\int_{-1}^{1}\frac{f(\boldsymbol{x},\theta)}{r^2}\,\phi(\xi,\eta)\,J(\xi,\eta)\,d\xi\,d\eta. \tag{43}$$

Here, $f(\boldsymbol{x},\theta)$ is the characteristic of the singular integral (42), so that $K(\boldsymbol{x},\xi,\eta)-f(\boldsymbol{x},\theta)/r^2$ is a nonsingular Kernel [8].

The first integral in (43) can then be evaluated by any standard numerical approach, while I can be decomposed into two iterated integrals as:

$$I_1=\frac{1}{2\,\mathrm{i}}\lim_{\varepsilon\to 0}\left\{\int_{-1}^{-\varepsilon}(I_2-I_3)\frac{d\eta}{\eta}+\int_{\varepsilon}^{1}(I_2-I_3)\frac{d\eta}{\eta}\right\} \tag{44}$$

with

$$I_2=\int_{-1}^{-\varepsilon}\frac{f(x,\theta)\,\phi(\xi,\eta)\,J(\xi,\eta)}{\xi-\mathrm{i}\,\eta}\,d\xi+\int_{\varepsilon}^{1}\frac{f(x,\theta)\,\phi(\xi,\eta)\,J(\xi,\eta)}{\xi-\mathrm{i}\,\eta}\,d\xi$$

$$I_3=\int_{-1}^{-\varepsilon}\frac{f(x,\theta)\,\phi(\xi,\eta)\,J(\xi,\eta)}{\xi+\mathrm{i}\,\eta}\,d\xi+\int_{\varepsilon}^{1}\frac{f(x,\theta)\,\phi(\xi,\eta)\,J(\xi,\eta)}{\xi+\mathrm{i}\,\eta}\,d\xi \tag{45}$$

The above reviewed numerical methods for one-dimensional singular integrals can then be applied for computing the integrals I_2, I_3, and I_1, which allow us to obtain the final result for the integral I.

4.6.1.3 Adaptive Numerical Integration Schemes

The possibility of evaluating a singular integral using a numerical approach which ignores the existence of the singularity has been studied by Davis and Rabinowitz [67] and El Tom [68], showing that such schemes can be convergent in certain circumstances. This fact, together with the necessity of integrating more accurately when the distance collocation point-element of integration is becoming shorter, and specially when this distance is zero, have recommended the use of standard quadratures of very high degree of accuracy. On the other hand, the use of these types of quadratures for all the elements would imply an excessive number of points of integration and the method becomes therefore uneconomical. The solution of this problem is the use of an adaptive numerical integration scheme (A.N.I.S.).

The first and less sophisticated A.N.I.S. for the B.E.M. arises as a consequence of some numerical analysis, which prove that, in order to compute the variables of the problem at internal points very close to the boundary, a more accurate integration is needed when the distance from the point to the boundary is about the length of the interval of integration on the closest boundary element.

An effective solution to the problem would be therefore to divide the element of integration into several subintervals with the same length as the distance from the element to the collocation point, and make the integration over each subelement by means of a standard quadrature (see 6.1), adding the contributions of all the intervals. The number of points of integration has been then increased according to an initial distance criterium. Different types of subintervals are shown in Fig. 17a for 2-D elements. For 3-D elements a similar criterium can be used, by changing the length of the interval by the maximum length of the sides of the subelement (Fig. 17b).

This heuristic criterium (some authors consider its analogy with some results corresponding to indirect methods) has been successfully used by different authors [7, 15], together with an analytical or semianalytical approach for the elements containing the collocation point. The main advantage of this method is, no doubt, its simplicity.

A more sophisticated A.N.I.S. for the B.E.M. was introduced by Lachat and Watson [69]. It consists in keeping the truncation error, for the numerical integration over an element, lower than a fixed upper bound. If the Gauss-Legendre quadrature is used, the truncation error is given by Davis and Rabinowitz [53] as:

$$E_n = \frac{f^{(2n)}(\xi)}{2n!\,2^n} \tag{46}$$

where n is the order of the quadrature.

To perform the integration (26) over a quadrilateral, for example (the extension to other situations is not difficult), so that the maximum truncation error is lower than a fixed bound (round off errors are not considered), a subdivision approach

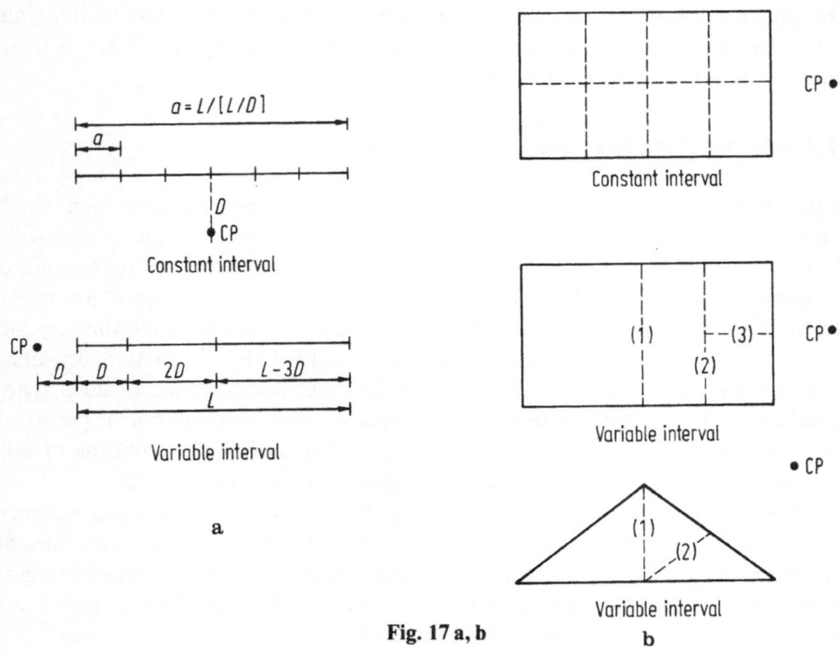

Fig. 17 a, b

is followed. The integral I to obtain is then approximated by a sum of several partial integrals in the form:

$$I = \sum_{i=1}^{N_1} \sum_{j=1}^{N_2} I_{ij}. \tag{47}$$

Here I_{ij} denotes the integral for the subelement i, j, and N_1, N_2 denote the number of subelements along each of the two directions ξ, η respectively (Fig. 18).

If normalized coordinates are used over each subelement (Fig. 18 b) the jacobian of the new mapping is given by $\frac{1}{N_1 N_2}$. From (26) the integral (47) can therefore

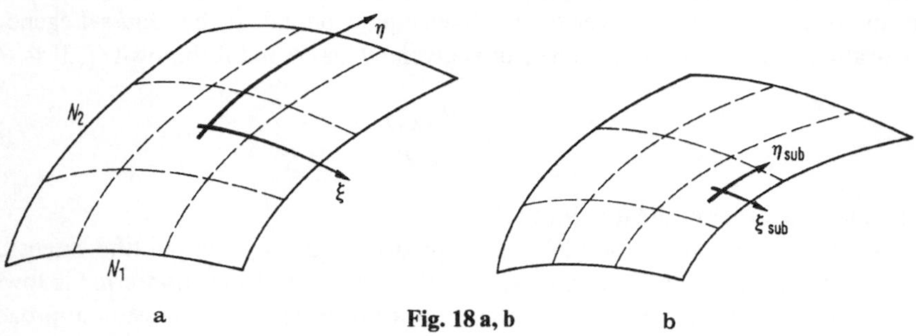

Fig. 18 a, b

be expressed as:

$$I \approx \frac{1}{N_1 N_2} \sum_{i=1}^{N_1} \sum_{j=1}^{N_2} \sum_{k=1}^{\lambda_1(i,j)} \sum_{l=1}^{\lambda_2(i,j)} w_k^{\lambda_1(i,j)} w_l^{\lambda_2(i,j)} f(\xi_k^{\text{sub}}, \eta_l^{\text{sub}}), \tag{48}$$

where $\lambda_1(i,j)$, $\lambda_2(i,j)$ denote the number of integration points along each direction for the subelement i, j and w_k, w_l, ξ_k^{sub}, η_l^{sub} are the weights and abscissas for each of those points in the chosen quadrature.

The number of subelements N_1, N_2 and integration points $\lambda_1(i,j)$, $\lambda_2(i,j)$ for each of them is chosen according to the premise of keeping the truncation error bounded. To get this, an estimation for the error is established using (46), and the $2\lambda_i$-th derivatives of the integrand must therefore be performed. In many cases however, the integrand is so complex that these derivatives cannot be easily performed, being necessary to employ some additional assumptions. For 3-D parabolic elements, for example, Lachat and Watson assume that:

(1) The truncation error for the singular term of the function $1/r$ is considered.

(2) The jacobian $J(\xi, \eta)$ in (26) is constant in the neighbourhood of the closest point, on the element of integration, to the collocation point.

(3) The variation of the distance r, between the collocation point and the element of integration, with respect to the tangential coordinates ξ, η is smaller than the variation of r with respect to r (this is only true for not very distorted elements).

From these assumptions it is possible to arrive [3] to an estimation of the truncation error for the integration over an element as:

$$E_n \leqq \frac{2GJ}{R^2} \sum_{i=1}^{2} (2\lambda_i + 1)\, \alpha_i^{2\lambda_i}$$

$$\alpha_i = \frac{\left(\left|\dfrac{\partial s}{\partial \xi_i}\right|\right)}{2R}, \quad \left|\frac{\partial s}{\partial \xi_i}\right| = \sqrt{\left(\frac{\partial x}{\partial \xi_i}\right)^2 + \left(\frac{\partial y}{\partial \xi_i}\right)^2 + \left(\frac{\partial z}{\partial \xi_i}\right)^2}. \tag{49}$$

Here, R is the minimum distance from the collocation point to the element of integration, J the jacobian $J(\xi, \eta)$ at the closest point P to the collocation point, G a constant and $\left|\dfrac{\partial s}{\partial \xi_i}\right|$ is computed at P.

With respect to the bound for this truncation error, the following assumptions are made:

(1) The bound is the same for both directions.

(2) The bound for each direction is proportional to the product $1/R$ by the area of the element, so that.

$$\text{Bound} = K \frac{1}{R^2} \cdot \text{Area}. \tag{50}$$

If we express now that the error must be lower than this bound, taking also in account that if the jacobian J is considered constant, it should be proportional to

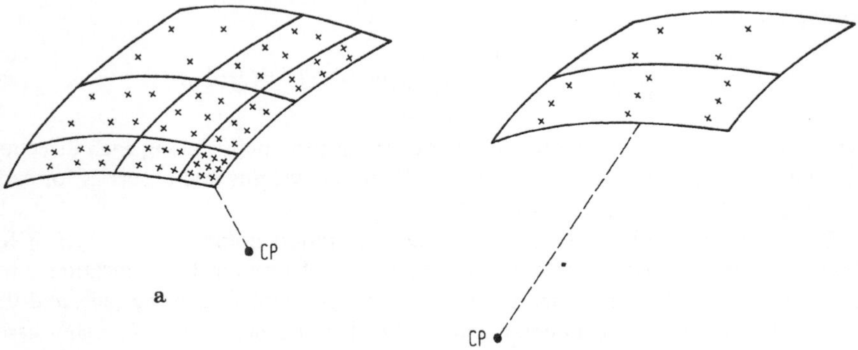

Fig. 19 a, b

the area of the element, we will have:

$$\sum_{i=1}^{2} (2\lambda_i + 1)\, \alpha_i^{2\lambda_i} \leqq 2C \tag{51}$$

and if it is assumed that the error for each direction ξ, η is the same, the number of points of integration for each subelement, along each direction will be given by

$$(2\lambda_i + 1)\, \alpha_i^{2\lambda_i} \leqq C. \tag{52}$$

The constant C is fixed according to previous numerical experiences, oscillating between 2^{-6} and 2^{-20}. Very accurate results with an unexcessive number of Gaussian points can be obtained with $C = 2^{-8}$.

The subdivision process for an element not containing the collocation point is shown by the scheme of Fig. 20, obtaining a distribution of points as shown in Fig. 19.

When the collocation point belongs to the element of integration, it is necessary to make some changes in this general approach, in order to get a greater number of points of integration in the neighborhood of the collocation point. This result is obtained by dividing the quadrilateral into several triangles as shown in Fig. 21 a. Each of these triangles is then subdivided using the previous scheme, by considering it as a degenerate quadrilateral with a zero length side, which coincides with the collocation point. Moreover, the distance R is taken as that between the collocation point and the opposite side of the triangle. The distribution of points which is obtained following this scheme is shown in Fig. 21 b.

This A.N.I.S. can be modified for triangles using a Hammer quadrature instead of the Gauss-Legendre quadrature. A different approach can be followed by considering the triangle as a degenerate quadrilateral, although in this case some special problems arise when the collocation point belongs to the integration element [22].

4.6.1.4 Computation of the Diagonal Terms of the Matrix B

We will finish this Sect. 4.6.1 with the study of the computation of the diagonal terms in (11), $B_{ik_{el}}$ with $\alpha = d(e, l)$.

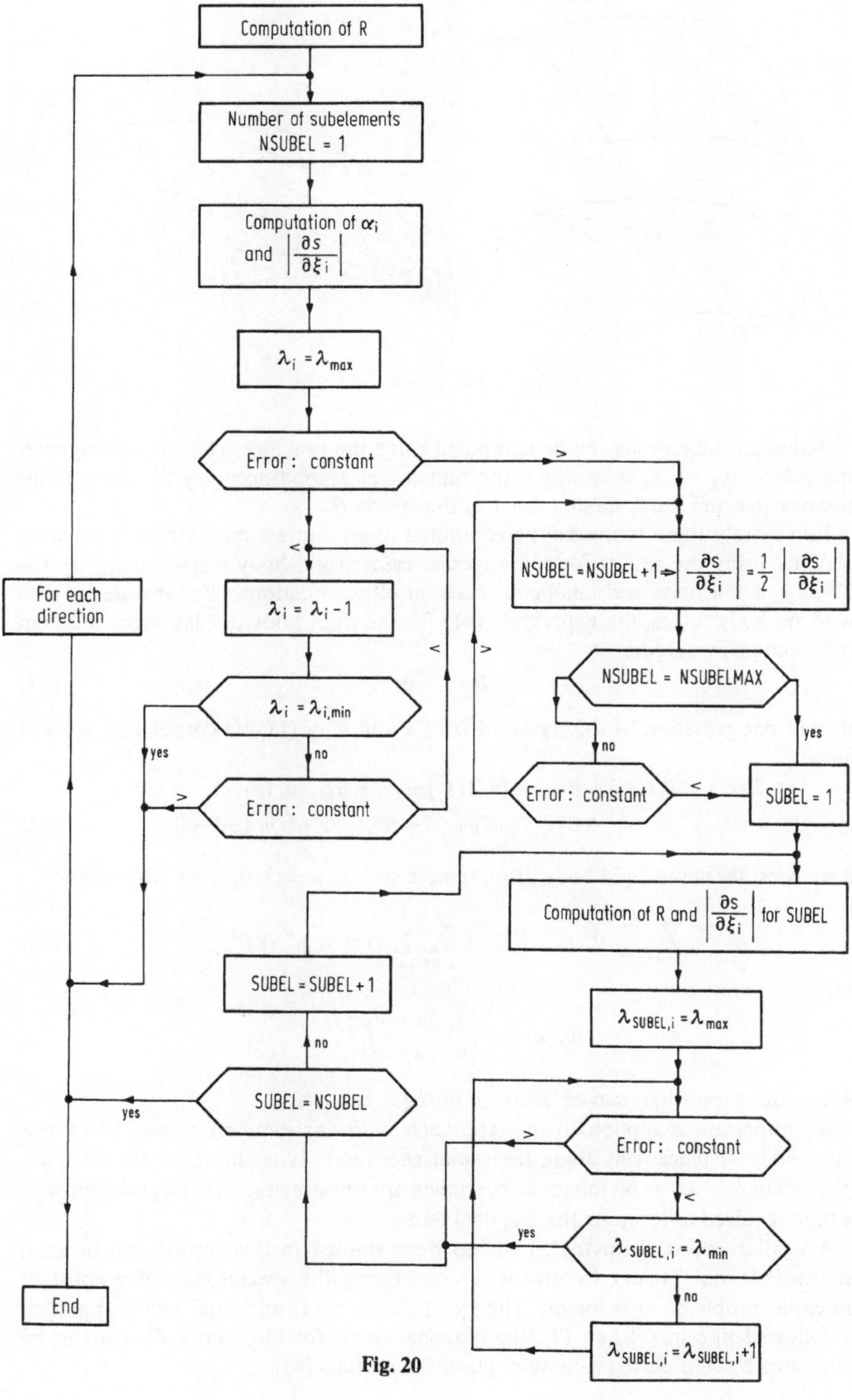
Computation of R
Number of subelements NSUBEL = 1
Computation of α_i and $\left|\frac{\partial s}{\partial \xi_i}\right|$
$\lambda_i = \lambda_{max}$
Error: constant
NSUBEL = NSUBEL + 1 ⇒ $\left|\frac{\partial s}{\partial \xi_i}\right| = \frac{1}{2}\left|\frac{\partial s}{\partial \xi_i}\right|$
For each direction
$\lambda_i = \lambda_i - 1$
NSUBEL = NSUBELMAX
$\lambda_i = \lambda_{i,min}$
yes
no
Error: constant
Error: constant
SUBEL = 1
Computation of R and $\left|\frac{\partial s}{\partial \xi_i}\right|$ for SUBEL
SUBEL = SUBEL + 1
$\lambda_{SUBEL,i} = \lambda_{max}$
SUBEL = NSUBEL
Error: constant
$\lambda_{SUBEL,i} = \lambda_{min}$
End
$\lambda_{SUBEL,i} = \lambda_{SUBEL,i}+1$

Fig. 20

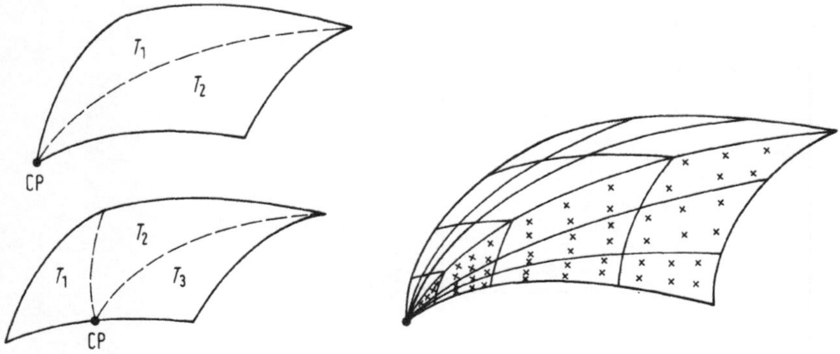

a b

Fig. 21 a, b

Although these terms can be computed using the previous schemes, adding later the matrix $\delta_{ik} - C_{ik}$, it is very important to get a good accuracy for these terms because they are the dominant terms of the matrix $\boldsymbol{B}$.

Fortunately these terms can be computed in an indirect but extremely accurate way, by using the results for some special cases (rigid body displacements in the Theory of Elasticity and analogous cases in other situations). For the elastic case with no body forces, the expression (11) for the rigid body displacement problem (no tractions) would be:

$$\boldsymbol{B} \cdot \boldsymbol{u} = \boldsymbol{0}. \tag{53}$$

If only one equation of this system [fixed α and k in (11)] is considered, we will have:

$$\begin{aligned} B^{\alpha}_{1k_{11}} u_1[d(1,1)] + B^{\alpha}_{2k_{11}} u_2[d(1,1)] + \ldots + B^{\alpha}_{1k_{el}} u_1(\alpha) \\ + B^{\alpha}_{2k_{el}} u_2(\alpha) + \ldots + B^{\alpha}_{3k_{pL_p}} u_3[d(p, L_p)] = 0. \end{aligned} \tag{54}$$

If we consider now a rigid body displacement such as $\boldsymbol{u} = (1, 0, 0)$ we will have:

$$\sum_{e=1}^{p} \sum_{l=1}^{L_p} \delta_{\alpha, d(e,l)} B^{\alpha}_{1k_{el}} = \sum_{e=1}^{p} \sum_{l=1}^{L_e} (1 - \delta_{\alpha, d(e,l)}) B^{\alpha}_{1k_{el}} \tag{55}$$

with

$$\delta_{\alpha, d(e,l)} = \begin{cases} 1, & \alpha = d(e,l) \\ 0, & \alpha \neq d(e,l) \end{cases}$$

Analogous expressions can be obtained for $B^{\alpha}_{2k_{el}}$ and $B^{\alpha}_{3k_{el}}$.

An important exception to this approach is the axisymmetric case, because a rigid body displacement along the radial coordinate is meaningless for this case (the strain $\varepsilon_\theta = u/r$ is no longer zero, so they are the stresses). An integral approach is then required to compute the diagonal terms.

A similar approach to that, that has been studied in this section can be used in the Potential Theory by means of considering the special case of a constant potential problem. Also for the Theory of Plasticity, if an initial strain approach is followed like in (6) and (7), the diagonal terms for the matrix $\boldsymbol{C}$ (13) can be computed by considering a constant plastic strain field [45].

```
      SUBROUTINE MATELA
C
C     COMPUTATION OF THE COEFFICIENTS MATRIX AND THE VECTOR OF INDEPENDENT
C     TERMS. THE MATRIX IS STORED IN FILE 1.
C
C
C                    DIMENSIONING AND DEFINITION OF ARRAYS
C
      COMMON /CIA/ IA(1)
      COMMON /CIAB/ IAB(1)
      COMMON /CONST/LEC,IMP,MEMO,MEMOB,NBP,NPER,NPEDP,NCON,NUDOSC,
     *              NPI,NUDOS,NELEM,NAPOYO,IPARA6,YOUNG,POIS,PORANA
      COMMON /WORK/ NDIM,RN1,ZN1,RN2,ZN2,RNI,ZNI,IER1,IER2,IER3,IER4,
     *              IER5,IER6,IER7,IER8,IER,NBUCLE,NELEM2,DIVI(201),
     *     U(2,2),T(2,2),UT(2,2),TT(2,2),CONSTA(5),CAREL(15)
      COMMON /CPUI/ TCPU,CPUT,CPU1,CPU2,CPU3,CPU4,CPU5,CPU6,CPU7,CPU8
      DIMENSION RIA(1)
      EQUIVALENCE(IA,RIA)
C
C                    CREATION OF THE ARRAYS AND LOCATION OF THE ONESALREADY
C                    CREATED.
C
      IER1=LOCA('COOR',NPER)
      IER2=LOCA('CONTOR',NPER)
      CALL CREAR('CAREL',15,NELEM,1,NPER,IER3)
      NDIM=4*NELEM
      NELEM2=2*NELEM
      CALL CREAR('VEC',NDIM,1,1,NPER,IER4)
      CALL CREARF('A',NELEM2,NELEM2,1,NPER,1,IER5)
C
C                    COMPUTATION OF THE CHARACTERISTICS OF EACH ELEMENT
C                    AND THE CONSTANTS NEEDED FOR THE INTEGRATION PROCESS
C
      DO 10 I=1,NELEM
           CALL COEX(RIA(IER1),I,RN1,ZN1,RN2,ZN2)
           IER=IER3+15*(I-1)
           CALL CARE(RN1,ZN1,RN2,ZN2,RIA(IER))
   10 CONTINUE
      CALL CONST1(YOUNG,POIS,CONSTA)
C
C                    INTEGRATION FROM EACH COLLOCATION POINT
C
      DO 1000 I=1,NELEM
C
C                     CO-ORDINATES OF THIS POINT
C
           JJ=15*(I-1)
           RNI=RIA(JJ+13+IER3)
           ZNI=RIA(JJ+14+IER3)
C
C                     INTEGRATION OVER EACH ELEMENT
C
           NBUCLE=IPARA6+1
           DO 500 J=1,NELEM
                CALL CEROS(UT,4)
                CALL CEROS(TT,4)
                IER=IER3+15*(J-1)
                CALL TRANS(RIA(IER),CAREL,15)
                DO 400 NB=1,NBUCLE
                     DO 20 NCAR=1,6
                          NPOS=2*NCAR
   20                CAREL(NPOS)=-CAREL(NPOS)*(2*NB-3)
                     CAREL(15)=-CAREL(15)*(2*NB-3)
C
C                     SEE IF THE INTEGRATION ELEMENT AND COLLOCATIONPOINT
C                     COINCIDE
C
                     IF(I.EQ.J.AND.NB.EQ.1) GO TO 31
C
C                     THEY DO NOT COINCIDE. SEE IF THE SUBDIVISION OF THE
C                     ELEMENT IS NEEDED,AND IF IT IS, IT IS DONE
C
                     CALL COEX(RIA(IER1),J,RN1,ZN1,RN2,ZN2)
                     ZN1=-ZN1*(2*NB-3)
                     ZN2=-ZN2*(2*NB-3)
                     CALL DIVEL(CAREL,RNI,ZNI,RN1,ZN1,RN2,ZN2,DIVI)
```

```
C
C                    NUMERICAL  INTEGRATION
C
                      CALL INTNU(CAREL,DIVI,CONSTA,RNI,ZNI,U,T)
                      GO TO 101
C
C                   THEY DO COINCIDE. A SEMIANALYTICAL INTEGRATION APPROACH
C                   IS USED
C
C                   COMPUTATION OF U(I,J) AND T(I,J)
C
   31                 CALL INTAN(CAREL,CONSTA,U,T)
  400            CONTINUE
                 CALL MONTAC(IA(IER2),IA(IER4),NDIM,CAREL,I,J,UT,TT)
C
C                   END OF THE INTEGRATION OVER THE ELEMENTS
C
  500       CONTINUE
C
C                   END OF THE INTEGRATION FROM EACH COLLOCATION POINT
C
 1000 CONTINUE
      RETURN
      END
```

```
      SUBROUTINE CARE(RN1,ZN1,RN2,ZN2,CAREL)
C
C     COMPUTATION OF THE CHARACTERISTICS OF EACH ELEMENT NEEDED FOR
C     THE INTEGRATION PROCESS
C
C********************DICTIONARY
C
C     RN1,ZN1              =CO-ORDINATES OF THE FIRST NODE OF THE ELEMENT
C     RN2,ZN2              =CO-ORDINATES OF THE SECOND NODE OF THEELEMENT
C     CAREL(1) A CAREL(10) =CARTESIAN CO-ORDINATES OF THE INTEGRATION POINTS
C     CAREL(11) A CAREL(12)=COMPONENTS OF THE NORMAL
C     CAREL(13)            =LENGTH OF THE ELEMENT
C     CAREL(14) A CAREL(15)=CO-ORDINATES OF THE C.O.G. OF THE ELEMENT
C*******************************
C
C                   DIMENSIONING AND DEFINITION OF ARRAYS
C
      COMMON /WORK/ A(25/),COINTR(5)
      DIMENSION CAREL(15)
C
C                   INTRINSIC CO-ORDINATES FOR GAUSSIAN INTEGRATION
C
      DATA COINTR/-.9061798459,-.5384693101,0.,.5384693101,.9061798459/
C
C
C                   COMPUTATION OF THE LENGTH OF THE ELEMENT
C
      CAREL(13)=SQRT((RN2-RN1)**2+(ZN2-ZN1)**2)
C
C                   COMPUTATION OF THE COMPONENTS OF THE NORMAL
C
      CAREL(11)=(ZN2-ZN1)/CAREL(13)
      CAREL(12)=(RN1-RN2)/CAREL(13)
C
C                   CO-ORDINATES OF THE CENTER OF GRAVITY OF THE ELEMENT
C
      CAREL(14)=(RN2+RN1)/2.
      CAREL(15)=(ZN2+ZN1)/2.
C
C                   CARTESIAN CO-ORDINATES OF THE INTEGRATION POINTS
C
      DO 10 II=1,5
            CAREL(2*II-1)=RN1*(1.-COINTR(II))/2.+RN2*(1.+COINTR(II))/2.
            CAREL(2*II)=ZN1*(1.-COINTR(II))/2.+ZN2*(1.+COINTR(II))/2.
   10 CONTINUE
C
C                   END OF THE PROCESS
C
      RETURN
      END
```

```
      SUBROUTINE INTNU(CAREL,DIVI,CONSTA,RNI,ZNI,U,T)
C
C     NUMERICAL COMPUTATION OF THE TENSORS U(I,J) AND T(I,J) WHEN
C     THE INTEGRATION ELEMENT AND COLLOCATION POINT DO NOT COINCIDE
C
C******************** DICTIONARY
C
C     CAREL(15)=CHARACTERISTICS OF THE INTEGRATION ELEMENT
C     DIVI(201)=NUMBER OF SUBDIVISIONS OF THE ELEMENT AND CARTESIAN
C               CO-ORDINATES OF THE INTEGRATION POINTS
C     CONSTA(5)=CONSTANTS OF THE INTEGRATION PROCESS
C     RNI,ZNI   =CARTESIAN CO-ORDINATES OF THE COLLOCATION POINT
C     U(2,2)    =TENSOR U(I,J)
C     T(2,2)    =TENSOR T(I,J)
C
C*******************************
C
C                    DIMENSIONING AND DEFINITION OF ARRAYS
C
      COMMON /CIA/ IA(1)
      COMMON /CONST/ LEC,IMP,MEMO,MEMOB,NBP,NPER,NPEDP,NCON,NUDOSC,
     *               NPI,NUDOS,NELEM,NAPOYO,IPARA6,YOUNG,POIS,PORANA
      COMMON /WORK/ A(250),PEGAUS(5),PII,NPGAUS,SIGMA(6)
      DIMENSION CAREL(15),DIVI(201),CONSTA(5),U(2,2),T(2,2)
C
C                    AUXILIARY FUNCTIONS
C
      DEFINE UURR(I)=URR(DIVI(2*I),DIVI(2*I+1),RNI,ZNI,CONSTA)
      DEFINE UUZR(I)=UZR(DIVI(2*I),DIVI(2*I+1),RNI,ZNI,CONSTA)
      DEFINE UURZ(I)=URZ(DIVI(2*I),DIVI(2*I+1),RNI,ZNI,CONSTA)
      DEFINE UUZZ(I)=UZZ(DIVI(2*I),DIVI(2*I+1),RNI,ZNI,CONSTA)
      DEFINE DURRR(I)=URRRP(DIVI(2*I),DIVI(2*I+1),RNI,ZNI,CONSTA)
      DEFINE DURRZ(I)=URRZP(DIVI(2*I),DIVI(2*I+1),RNI,ZNI,CONSTA)
      DEFINE DUZRR(I)=UZRRP(DIVI(2*I),DIVI(2*I+1),RNI,ZNI,CONSTA)
      DEFINE DUZRZ(I)=UZRZP(DIVI(2*I),DIVI(2*I+1),RNI,ZNI,CONSTA)
      DEFINE DURZR(I)=URZRP(DIVI(2*I),DIVI(2*I+1),RNI,ZNI,CONSTA)
      DEFINE DURZZ(I)=URZZP(DIVI(2*I),DIVI(2*I+1),RNI,ZNI,CONSTA)
      DEFINE DUZZR(I)=UZZRP(DIVI(2*I),DIVI(2*I+1),RNI,ZNI,CONSTA)
      DEFINE DUZZZ(I)=UZZZP(DIVI(2*I),DIVI(2*I+1),RNI,ZNI,CONSTA)
      DATA PEGAUS/.236926885,.4786286704,.5688888889,.4786286704,
     *           .236926885/
      G=CONSTA(5)
      PII=CAREL(13)*4.*ATAN(1.)
      CALL CEROS(U,4)
      CALL CEROS(SIGMA,8)
C
C                    COMPUTATION OF THE TENSOR U(I,J)
C
      NPGAUS=DIVI(1)*5
      DO 100 I=1,NPGAUS
         J=(I-1)/5
         J=I-5*J
         U(1,1)=U(1,1)+UURR(I)*PEGAUS(J)*PII*DIVI(2*I)
         U(1,2)=U(1,2)+UUZR(I)*PEGAUS(J)*PII*DIVI(2*I)
         U(2,1)=U(2,1)+UURZ(I)*PEGAUS(J)*PII*DIVI(2*I)
         U(2,2)=U(2,2)+UUZZ(I)*PEGAUS(J)*PII*DIVI(2*I)
C
C                    COMPUTATION OF THE TENSOR SIGMA(I,J,K)
C
         SIGMA(1)=SIGMA(1)+2.*G*(CONSTA(3)*DURRR(I)+CONSTA(4)*
     *       (UURR(I)/DIVI(2*I)+DUZRZ(I)))*PEGAUS(J)*PII*DIVI(2*I)
         SIGMA(2)=SIGMA(2)+G*(DURRZ(I)-DUZRR(I))*PEGAUS(J)*PII*
     *       DIVI(2*I)
         SIGMA(3)=SIGMA(3)+2.*G*(CONSTA(3)*DUZRZ(I)+CONSTA(4)*
     *       (UURR(I)/DIVI(2*I)+DURRR(I)))*PEGAUS(J)*PII*DIVI(2*I)
         SIGMA(4)=SIGMA(4)+2.*G*(CONSTA(3)*DURZR(I)+CONSTA(4)*
     *       (UUZR(I)/DIVI(2*I)+DUZZZ(I)))*PEGAUS(J)*PII*DIVI(2*I)
         SIGMA(5)=SIGMA(5)+G*(DURZZ(I)+DUZZR(I))*PEGAUS(J)*PII*
     *       DIVI(2*I)
         SIGMA(6)=SIGMA(6)+2.*G*(CONSTA(3)*DUZZZ(I)+CONSTA(4)*
     *       (UUZR(I)/DIVI(2*I)+DURZR(I)))*PEGAUS(J)*PII*DIVI(2*I)
  100 CONTINUE
C
```

```
C                       COMPUTATION OF THE TENSOR T(I,J)
C
      T(1,1)=SIGMA(1)*CAREL(11)+SIGMA(2)*CAREL(12)
      T(2,1)=SIGMA(2)*CAREL(11)+SIGMA(3)*CAREL(12)
      T(1,2)=SIGMA(4)*CAREL(11)+SIGMA(5)*CAREL(12)
      T(2,2)=SIGMA(5)*CAREL(11)+SIGMA(6)*CAREL(12)
C
C                       END OF THE PROCESS
C
      RETURN
      END
```

```
      SUBROUTINE DIVEL[CAREL,RNI,ZNI,RN1,ZN1,RN2,ZN2,DIVI]
C
C     SEES IF IT IS NECESSARY THE SUBDIVISION OF AN ELEMENT AND COMPUTES
C     THE NUMBER OF NEEDED SUBDIVISIONS. ALSO COMPUTES THE CARTESIAN
C     CO-ORDINATES OF THE INTEGRATION POINTS, FOR A MAXIMUM OF
C     20 SUBDIVISIONS
C
C*******************DICTIONARY
C
C     CAREL[15]      =CHARACTERISTICS OF THE INTEGRATION ELEMENT
C     RNI,ZNI        =CARTESIAN CO-ORDINATES OF THE COLLOCATION POINT
C     RN1,ZN1        =CARTESIAN CO-ORDINATES OF THE FIRST NODE OF THE ELEMENT
C     RN2,ZN2        =CARTESIAN CO-ORDINATES OF THE SECOND NODE OF THE ELEMENT
C     DIVI[1]        =NUMBER OF SUBDIVISIONS
C     DIVI[2 A 200]=CARTESIAN CO-ORDINATES OF THE INTEGRATION POINTS
C******************************
C
C
C
C
                     DIMENSIONING AND DEFINITION OF ARRAYS

      COMMON /WORK/ AN[250],RP,ZP,CUR[15],NSUBEL,R1,Z1,R2,Z2,DMIN,ANEL
      DIMENSION CAREL[15],DIVI[201]
C
C                    COMPUTATION OF THE DISTANCE BETWEEN THE INTEGRATION
C                    ELEMENT AND THE COLLOCATION POINT, AND ALSO THE
C                    CARTESIAN CO-ORDINATES OF THE POINT OF THE ELEMENT
C                    FOR WHICH THIS DISTANCE IS COMPUTED
C
      IF[CAREL[11].GT.2.E-6]GO TO 1
      RP=RNI
      GO TO 21
    1 RP=[ZNI-ZN1+RN1*[ZN2-ZN1]/[RN2-RN1]-CAREL[12]/CAREL[11]*
     *RNI]/[[ZN2-ZN1]/[RN2-RN1]-CAREL[12]/CAREL[11]]
   21 IF[RP.GT.MAX0[RN1,RN2]] RP=MAX0[RN1,RN2]
      IF[RP.LT.MIN0[RN1,RN2]] RP=MIN0[RN1,RN2]
      ZP=ZN1+[RP-RN1]*[ZN2-ZN1]/[RN2-RN1]
      DMIN=SQRT[[RNI-RP]**2+[ZNI-ZP]**2]
C
C                    IF THE DISTANCE IS BIGGER THAN THE LENGTH OF THE ELEMENT
C                    THERE IS NO SUBDIVISION AND ONLY 5 INTEGRATION POINTS
C                    ARE USED
C
      IF[DMIN.LT.CAREL[13]] GO TO 11
      DIVI[1]=1
      CALL TRANS[CAREL[1],DIVI[2],10]
      RETURN
C
C                    IF THE DISTANCE IS SMALLER THAN THE LENGTH OF THE ELEMENT
C                    THE NUMBER OF SUBDIVISIONS IS COMPUTED
C
   11 NSUBEL=CAREL[13]/DMIN+1
      IF[NSUBEL.GT.20] NSUBEL=20
      DIVI[1]=NSUBEL
C
C                    COMPUTATION OF THE CARTESIAN CO-ORDINATES OF THE
C                    INTEGRATION POINTS
C
      R1=RN1
      Z1=ZN1
      ANEL=CAREL[13]/NSUBEL
```

```
      DO 100 I=1,NSUBEL
           R2=R1-ANEL*CAREL(12)
           Z2=Z1+ANEL*CAREL(11)
           IF(I.EQ.NSUBEL) R2=RN2
           IF(I.EQ.NSUBEL) Z2=ZN2
           CALL CARE(R1,Z1,R2,Z2,COR)
           II=(I-1)*10+2
           CALL TRANS(COR,DIVI(II),10)
           R1=R2
           Z1=Z2
  100 CONTINUE
C
C                 END OF THE PROCESS
C
       RETURN
       END
```

```
      SUBROUTINE INTAN(CAREL,CONSTA,U,T)
C
C     COMPUTATION OF THE TENSORS U(I,J) Y T(I,J) WHEN THE COLLOCATION
C     POINT AND THE INTEGRATION ELEMENT COINCIDE
C
C*******************DICTIONARY
C
C     CAREL(15)-CHARACTERISTICS OF THE ELEMENT
C     CONSTA(5)-CONSTANTS OF THE INTEGRATION PROCESS
C     U(2,2)   =TENSOR U(I,J)
C     T(2,2)  =TENSOR T(I,J)
C*******************************
C
C                 DIMENSIONING AND DEFINITION OF ARRAYS
C
      COMMON /CONST/ LEC,IMP,MEMO,MEMOB,NBP,NPER,NPEDP,NCON,NUDOSC,
     *           NPI,NUDOS,NELEM,NAPOYO,IPARA6,YOUNG,POIS,PORANA
      COMMON /WORK/ AN(18),DIVI(201),AN1(36),SCARE(15),UN(2,2),
     *           TN(2,2),SIGMA(6)
      DIMENSION CAREL(15),CONSTA(5),U(2,2),T(2,2)
C
C                 DEFINITION OF SOME AUXILIAR FUNCTIONS
C
      DEFINE URR=AURR(CAREL,PORANA,CONSTA)
      DEFINE URZ=AURZ(CAREL,PORANA,CONSTA)
      DEFINE UZR=AUZR(CAREL,PORANA,CONSTA)
      DEFINE UZZ=AUZZ(CAREL,PORANA,CONSTA)
      DEFINE DURRR=ADURRR(CAREL,PORANA,CONSTA)
      DEFINE DURRZ=ADURRZ(CAREL,PORANA,CONSTA)
      DEFINE DURZR=ADURZR(CAREL,PORANA,CONSTA)
      DEFINE DURZZ=ADURZZ(CAREL,PORANA,CONSTA)
      DEFINE DUZRR=ADUZRR(CAREL,PORANA,CONSTA)
      DEFINE DUZRZ=ADUZRZ(CAREL,PORANA,CONSTA)
      DEFINE DUZZR=ADUZZR(CAREL,PORANA,CONSTA)
      DEFINE DUZZZ=ADUZZZ(CAREL,PORANA,CONSTA)
      DEFINE URRDR=AURRDR(CAREL,PORANA,CONSTA)
      DEFINE URZDR=AURZDR(CAREL,PORANA,CONSTA)
C
C                 ANALYTICAL COMPUTATION OF THE TENSOR U(I,J)
C
      G=CONSTA(5)
      U(1,1)=URR
      U(1,2)=UZR
      U(2,1)=URZ
      U(2,2)=UZZ
C
C                 ANALYTICAL COMPUTATION OF THE TENSOR SIGMA(I,J,K)
C
      SIGMA(1)=2.*G*(CONSTA(3)*DURRR+CONSTA(4)*(URRDR+DUZRZ))
      SIGMA(2)=G*(DURRZ+DUZRR)
      SIGMA(3)=2.*G*(CONSTA(3)*DURZR+CONSTA(4)*(URRDR+DURRR))
      SIGMA(4)=2.*G*(CONSTA(3)*DURZR+CONSTA(4)*(URZDR+DUZZZ))
      SIGMA(5)=G*(DURZZ+DUZZR)
      SIGMA(6)=2.*G*(CONSTA(3)*DUZZZ+CONSTA(4)*(URZDR+DURZR))
C
C                 ANALYTICAL COMPUTATION OF THE TENSOR T(I,J)
C
      T(1,1)=SIGMA(1)*CAREL(11)+SIGMA(2)*CAREL(12)
      T(2,1)=SIGMA(2)*CAREL(11)+SIGMA(3)*CAREL(12)
```

```
      T(1,2)=SIGMA(4)*CAREL(11)+SIGMA(5)*CAREL(12)
      T(2,2)=SIGMA(5)*CAREL(11)+SIGMA(6)*CAREL(12)
C
C                    NUMERICAL PART OF THE PROCESS. DIVISION OF THE TWO
C                    HALVES OF THE ELEMENT AND INTEGRATION PROCESS
C
      IC=-1
      DO 100 I=1,2
           RN2=CAREL(14)-CAREL(12)*CAREL(13)*IC/2
           ZN2=CAREL(15)+CAREL(11)*CAREL(13)*IC/2
           RN1=CAREL(14)-CAREL(12)*CAREL(13)*PORANA*IC/2
           ZN1=CAREL(15)+CAREL(11)*CAREL(13)*PORANA*IC/2
           CALL DIVAN(RN1,ZN1,RN2,ZN2,PORANA,DIVI)
           CALL INTNU(CAREL,DIVI,CONSTA,CAREL(14),CAREL(15),UN,TN)
           DO 20 J=1,2
                DO 10 K=1,2
                     U(J,K)=U(J,K)+UN(J,K)
                     T(J,K)=T(J,K)+TN(J,K)
   10           CONTINUE
   20      CONTINUE
           IC=1
  100 CONTINUE
C
C                    INCREASING THE TENSOR T(I,J) WITH C(I,J)
C
      T(1,1)=T(1,1)+0.5
      T(2,2)=T(2,2)+0.5
C
C                    END OF THE PROCESS
C
      RETURN
      END
```

```
       SUBROUTINE DIVAN(RN1,ZN1,RN2,ZN2,PORANA,DIVI)
C
C     SUBDIVISION OF EACH HALF OF THE ELEMENT WHEN THE INTEGRATION
C     ELEMENT AND COLLOCATION POINT COINCIDE
C
C******************DICTIONARY
C
C     RN1,ZN1   =CO-ORDINATES OF THE FIRST NODE OF THE ELEMENT
C     RN2,ZN2   =CO-ORDINATES OF THE SECOND NODE OF THE ELEMENT
C     PORANA    =PERCENTAGE OF THE LENGTH OF THE ELEMENT FOR ANALYTICAL
C                INTEGRATION
C     DIVI(201)=NUMBER OF SUBDIVISIONS AND CO-ORDINATES OF THE
C                INTEGRATION POINTS
C******************************
C
C                    DIMENSIONING AND DEFINITION OF ARRAYS
C
       COMMON /WORK/ AN(249),N,R1,Z1,R2,Z2,SCAREL(15)
       DIMENSION DIVI(201)
C
C     COMPUTATION OF THE NUMBER OF SUBDIVISIONS
C
       N=-ALOG(PORANA)/ALOG(2)+1
       DIVI(1)=N
       AL=SQRT((RN2-RN1)**2+(ZN2-ZN1)**2)
       TR=(RN2-RN1)/AL
       TZ=(ZN2-ZN1)/AL
C
C                    COMPUTATION OF THE CO-ORDINATES OF THE INTEGRATION POINTS
C
      R1=RN1
      Z1=ZN1
      DO 100 I=1,N
            BL=2*AL*(PORANA+1)
            R2=R1+2**(I-2)*PORANA*TR*BL
            Z2=Z1+2**(I-2)*PORANA*TZ*BL
            IF(I.NE.N) GO TO 11
            R2=RN2
            Z2=ZN2
   11       CALL CARE (R1,Z1,R2,Z2,SCAREL)
            K=10*(I-1)+2
            CALL TRANS(SCAREL,DIVI(K),10)
             R1=R2
             Z1=Z2
```

```
  100 CONTINUE
C
C                    END OF THE PROCESS
C
      RETURN
      END
```

```
      SUBROUTINE COEX(COOR,NEL,RN1,ZN1,RN2,ZN2)
C
C     COMPUTATION OF THE CARTESIAN CO-ORDINATES OF THE NODES OF
C     A GIVEN ELEMENT
C
C*******************DICTIONARY
C
C     COOR(2,NUDOS)-CO-ORDINATES OF THE NODES OF THE PROBLEM
C     NEL          =NUMBER OF THE ELEMENT UNDER CONSIDERATION
C     RN1,ZN1      =CO-ORDINATES OF THE FIRST NODE OF THE ELEMENT
C     RN2,ZN2      =CO-ORDINATES OF THE SECOND NODE OF THE LEMENT
C     N1,N2        =FIRST AND SECOND NODES OF THE ELEMENT
C******************************
C
C                    DIMENSIONING AND DEFINITION OF ARRAYS
C
      COMMON /CIA/ IA(1)
      COMMON /CONST/ LEC,IMP,MEMO,MEMOB,NBP,NPER,NPEDP,NCON,NUDOSC,
     *               NPI,NUDOS,NELEM,NAPOYO,IPARA6,YOUNG,POIS,PORANA
      COMMON /WORK/ A(255),IER1,NCONT
      DIMENSION COOR(2,NUDOS)
C
C                    LOCATION OF THE BOUNDARY TO WHICH THIS ELEMENT BELONGS
C
      IER1=LOCA('NELE',1)
      NCONT=NCONEL(IA(IER1),NEL)
C
C                    NODES OF THE ELEMENT AND THEIR CO-ORDINATES
C
      N1=NEL+NCONT-1
      N2=N1+1
      RN1=COOR(1,N1)
      ZN1=COOR(2,N1)
      RN2=COOR(1,N2)
      ZN2=COOR(2,N2)
      RETURN
      END
```

```
      SUBROUTINE CONST1(YOUNG,POIS,CONSTA)
C
C     COMPUTATION OF THE CONSTANTS NEEDED FOR THE INTEGRATION PROCESS
C
C*******************DICTIONARY
C
C     YOUNG      =ELASTIC MODULUS OF THE MATERIAL
C     POIS       =POISSON RATIO OF THE MATERIAL
C     CONSTA(5)=COMPUTED CONSTANTS
C******************************
C
C                    DIMENSIONING AND DEFINITION OF ARRAYS
C
      DIMENSION CONSTA(5)
C
C                       COMPUTATION OF THE CONSTANTS
C
      CONSTA(5)=YOUNG/2./(1.+POIS)
      PI=4.*ATAN(1.)
      CONSTA(1)=16.*PI*PI*CONSTA(5)*(1.-POIS)
      CONSTA(2)=3.-4.*POIS
      CONSTA(3)=(1.-POIS)/(1.-2*POIS)
      CONSTA(4)=CONSTA(3)-1.
      CALL IMCONS(CONSTA)
      RETURN
      END
      FUNCTION NCONEL(NELE,NEL)
C
C     DETECTION OF THE BOUNDARY TO WHICH AN ELEMENT BELONGS
C
```

```
C*******************DICTIONARY
C
C     NELE(NCON)=NUMBER OF ELEMENTS OF EACH BOUNDARY
C     NEL       =NUMBER OF THE ELEMENT UNDER CONSIDERATION
C******************************
C
C
C                  DIMENSIONING AND DEFINITION OF ARRAYS
C
      COMMON /CONST/LEC,IMP,MEMO,MEMOB,NRP,NPER,NPEDP,NCON
      COMMON /WORK/ A(227),NELMAX,NCONM1
      DIMENSION NELE(NCON)
C
C                  DETECTION  OF THE BOUNDARY
C
      NELMAX=NELE(1)
      DO 10 I=1,NCON
           IF(NEL.LE.NELMAX) GO TO 11
           NELMAX=NELMAX+NELE(I+1)
   10 CONTINUE
C
C                  THE ELEMENT DOES NOT BELONG TO ANY BOUNDARY
C
      WRITE(IMP,20000)NEL
      NCONEL=0
      RETURN
C
C                  THE ELEMENT BELONGS TO BOUNDARY I
C
   11 NCONEL=I
      RETURN
20000 FORMAT(10X,'THE ELEMENT',I4,' DOES NOT BELONG TO ANY BOUNDARY'/)
      END
```

```
      SUBROUTINE CEROS(XA,JA)
C
C     PUT TO ZERO JA ELEMENTS OF THE ARRAY XA
C
C*****************DICTIONARY
C
C     XA=ARRAY TO BE PUT TO ZERO
C     JA=NUMBER OF ELEMENTS TO BE PUT TO ZERO
C*****************************
C
C
C                  DIMENSIONING AND DEFINITION OF ARRAYS
C
      DIMENSION XA(1)
C
C
C                  THE ARRAY IS PUT TO ZERO
C
      DO 100 IA=1,JA
         XA(IA)=0.0
  100 CONTINUE
      RETURN
      END
```

4.6.2 Boundary Conditions and Assemblage of the Matrices

The imposition of boundary conditions (b.c.) is, as was deduced in (12), other essential point of this method, in order to get the same number of equations than unknowns.

There are three different types of boundary conditions for each element: Dirichlet conditions (essential b.c. are given); Neumann conditions (natural b.c. are known); mixed conditions. It is necessary to remember now that, as it is well known, a pure Dirichlet problem all over the boundary has infinite solutions. In order to avoid this problem there are two different approaches from a practical point of view. The first consists in replacing some of the natural b.c. by essential

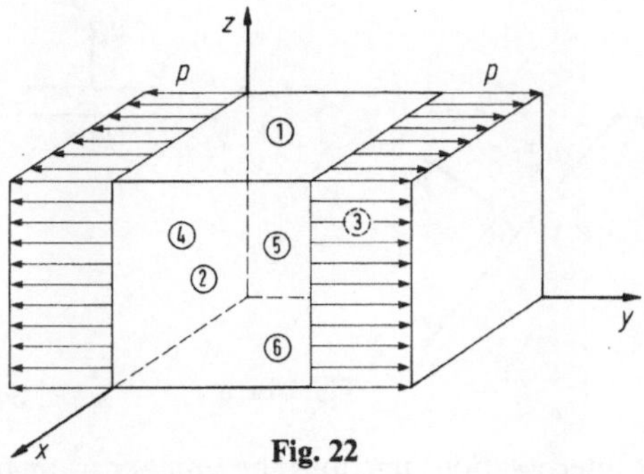

Fig. 22

ones that do not alterate the final results. With this, one of the solutions of the problem is obtained, being possible to get any other by adding a rigid body movement to the above first solution (constant potential solution for Potential Theory). For the hexahedron in Fig. 22, for instance, with constant tractions over two faces (3) and (4), a solution can be obtained by fixing the normal displacements of faces (4) – (6) or any other sequence of faces normal to the global axes.

The second method lies in applying the b.c. as in a Neumann problem and once the system of linear equations has been obtained, some of the displacements of certain points of the boundary are put to zero, so that as in the former solution the deformed configuration is not changed. In Fig. 22 it could be the normal displacements of the center of gravity of the faces (4) – (6). The way it is accomplished is the same that boundary conditions are prescribed in the F.E.M.

For the rest of situations, the complexity of the treatment is closely related to the interpolation functions for the displacements and tractions, being possible to differentiate two cases: constant functions and non-constant ones. This is due to the fact that a special treatment is needed when there are several elements with a common node which is a feature of the non-constant interpolation functions.

For the case of constant approximation there is no problem. The number of unknowns for each element is always the same than the number of equations that can be obtained for that element as a collocation point. Also the b.c. can be defined element by element or node by node, so that both concepts coincide. For the triangle in Fig. 23 three of the six possible variables ($u, v, w, \sigma, \tau_1, \tau_2$) will be known, whatever b.c. are prescribed. The rest three are the unknowns for this element, which is therefore completely determined.

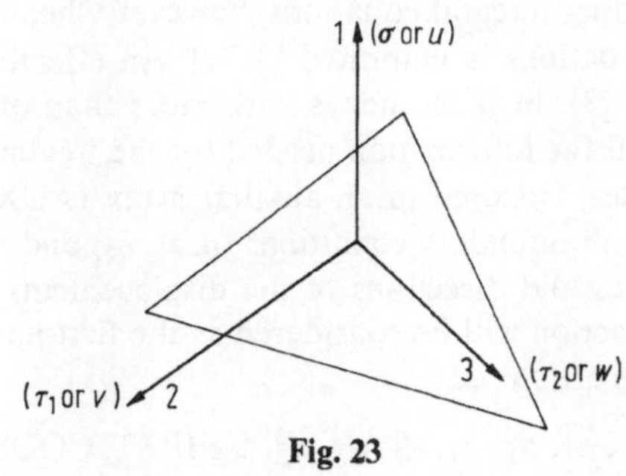

Fig. 23

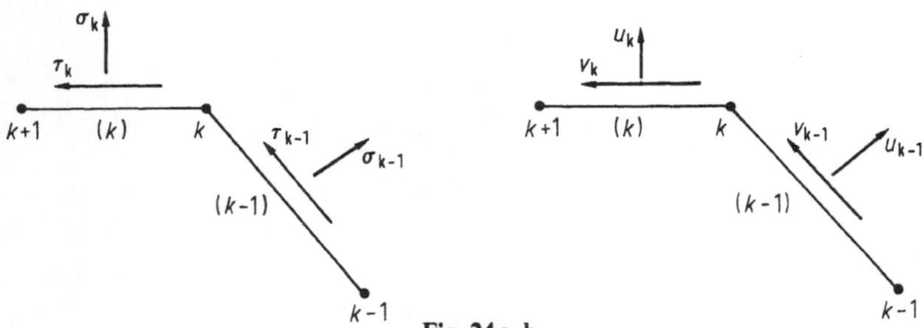

a **Fig. 24 a, b** b

For non-constant interpolation functions the number of unknowns for each node is not determined being possible that this number is greater than the number of equations that can be established for this collocation point. In order to fix these ideas we will consider the two situations in Fig. 24 for Plane Elasticity.

In the first case, the tractions on the elements $(k-1)$ and (k) are known, so that, the unknowns for the node k are the displacements (2 unk.). Therefore, and because two integral equations can be written for that node, the problem is well established. In case (b), the displacements are known for both elements (k) and $(k-1)$, the tractions are the unknowns (4 for node k: $\sigma_{(k-1)}$, $\sigma_{(k)}$, $\tau_{(k-1)}$, $\tau_{(k)}$, where the index in brackets represents the element to which the traction is associated). Some approaches to solve this indeterminancy (not consisting in incrementing the number of equations by writing the integral equations for additional collocation points, Brebbia [17]) are shown in Chaudonneret [70], Paris and Alarcon [71] and Paris [38].

In the last, the two principal stresses at the node k are considered as the unknowns associated to it, in the same way that it is done in the tree-dimensional case that will be studied along the next paragraphs. The same scheme can be followed in Potential Theory changing the concept of principal stresses by the module of the gradient of the potential function.

In the 3-D case the situation is much more complex, not only due to the greater number of dimensions but mainly, because the number of faces that can intersect at a specific node is not bounded, so it is the number of unknowns associated to it. Since now a "face" will be understood as the set of elements which have the same normal and the same type and value of boundary conditions at the node under study.

When the number of unknowns for a node is three, the problem is determined only with its corresponding integral equations, however when it is not, the number of different possible situations is unlimited [3, 72]. An effective way to solve this problem is included in [3]. In it the nodes with more than one face are detected ("critical nodes") and all the information needed for the treatment of the boundary conditions of these nodes, is stored in an auxiliar array (AUXCRI). For example for a node with faces with boundary conditions (u, τ_1, τ_2) and any number of faces with (σ, τ_1, τ_2) the values and directions of the displacements u_1, u_2, the number of the face where the traction will be considered as the first unknown and the code for this specific situation are stored.

$$u_1,\ d_1^1,\ d_1^2,\ d_1^3,\ u_2,\ d_2^1,\ d_2^2,\ d_2^3,\ \text{NFIRST},\ \text{CODE}.$$

Two situations may be considered:

1. Number of unknowns tractions is less than $m = a + 3 - d$, where d is the number of unknown displacements and:

$$a = \binom{n}{2} - \binom{n-3}{2} \qquad n > 4$$

$$a = \binom{n}{2} \qquad n \leqq 4 \tag{56}$$

n – number of faces.

2. Number of unknowns is greater than m.

In the first case "a" tractions can be expressed as a function of the rest using the same number of equilibrium equations in the node (Cauchy's relations). With this, in the worst case, only $m - a = 3 - d$ unknown tractions there will appear, and therefore three will be the total number of remaining unknowns, so that the problem is again determined. The equilibrium relations are established between two faces (considering the continuity of the stress tensor) as:

$$\boldsymbol{T}^{(1)} \cdot \boldsymbol{\nu}^{(2)} = \boldsymbol{T}^{(2)} \cdot \boldsymbol{\nu}^{(1)} \tag{57}$$

where $\boldsymbol{T}^{(i)}$ is the traction vector on the face i at the node and is the normal on the face i at the same point. A complete study of these cases, including the different typology that can appear may be found in [3].

For computational purposes, it is also convenient to store in AUXCRI the data needed to apply the boundary conditions corresponding to that node.

In the second case, when the number of unknown tractions is greater than m it is not possible to use (57). In that case the principal stresses are again chosen as the unknowns for the node. Consequently in those cases, the displacements must always be known (the continuity of the stress tensor is again assumed). In order to express the rest of the tractions as a function of the principal stresses, the principal directions have to be computed, which is only possible if the variation of the three displacements along three linearly independent directions (1, 2, 3) is known, at the node under study. With a change of coordinates, the variation of these displacements along the global directions (x, y, z) is computed. From this, the strain tensor is obtained as they are the principal directions. (A detailed explanation may be found in [3].) Two of these principal directions are stored in AUXCRI.

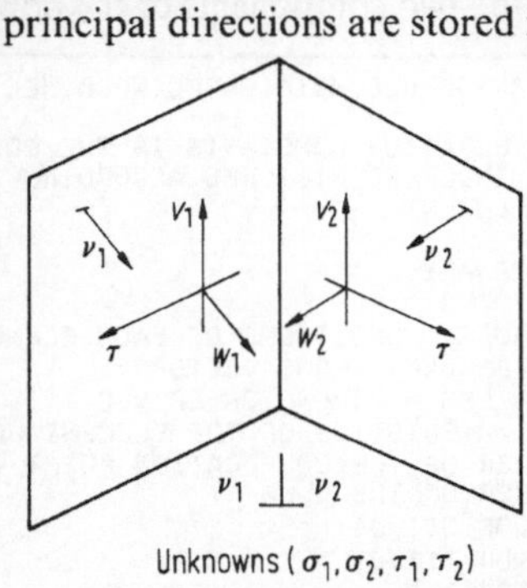

Fig. 25

In practice, however, some particular situations can appear for which some simplifications must be done. For example, when the number of unknown tractions in m, but two faces are orthogonal and the corresponding tangential tractions are known (Fig. 25). Of course, it is not possible at this moment to study all the possible examples, referring the reader to [3].

Once the unknowns have been chosen, it is only needed to define the most convenient sequence of operations to get the system of equations using the above processes when necessary. There are two main possibilities for that sequence:

(1) To integrate from collocation point to collocation point over all the elements. (Obtention of the matrix row by row.)
(2) To integrate over each element from all the C.P. (Obtention of the matrix column by column.)

The second one has advantages vis-a-vis the former, mainly because much more data for each element are needed during the formation of the matrix (coordinates, points of integration, components of the normal, boundary conditions) than those of the nodes (only coordinates) so that those data are computed only once for each element, reducing drastically the computation time. The only disadvantage occurs when, due to the dimension of the problem, the matrix has to be stored in auxiliary memory, because the majority of the solution schemes work with row blocks and not with column ones. In order to avoid this problem three different approaches can be followed:

(a) A reordering process to a row block scheme [22], which means an important loss of time.
(b) To use a column block solution scheme [3], which seems to be the correct approach.
(c) To follow the process (1) above, but computing and storing only once all the needed data for each element. This is a good method for time purposes, but instead it needs more memory, being the best for small problems, and also when these data are used in another part of the program (Plasticity) [7].

In the attached program this last scheme has been used, computing and storing (array CAREL) all the data for each element (integration point coordinates, length, coordinates of the C.O.G., the two components of the normal).

```
      SUBROUTINE MONTAC(CONTOR,VEC,NDIM,CAREL,NUDO,NEL,U,T)
C
C     IT ASSEMBLES THE INTEGRATION CONSTANTS IN THE COEFFICIENTS MATRIX
C     OR IN THE VECTOR OF INDEPENDENT TERMS ACCORDING TO THE BOUNDARY
C     CONDITIONS OF EACH ELEMENT
C
C*******************DICTIONARY
C
C     CONTOR(3,NELEM)=BOUNDARY CONDITIONS OF EACH ELEMENT
C     VEC(NDIM)       =INDEPENDENT TERMS VECTOR
C     NDIM            = 2*NELEM = DIMENSION OF VEC
C     CAREL(15)       =CHARACTERISTICS OF THE ELEMENT UNDER CONSIDERATION
C     NUDO            =NUMBER OF THE COLLOCATION POINT
C     NEL             =NUMBER OF THE ELEMENT
C     U(2,2)          =TENSOR U(I,J)
C     T(2,2)          =TENSOR T(I,J)
C*******************************
C
C                  DIMENSIONING AND DEFINITION OF ARRAYS
C
```

```
      COMMON /CIAB/ IAB(1)
      COMMON /CONST/ LEC,IMP,MEMO,MEMOB,NBP,NPER,NPEDP,NCON,NUDOSC,
     *               NPI,NUDOS,NELEM,NAPOYO,IPARA6,YOUNG,POIS,PORANA
      COMMON /WORK/ AA(227),UG(4),TG(4),ICODI(4),NFILA,NCOL
      DIMENSION CONTOR(3,NELEM),VEC(NDIM),CAREL(15),U(2,2),T(2,2)
      DIMENSION RIAB(1)
      EQUIVALENCE (IAB,RIAB)
      DEFINE RIABI(J)=RIAB(J)
      DEFINE A(I,J)=RIABI(IOB('A',I,J,1))
C
C                    DECODIFICATION OF THE BOUNDARY CONDITIONS
C
      CALL DECODI(4,CONTOR(1,NEL),ICODI)
C
C                    CHANGE OF CO-ORDINATES WHEN NEEDED
C
      IF(ICODI(3).EQ.1) CALL TRACOR(U,CAREL(11))
      IF(ICODI(4).EQ.1) CALL TRACOR(T,CAREL(11))
C
C                    ASSEMBLING PROCESS
C
      DO 100 I=1,2
           NFILA=2*(NUDO-1)+I
           DO 90 J=1,2
                NCOL=2*(NEL-1)+J
                IF(ICODI(3-J).LT.3) GO TO 11
C
C                    THE DISPLACEMENT IN DIRECTION J IS NOT KNOWN
C
                A(NFILA,NCOL)=A(NFILA,NCOL)+T(I,J)
                GO TO 21
C
C                    THE DISPLACEMENT IN DIRECTION J IS KNOWN
C
   11           VEC(NFILA)=VEC(NFILA)-T(I,J)*CONTOR(J+1,NEL)
                GO TO 31
C
C                    THE TRACTION IS KNOWN
C
   21           VEC(NFILA)=VEC(NFILA)+U(I,J)*CONTOR(J+1,NEL)
                GO TO 90
C
C                    THE TRACTION IS NOT KNOWN
C
   31           A(NFILA,NCOL)=A(NFILA,NCOL)-U(I,J)
C
C                     END OF THE PROCESS
C
   90      CONTINUE
  100 CONTINUE
      RETURN
      END
```

```
      SUBROUTINE DECODI(NCIFRA,PALA,JCIFRA)
C
C     DECODIFICATION OF A WORD
C
C*******************DICTIONARY
C
C      NCIFRA          =NUMBER OF DIGITS TO DECODE
C      PALA            =WORD TO DECODE
C      JCIFRA(NCIFRA)=ARRAY OF DECODED DIGITS
C*****************************
C
C                    DIMENSIONING AND DEFINITION OF ARRAYS
C
      DIMENSION JCIFRA(NCIFRA)
C
C
C                    DECODIFICATION OF THE WORD
C
      PAL=PALA
      IPALA=PAL/10.**(NCIFRA)
      PAL-PAL-IPALA*10**(NCIFRA)
      DO 100 I=1,NCIFRA
           II=I-1
```

```
          JCIFRA(NCIFRA-I+1)=PAL/10.**(NCIFRA-I)
          IF(I.EQ.1)GO TO 100
          DO 10 J=1,II
               JCIFRA(NCIFRA-I+1)=JCIFRA(NCIFRA-I+1)-
     *         JCIFRA(NCIFRA-J+1)*10.**(I-J)
 10       CONTINUE
100    CONTINUE
       RETURN
       END
```

```
      SUBROUTINE TRACOR(XM,V)
C
C     TRANSFORMATION OF THE MATRIX XM BY CHANGING THE CO-ORDINATES
C
C********************DICTIONARY
C
C     XM(2,2) = MATRIX TO TRANSFORM
C     V(2)    = COMPONENTS OF THE NORMAL TO THE ELEMENT
C*******************************
C
      DIMENSION XM(2,2),V(2)
      DIMENSION B(2,2)
C
C                    CHANGE OF CO-ORDINATES
C
      DO 10 J=1,2
            B(J,1)=-XM(J,1)*V(2)+XM(J,2)*V(1)
            B(J,2)=XM(J,1)*V(1)+XM(J,2)*V(2)
   10 CONTINUE
      CALL TRANS(B,XM,4)
      RETURN
      END
```

```
      SUBROUTINE TRANS(XA,XB,JA)
C
C     IT TRANSFERS JA ELEMENTS OF THE ARRAY XA TO THE ARRAY XB
C
C********************DICTIONARY
C
C     XA=ARRAY FROM WHICH THE ELEMENTS ARE TRANSFERRED
C     XB=ARRAY TO WHICH THE ELEMENTS ARE TRANSFERRED
C     JA=NUMBER OF ELEMENTS TO BE TRANSFERED
C*******************************
C
C
C                    DIMENSIONING AND DEFINITION OF ARRAYS
C
      DIMENSION XA(1),XB(1)
C
C
C                    TRANSFER OF THE ELEMENTS
C
      DO 100 IA=1,JA
           XB(IA)=XA(IA)
  100 CONTINUE
      RETURN
      END
```

4.6.3 Symmetries

In this epigraph we mean by "symmetry" only planar symmetry with respect to the global coordinate planes or sectorial symmetry. On the other hand, axial symmetry is not considered since a different approach is followed in this case (an alternative fundamental solution is used). The final goal of using symmetry conditions will be the reduction of the number of elements needed to define the problem, and therefore the number of equations needed to solve it, with an important gain of time and memory.

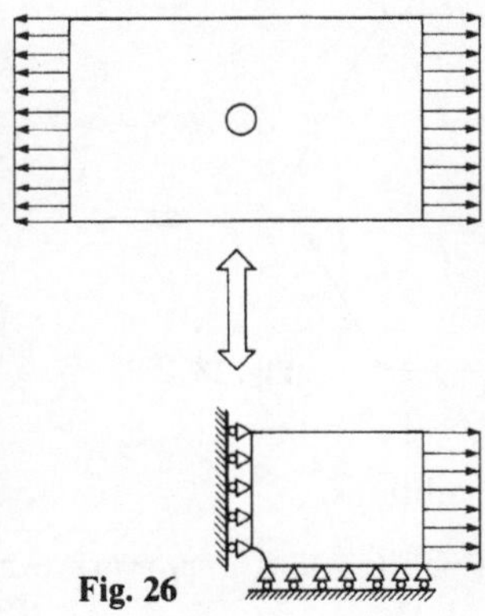

Fig. 26

A first approach corresponds to employ the knowledge we have on the traction and displacement distributions over the planes of symmetry, substituting the initial problem by a smaller one, with the same solution (Fig. 26).

The second method consists in introducing the symmetry conditions over the coordinates and boundary conditions, so as in the final results. The element 1, for example, in Fig. 27a has as "reflected" the elements 2, 3, and 4, whose coordinates and boundary conditions are known as a function of those of the element 1. With this, only the data of this element are needed to define the problem, and of course the symmetry conditions (Fig. 27b, c, d). Also the relation between the unknowns of those elements is considered in the assemblage of the matrix. If we imagine an integral equation corresponding any collocation point, this can be expressed (considering constant approximation to simplify) as:

$$\begin{aligned} &a_{i1} u_1 + a_{i2} v_1 + a_{i3} u_2 + a_{i4} v_2 + \ldots + a_{i2n-1} u_n + a_{i2n} v_n \\ &\quad = b_{i1} \sigma_1 + b_{12} \tau_1 + \ldots + b_{i2n-1} \sigma_n + b_{i2n} \tau_n + p_i . \end{aligned} \tag{58}$$

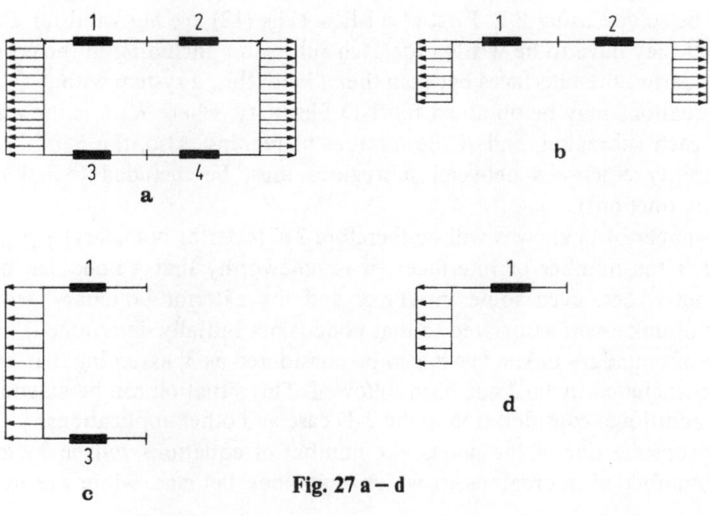

Fig. 27 a – d

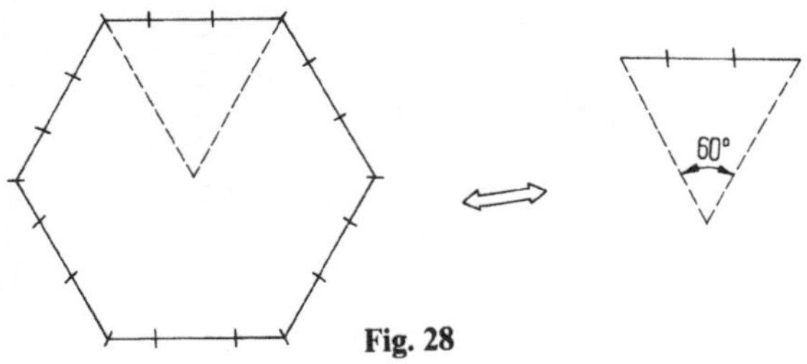

Fig. 28

But using the symmetry conditions:

$$u_1 = -u_2 = -u_3 = u_4, \qquad v_1 = v_2 = -v_3 = -v_4$$

$$\sigma_1 = \sigma_2 = -\sigma_3 = -\sigma_4, \qquad \tau_1 = -\tau_2 = -\tau_3 = \tau_4$$

and including these in (58):

$$(a_{i1} - a_{i3} - a_{i5} - a_{i7})\, u_1 + (a_{i2} + a_{i4} - a_{i6} - a_{i8})\, v_i + \dots$$
$$= (b_{i1} + b_{i3} - b_{i5} - b_{i7})\, \sigma_1 + (b_{i2} - b_{i4} - b_{i6} + b_{i8})\, \tau_1 + \dots + p_i. \qquad (59)$$

With this, the system of equations has been reduce drastically, since only the C.P. of Fig. 27 a are used.

This method, however, has a disadvantage compared with the former that is the greater number of integrations that have to be carried out, although the number of equations to solve is smaller. For 2-D the second approach is better, but for 3-D discretization is not clear. An additional advantage of the second is its greater accuracy. This last approach can be used effectively for sectorial symmetries as in Fig. 28 in exactly the same way that explained above [73].

4.6.4 Piecewise Heterogeneous Bodies

Complex computational problems appear when piecewise heterogeneous bodies have to be solved using B.E. First of all Eqs. (11), (12) are not valid for all the domain, but they have to be written for each subregion, including in the corresponding boundaries, the interfaces between them. With this, a system with $\sum_{r=1}^{R} 3 \times K(r)$ linear equations may be obtained for 3-D Elasticity, where $K(r)$ is the number of C.P. of each subregion, and R the number of regions. Also, the equilibrium and compatibility equations between subregions must be included (equal displacements and tractions).

The number of unknowns will be therefore $3K$ (exterior boundary) $+ \sum_{i=1}^{I} 6K(i)$, where I is the number of interfaces (it is noteworthy that a node can belong to several interfaces, even some interfaces and the exterior boundary, so that the number of unknowns associated to that node is not initially determined). Here, the number of boundary unknowns has to be considered as 3, assuming that any of the processes included in 4.6.2 has been followed. This situation can be extended without any additional consideration to the 2-D case and other applications.

If we consider one of the nodes, the number of equations will be $3 \times R_q$, being R_q the number of subregions to which that node belongs, while the number of unknowns is $6 I_q$ (or $6 I_q + 3$ if the node also belongs to the exterior boundary) with I_q the number of interfaces. This number can be bigger than the former, making it necessary to compute these two numbers and remove the excess of unknowns using some kind of approximation (condensation process).

The second problem to solve in this case is to arrive to an appropriate structure for the coefficient matrix, in order to make possible effective assemblage and solving processes.

Finally it has to be noticed that the computation of the diagonal submatrices (see 6.1.4) has to be done for each subregion, being convenient, in this case, from a computational point of view, to store all of these matrices in an additional array.

4.6.4.1 Condensation Process

The first question to answer in this case is: which are the unknowns to remove in order to get a systematic process? It seems clear that the interface tractions have to be these ones. Now, which interfaces? We have chosen the last ones, that is, the interfaces between the subregions $1, n$; $n, n-1$; $n-1, n-2$; ..., being n the number of regions connected at that specific node. Of course, we have to assume that those interfaces do exist, which implies that the numbering of the regions is not arbitrary. It has to be a cyclic one at the condensation nodes, which is not difficult to obtain as is shown in Fig. 29.

Once the unknowns have been chosen we have to develop a convenient approximation process. The approach proposed here [3] consists in writing the unknowns to be eliminated as a function of others to be removed in such a way that the general solution and the computation process is not very distorted and also it is easy to apply.

The first possibility is again to consider the principal stresses as the unknowns and write the rest of tractions as a function of them. There are two main disadvantages in this method. First the continuity of the stress tensor has to be assumed, which is not the case at a point belonging to two regions with different material properties. Second the principal directions have to be computed and there is no simple way to do it.

A second approach uses the relations (57) to remove the excess of unknowns. This is possible only in some cases (see 4.6.2), therefore its application is not effective. Also this method represents an important difficulty in the assemblage process, because there appear new terms in the columns corresponding to the tractions of the node, not only in the equations corresponding to the regions which forms the

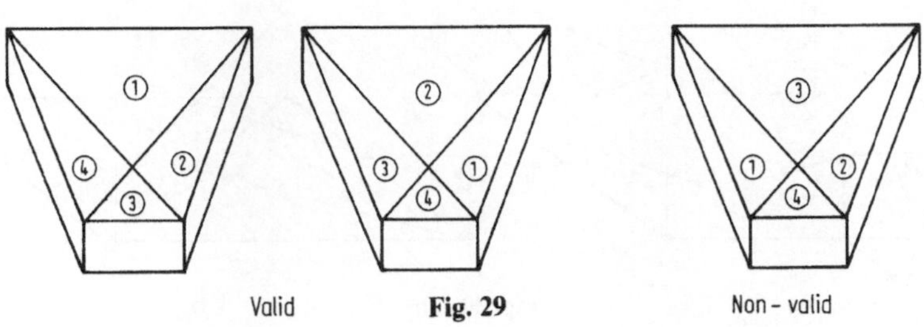

Fig. 29

interface to which those tractions are associated, but also with other regions that do not form that interface but coincide at the node, which is specially important (see 4.6.4.2).

A third method is to choose as unknowns the six independent components of the stress tensor, changing the unknowns of the first two interfaces by these last. The rest of interfaces not removed, preserve their own unknowns, while the rest can be expressed as a function of the first. It is clear that there exist a certain redundancy, which can be used as comprobation. The disadvantages of this method are again the necessity of assuming the continuity of the stress tensor and the problems that appear during the assemblage process.

The chosen approach consists in relating the tractions to be eliminated with those of the rest of the nodes of the element under consideration. The only difficulty is the selection of the nodes needed for this relation, so that in an element can exist more than one node with the same tractions to condense. In the most common case, with only one node to condense in the element, we have used an approximation function for the tractions of the node of the form:

$$t_i = N^n \, t_i^n \tag{60}$$

where n corresponds to the nodes where the tractions are known or are not removing unknowns of the problem. With this it is possible to express the condensing tractions as a function of the rest (Fig. 30a). For the case of a parabolic quadrilateral (60) can be written:

$$t_i = a_{0i} + a_{1i}\,\xi + a_{2i}\,\eta + a_{3i}\,\xi^2 + a_{4i}\,\eta^2 + a_{5i}\,\xi\,\eta^2 + a_{6i}\,\xi^2\,\eta \tag{61}$$

where the term $\xi\,\eta$ has been eliminated from the original formulation, in order to save the symmetry. With this it is possible to express the constants a_{0i}, a_{1i}, a_{2i}, a_{3i}, a_{4i}, a_{5i}, a_{6i}, and the traction t_i as a function of the rest $t_i^1, \ldots t_i^7$ (see Fig. 30a).

For node 1, for example, we have:

$$t_i^1 = t_i^2 + t_i^4 - t_i^3 \tag{62}$$

which corresponds to the linear extrapolation of those values as can be seen in Fig. 30b.

A slightly different approach has to be followed when some of those tractions are also removing ones, being possible to study the different possibilities in Doblaré [3].

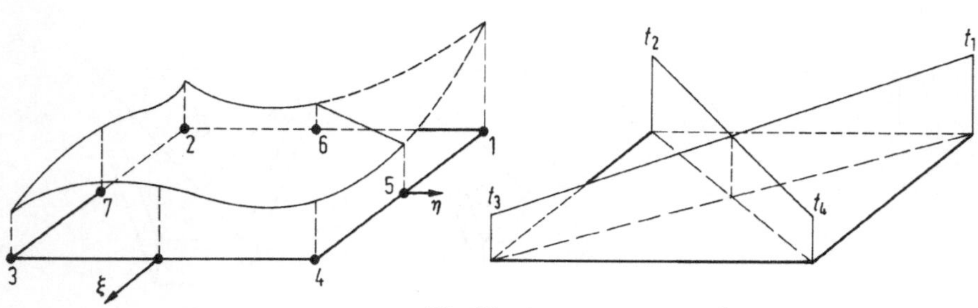

a **Fig. 30 a, b** b

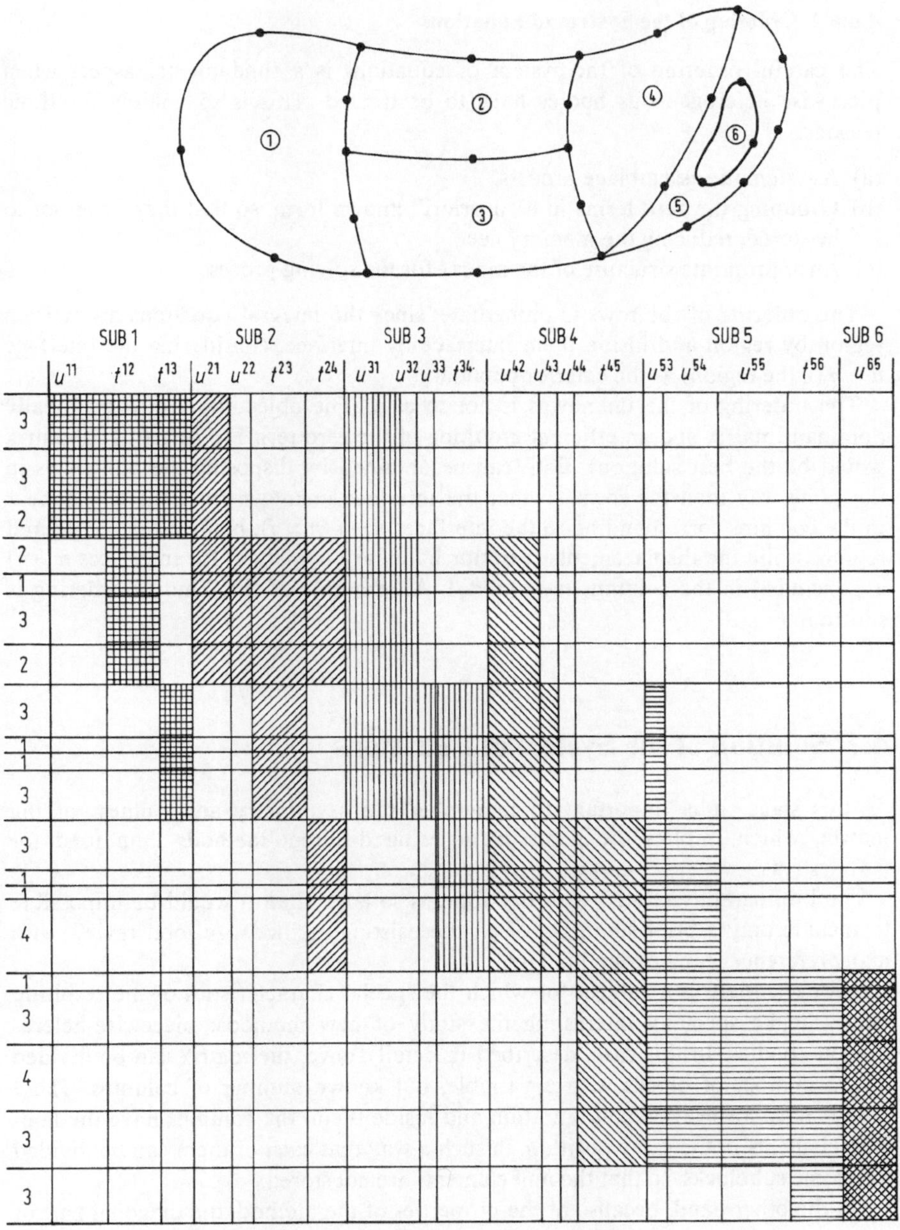
①
②
③
④
⑤
⑥
SUB 1
SUB 2
SUB 3
SUB 4
SUB 5
SUB 6
u11 t12 t13 u21 u22 t23 t24 u31 u32 u33 t34 u42 u43 u44 t45 u53 u54 u55 t56 u65
3
3
2
2
1
3
2
3
1
1
3
3
1
1
4
1
3
4
3
3

Fig. 31

4.6.4.2 Ordering of the System of Equations

The careful ordering of the system of equations is a fundamental aspect when piecewise heterogeneous bodies have to be treated. This is so, mainly for three reasons:

(a) A systematic assemblage process.
(b) Grouping the zero terms in a "a priori" known form, so that they have not to be stored, reducing the memory need.
(c) An appropriate structure of the matrix for the solving process.

The ordering of the rows is immediate, since the integral equations are written region by region and inside them interface by interface, considering the interface $n - n$ (n the region) as the exterior boundary.

The ordering of the unknowns is not so easy. The objectives are: a diagonally dominant matrix and an efficient grouping of the zero terms (a quasi-band matrix would be the best solution). Both can be obtained by disposing the unknowns in the same way than the rows, so that the interface unknown tractions will appear in the columns corresponding to the interfaces $n - i$ $(n > i)$, being n the considered region, while the displacements (exterior boundary unknowns for interfaces $n - n$) are included in the columns $n - j$ $(n < j)$. An example of this kind of ordering is shown in Fig. 31.

4.7 Solution of the System of Equations

At this stage, it is important to remember the asymmetry and fullness of the matrix, which implies the necessity of using different methods than used for F.E. systems.

The bibliography related with this topic is so large than it would be impossible to include only a brief summary of all the existing methods. A good review with many references can be found in [74].

However, there is a subject for which the special characteristics of the resulting matrix make specially interesting the study of new methods: piecewise-heterogeneous bodies. In this case, described in detail above, the matrix can be divided into a number of blocks with a variable, but known number of columns. These groups have a specified configuration and inside them, the columns have the non-zero elements in the same position, in such a way that each of them can be divided into some subblocks, so that the null elements are not stored.

On the other hand, because of the properties of the method, the diagonal part of the matrix always has a full band. So we are interested in a sparse matrix composed by different blocks and subblocks which seem to recommend an hyper-matrix solving approach.

There are some already developed methods to solve this kind of problems using elimination solvers [75, 76] (with additional references). However there exist some disadvantages in these approaches. First of all they need to foresee some additional storage for the new non-zero elements that appear during the elimination process, wich, even when using an optimal pivotting selection, represent an important additional storage need.

An other problem consists in the fact that the matrix (as was seen in 4.6.4.4) is stored column by column, so that we require an elimination solver which uses this configuration or, on the other hand, a reordering preprocessor is needed. Both situations are not advantageous.

For all of this, we recommend the use of an iterative solver for these cases, specially for large systems which cannot be solved using an in-core method. Thinking of those methods, we have to start saying that the Gauss-Seidel, Southwell or Jacobi ones work with a symmetric matrix, which in this case represents the need of making the product $\boldsymbol{A}^T \cdot \boldsymbol{A}$, which is not possible except destroying the initial configuration of the system. The approach we will develop here is a modified Conjugate Gradient method, changing its standard sequence in order to keep the same form for the matrix along all the process [3].

The C.G.M. is a well-known method introduced by Hestenes and Seifel for solving systems of linear equations:

$$\boldsymbol{A}\,\boldsymbol{x} = \boldsymbol{b}$$

where $\boldsymbol{A}$ is an $n \times n$ symmetric matrix.

Some special characteristics of this method are that, with no roundoff errors, it obtains the exact solution in less than $n+1$ iterations, and also that the matrix have not to be in in-core storage but only to be multiplied by a vector, as will be seen.

Other important properties are: it does not need an "a priori" estimate of any parameters, it uses the distribution of eigenvalues of the operator and finally it requires few restrictions for the matrix in order to have an optimum behavior. These are some of the reasons that have justified the selection of this method that has been profusely applied by other authors [77–79], etc.

As a summary of the computation process we have:

$$\beta_i = -\frac{\boldsymbol{v}_{i-1}^T \boldsymbol{A}\,\boldsymbol{r}_i}{\boldsymbol{v}_{i-1}^T \boldsymbol{A}\,\boldsymbol{v}_{i-1}}$$

$$\boldsymbol{v}_i = \boldsymbol{r}_i + \beta_i\,\boldsymbol{v}_{i-1} \tag{63}$$

$$\alpha = \frac{\boldsymbol{v}_i^{\mathrm{T}} \boldsymbol{r}_i}{\boldsymbol{v}_i^T \boldsymbol{A}\,\boldsymbol{v}_i}$$

where $\boldsymbol{x}_i$ is the solution vector at iteration i, and $\boldsymbol{r}_i$ is the residual vector $(\boldsymbol{b} - \boldsymbol{A}\,\boldsymbol{x}_i)$ at that iteration. Now $\boldsymbol{x}_{i+1}$ and $\boldsymbol{r}_{i+1}$ can be computed using:

$$\boldsymbol{x}_{i+1} = \boldsymbol{x}_i + \alpha\,\boldsymbol{v}_i$$

$$\boldsymbol{r}_{i+1} = \boldsymbol{r}_i - \alpha\,\boldsymbol{A}\,\boldsymbol{v}_i. \tag{64}$$

In order to apply this method of solution to B.E. systems, two main aspects have to be considered: the optimum sequence of operations and the selection of the initial iteration vector $\boldsymbol{x}_0$.

Concerning the first, it has to be noticed that the sequence included in (63–64) is only valid for symmetric matrices, but the B.E. one is not. A first possibility to save this problem would be the premultiplication of the system by the transponse matrix $\boldsymbol{A}^T$.

$$\boldsymbol{A}^T \boldsymbol{A}\, \boldsymbol{x} = \boldsymbol{A}^T \boldsymbol{b} \tag{65}$$

where $\boldsymbol{A}^T\boldsymbol{A}$ is already symmetric. Now it is possible to employ the above sequence; however, as will be seen in a moment, this is not the optimum one for the problem we are concerned.

If we make:

$$\boldsymbol{v}_{i-1}^T \boldsymbol{A}\, \boldsymbol{v}_{i-1} = \frac{\boldsymbol{v}_{i-1}^T \boldsymbol{r}_{i-1}}{\alpha_{i-1}}$$

we have:

$$\beta_i = -\frac{\alpha_{i-1} \boldsymbol{v}_{i-1}^T \boldsymbol{A}\, \boldsymbol{r}_i}{\boldsymbol{v}_{i-1}^T \boldsymbol{r}_{i-1}} = -\frac{\alpha_{i-1} \boldsymbol{r}_{i-1}^T \boldsymbol{A}\, \boldsymbol{v}_{i-1}}{\boldsymbol{v}_{i-1}^T \boldsymbol{r}_{i-1}}$$

and as

$$-\alpha_{i-1} \boldsymbol{A}\, \boldsymbol{v}_{i-1} = \boldsymbol{r}_i - \boldsymbol{r}_{i-1}$$

β_i can be written as

$$\beta_i = \frac{\boldsymbol{r}_i^T (\boldsymbol{r}_i - \boldsymbol{r}_{i-1})}{\boldsymbol{v}_{i-1}^T \boldsymbol{r}_{i-1}}.$$

It can be shown that:

$$\boldsymbol{r}_i^T \boldsymbol{r}_{i-1} = \boldsymbol{0}$$
$$\boldsymbol{v}_{i-1}^T \boldsymbol{r}_{i-1} = \boldsymbol{r}_{i-1}^T \boldsymbol{r}_{i-1}$$

and with this

$$\beta_i = \frac{\boldsymbol{r}_i^T \boldsymbol{r}_i}{\boldsymbol{r}_{i-1}^T \boldsymbol{r}_{i-1}}$$

so the sequence can be dispose as:

$$\beta_i = \frac{\boldsymbol{r}_i^T \boldsymbol{r}_i}{\boldsymbol{r}_{i-1}^T \boldsymbol{r}_{i-1}}$$
$$\boldsymbol{v}_i = \boldsymbol{r}_i + \beta_i \boldsymbol{v}_{i-1}$$
$$\alpha_i = \frac{\boldsymbol{r}_i^T \boldsymbol{r}_i}{\boldsymbol{v}_i^T \boldsymbol{A}\, \boldsymbol{v}_i}$$
$$\boldsymbol{x}_{i+1} = \boldsymbol{x}_i + \alpha\, \boldsymbol{v}_i$$
$$\boldsymbol{r}_{i+1} = \boldsymbol{r}_i - \alpha_i \boldsymbol{A}\, \boldsymbol{v}_i \tag{66}$$

It has to be noticed here that only the product $\boldsymbol{A} \cdot \boldsymbol{v}_i$ has to be made for each iteration. In our case changing $\boldsymbol{A}$ by $\boldsymbol{A}^T\boldsymbol{A}$ we have:

$$\beta_i = \frac{\boldsymbol{r}_i^T \boldsymbol{r}_i}{\boldsymbol{r}_{i-1}^T \boldsymbol{r}_{i-1}}$$
$$\boldsymbol{v}_i = \boldsymbol{r}_i + \beta_i \boldsymbol{v}_{i-1}$$
$$\alpha_i = \frac{\boldsymbol{r}_i^T \boldsymbol{r}_i}{\boldsymbol{v}_i^T \boldsymbol{A}^T \boldsymbol{A}\, \boldsymbol{v}_i}$$

$$\begin{aligned} x_{i+1} &= x_i + \alpha_i v_i \\ r_{i+1} &= r_i - \alpha_i A^T A v_i. \end{aligned} \tag{67}$$

With this, no additional previous operation is required, and only two matrix by vector products have to be made for each iteration.

$$A v_i, A^T (A v_i).$$

This sequence is more efficient than the first explained only when the number of iterations is less than the number of equations, which is the usual case for the C.G.M. as was mentioned before.

Referring the second aspect, the selection of x_0, we have studied two different possibilities:

$- x_0 = \mathbf{0}$.

The main advantage of this method is: the simplicity. On the contrary the number of iterations is in most cases greater than those of other methods.

$- x_0 =$ the resulting vector of dividing the independent terms vector b by the diagonal of the matrix A.

The number of iterations is now about 4 to 5% less than that of the above situation. Also the obtention of x is immediate, even when the matrix is stored in auxiliary memory, because (as was explained in 4.6.1.4) the diagonal submatrices are stored in an additional easy-access array. The main disadvantage is that the matrix in B.E. problems is not diagonally dominant and the initial vector is not very close to the exact solution.

```
      SUBROUTINE SOLVE
C
C  SOLUTION OF THE SYSTEM OF EQUATIONS
C
C
C ********************  DICTIONARY
C
C       VEC = VECTOR OF INDEPENDENT TERMS
C ****************************************************
C
C                  DIMENSIONING AND DEFINITION OF ARRAYS
C
      COMMON /CIA/ IA(1)
      COMMON /CIAB/ IAB(1)
      COMMON /CONST/ LEC,IMP,MEMO,MEMOB,NRP,NPER,NPEDP,NCON,NUDOSC,
     *               NPI,NUDOS,NELEM
      COMMON /WORK/ IER,NDIM,IER1
C
C                  STARTING OF THE PROCESS
C
      IER=LOCA('VEC',NPER)
      NDIM=2*NELEM
      IER1=IER+NDIM
      CALL SOLUG (NDIM,IA(IER),IA(IER1))
C
C                  END OF THE PROCESS
C
      RETURN
      END
```

```
      SUBROUTINE SOLUG(N,C,IPIVO)
C
C  SOLUTION OF A SYTEM OF LINEAR EQUATIONS BY A STANDARD GAUSS ELIMINATION
C  PROCESS
C
C
C ********************  DICTIONARY
C
C        A     = COEFFICIENTS MATRIX
C        C     = VECTOR OF INDEPENDENT TERMS
C        IPIVO = PIVOTS VECTOR
C        N     = DIMENSION OF THE SYSTEM
C ***************************************************
C
C                   DIMENSIONING AND DEFINITION OF ARRAYS
C
      COMMON /CIAB/ IAB(1)
      COMMON /CONST/ LEC,IMP
      DIMENSION IPIVO(N),C(N)
      DIMENSION RIAB(1)
      EQUIVALENCE(RIAB,IAB)
      DEFINE RIABI(I)=RIAB(I)
      DEFINE A(I,J)=RIABI(IOB('A',I,J,1))
C
C                   STARTING OF THE PROCESS
C
      CALL CEROS(IPIVO,N)
      DO 50 J=1,N
          DMAX=0.0
          DO 40 K=1,N
              IF(IPIVO(K)-1)31,40,31
   31         IF(ABS(DMAX)-ABS(A(K,K)))41,40,40
   41         DMAX=A(K,K)
              IR=K
   40     CONTINUE
          IF(ABS(DMAX)-2.0E-8)51,51,61
   51     WRITE(IMP,20000)
          GO TO 71
   61     IPIVO(IR)=IPIVO(IR)+1
          A(IR,IR)=1.0
          DO 30 I=1,N
              A(IR,I)=A(IR,I)/DMAX
   30     CONTINUE
          C(IR)=C(IR)/DMAX
          DO 20 K=1,N
              IF(K-IR)81,20,81
   81         R=A(K,IR)
              A(K,IR)=0.0
              DO 10 L=1,N
                  A(K,L)=A(K,L)-A(IR,L)*R
   10         CONTINUE
              C(K)=C(K)-C(IR)*R
   20     CONTINUE
   50 CONTINUE
C
C                   END OF THE PROCESS
C
   71 RETURN
20000 FORMAT(1H1,'******  SINGULAR MATRIX  ******')
      END
```

```
      SUBROUTINE ELACON
C
C     THIS ROUTINE PRINTS THE RESULTS ALONG THE BOUNDARY
C
C                   DIMENSIONING AND DEFINITION OF ARRAYS
C
      COMMON /CIA/ IA(1)
      COMMON /CONST/ LEC,IMP,MEMO,MEMOB,NBP,NPER,NPEDP,NCON,NUDOSC,
     *               NPI,NUDOS,NELEM
      COMMON /WORK/ NDIM,IER1,IER2
      IER1=LOCA('CONT',NPER)
      IER2=LOCA('VEC',NPER)
      NDIM=4*NELEM
```

```
C
C                 START OF THE PROCESS
C
         CALL MODIFY(IA(IER1),IA(IER2),NDIM)
         CALL IMRES
C
C                 END OF THE PROCESS
C
      RETURN
      END
```

```
      SUBROUTINE MODIFY (CONTOR,VEC,NDIM)
C
C     MODIFICATION OF THE VECTOR OF RESULTS
C
C ********************DICTIONARY
C
C          CONTOR(3,NELEM) = BOUNDARY CONDITIONS
C          VEC             = VECTOR OF RESULTS
C          NDIM            = 4*NELEM
C ***************************************************
C
C
C                 DIMENSIONING AND DEFINITION OF ARRAYS
C
      COMMON /CIA/ IA(1)
      COMMON /CONST/ LEC,IMP,MEMO,MEMOB,NBP,NPER,NPEDP,NCON,NUDOSC,
     *               NPI,NUDOS,NELEM
      COMMON /WORK/ N(3),ICODI(4),IER,IER1
      DIMENSION VEC(NDIM),CONTOR(3,NELEM)
C
C
C                 START OF THE PROCESS
C
      DO 100 I=1,NELEM
         M=2*(NELEM+I)
         J=M-1
         N=2*I
         K=N-1
         CALL DECODI(4,CONTOR(1,I),ICODI)
         IF(ICODI(2).GE.2) GO TO 21
C
C     THE RADIAL DISPLACEMENT IS KNOWN
C
         IF(ICODI(1).EQ.4) GO TO 11
         VEC(J)=VEC(K)
         VEC(K)=CONTOR(2,I)
         GO TO 41
   11    VEC(J)=VEC(N)
         VEC(N)=VEC(K)
         VEC(K)=CONTOR(2,I)
         GO TO 41
C
C     THE RADIAL TRACTION IS KNOWN
C
   21    IF(ICODI(2).NE.2) GO TO 31
         VEC(J)=CONTOR(3,I)
         GO TO 41
   31    VEC(J)=CONTOR(2,I)
   41    IF(ICODI(1).EQ.4) GO TO 51
C
C     THE AXIAL DISPLACEMENT IS KNOWN
C
         VEC(M)-VEC(N)
         IF(ICODI(1).EQ.2) GO TO 81
         VEC(N)=CONTOR(2,I)
         GO TO 71
   81    VEC(N)=CONTOR(3,I)
         GO TO 71
C
C     THE AXIAL TRACTION IS KNOWN
C
   51   IF(ICODI(2).EQ.2) GO TO 61
        VEC(M)=CONTOR(3,I)
        GO TO 71
```

```
   61   VEC(J)=CONTOR(3,I)
   71   IER=LOCA('CAREL',NPER)
        IER1=IER+15*(I-1)+10
C       IF(ICODI(4).NE.0) CALL TRAC(VEC(K),IA(IER1))
C       IF(ICODI(3).EQ.0) CALL TRAC(VEC(J),IA(IER1))
  100 CONTINUE
C
C                    END OF THE PROCESS
C
      RETURN
      END
```

```
      SUBROUTINE IMCOCO
C
C     PRINTING OF THE BOUNDARY NODE CO-ORDINATES
C
C
C                     DIMENSIONING AND DEFINITION OF ARRAYS
C
      COMMON/CONST/LEC,IMP,MEMO,MEMOB,NGP,NPER,NPEDP,NCON,NUDOSC
      COMMON/C1A/IA(1)
      DIMENSION NELE(1)
      DIMENSION COOR(1)
      EQUIVALENCE (COOR,IA,NELE)
      N1=LOCA('COOR',NPER)
      N2=LOCA('NELE',1)
C
C     PRINTING OF THE BOUNDARY NODE CO-ORDINATES
C
            WRITE(IMP,20000)
20000 FORMAT(1H1,7X,48('*')/8X,'* * *  CO-ORDINATES OF THE'
     *              ' BOUNDARY NODES  * * *'/8X,48('*')//)
      NF=0
      DO 10 I=1,NCON
          WRITE(IMP,20001)I
20001     FORMAT(////5X,'BOUNDARY',2X,I3/5X,13('=')/)
          WRITE(IMP,20002)
20002     FORMAT(3(11X,'NODE',2X,'RADIAL COOR',2X,'AXIAL COOR')/3(11X,
     *           4('-'),2X,11('-'),2X,10('-')))
          NI=1+NF
          NF=NI+NELE(I+N2-1)
          WRITE(IMP,20003)(J,COOR(2*J-2+N1),COOR(2*J-1+N1),J=NI,NF)
20003     FORMAT(3(12X,I3,3X,F10.4,2X,F10.4))
   10 CONTINUE
      RETURN
      END
```

```
      SUBROUTINE IMCCON
C
C     PRINTING OF THE BOUNDARY CONDITIONS
C
C******************** DICTIONARY
C
C     NCON                 = NUMBER OF DISJOINT BOUNDARIES
C     NELE(NCON)           = NUMBER OF BOUNDARY ELEMENTS OF EACH BOUNDARY
C     CONTOR(1,NCLEM)      = BOUNDARY CONDITIONS CODE FOR EACH ELEMENT
C                  CODE    = NUMBER WITH FOUR DIGITS ABCD
C                    * A : TYPE OF CO-ORDINATES FOR WHICH THE DISPLACEMENTS
C                          ARE GIVEN
C                          0 GLOBAL CO-ORDINATES
C                          1 LOCAL CO-ORDINATES
C                    * B : TYPE OF CO-ORDINATES FOR THE TRACTIONS
C                          0 GLOBAL CO-ORDINATES
C                          1 LOCAL CO-ORDINATES
C                    * C : THE DISPLACEMENTS THAT ARE KNOWN
C                          1 RADIAL DISPLACEMENT
C                          2 AXIAL DISPLACEMENT
C                    * D : THE TRACTIONS THAT ARE KNOWN
C                          3 RADIAL TRACTION
C                          4 AXIAL TRACTION
C     NAPOYO               = NUMBER OF NODES WITH RESTRICTED DEGREES OF FREEDOM
C     NANU(NAPOYO)         = NUMBER OF THOSE DEGREES OF FREEDOM
C********************************
```

```
C
C                          DIMENSIONING AND DEFINITION OF ARRAYS
C
      COMMON/CONST/LEC,IMP,MEMO,MEMOB,NBP,NPER,NPFDP,NCON,
     *          NUDOSC,NPI,NUDOS,NELEM,NAPOYO
      COMMON/CIA/IA(1)
      DIMENSION NELE(1)
      DIMENSION CONTOR(1)
      DIMENSION NANU(1)
      EQUIVALENCE (CONTOR,IA,NELE,NANU)
      N1=LOCA('NELE',1)
      N2=LOCA('CONTOR',NPER)
      N3=LOCA('NANU',1)
C
C     PRINTING OF BOUNDARY CONDITIONS
C
            WRITE(IMP,20000)
20000       FORMAT(1H1,7X,37('*')//8X,'* * *     BOUNDARY '
     *       'CONDITIONS    * * *'//8X,37('*')///)
      NEF=0
      DO 10 I=1,NCON
            WRITE(IMP,20001)I
20001       FORMAT(////5X,'BOUNDARY',2X,I3/5X,13('=')//)
            WRITE(IMP,20002)
20002       FORMAT(/5X,'ELEMENT',2X,' CODE ',4X,'RADIAL DIS',4X,
     *              'AXIAL DIS',4X,'RADIAL TRA',4X,'AXIAL TRA'/5X,
     *       7('-'),2X,6('-'),4(4X,10('-')))
      NEI=1+NEF
      NEL=NELE(N1+I-1)
      NEF=NEI+NEL-1
            DO 20 J=NEI,NEF
              M=3*J-1+N2
              NCOD=CONTOR(M-2)
              N=NCOD-INT(NCOD/100.)*100
              IF(N.EQ.14)GOTO 11
              IF(N.EQ.23)GOTO 21
              IF(N.EQ.12)GOTO 31
              IF(N.EQ.34)GOTO 41
              GOTO 20
   11      WRITE(IMP,20003)(J,NCOD,CONTOR(M-1),CONTOR(M))
20003      FORMAT(7X,I3,6X,I4,5X,F10.4,4X,' UNKNOWN ',4X,F10.4,
     *             4X,' UNKNOWN ')
              GOTO 20
   21      WRITE(IMP,20004)(J,NCOD,CONTOR(M-1),CONTOR(M))
20004      FORMAT(7X,I3,6X,I4,5X,' UNKNOWN ',4X,F10.4,4X,F10.4
     *             ,4X,' UNKNOWN ')
              GOTO 20
   31      WRITE(IMP,20005)(J,NCOD,CONTOR(M-1),CONTOR(M))
20005      FORMAT(7X,I3,6X,I4,5X,F10.4,4X,F10.4,4X,' UNKNOWN '
     *             ,4X,' UNKNOWN ')
              GOTO 20
   41      WRITE(IMP,20006)(J,NCOD,CONTOR(M-1),CONTOR(M))
20006      FORMAT(7X,I3,6X,I4,5X,' UNKNOWN ',4X,' UNKNOWN ',4X
     *             ,F10.4,4X,F10.4)
   20       CONTINUE
   10 CONTINUE
C
C      PRINTING OF THE RESTRICTED DEGREES OF FREEDOM
C
      WRITE(IMP,20007)
20007 FORMAT(///5X,'NODE',3X,'RESTRICT D.O.F.'/5X,4('-'),3X,14('-')/)
      DO 30 I=1,NAPOYO
            NAP=NANU(I+N3-1)
            NUDO=INT(NAP/2.)
            IRES=NAP-NUDO*2
            IF(IRES.EQ.0)GO TO 51
            NN=NUDO+1
            WRITE(IMP,20008)NN
20008       FORMAT(6X,I3,5X,'RADIAL DISP')
            GOTO 30
   51       WRITE(IMP,20009)NUDO
20009       FORMAT(6X,I3,5X,'AXIAL DISP')
   30 CONTINUE
      RETURN
      END
```

```
      SUBROUTINE IMCONS (CONSTA)
C
C     PRINTING OF THE ARRAY 'CONSTA' WHICH CONTAINS THE CONSTANTS
C     NEEDED FOR THE INTEGRATION PROCESS
C
C******************** DICTIONARY
C
C     CONSTA(1) = 16*(PI**2)*G*(1-NU)
C     CONSTA(2) = 3-4*NU
C     CONSTA(3) = (1-NU)/(1-2*NU)
C     CONSTA(4) = CONSTA(3)-1
C     CONSTA(5) = G = E/2(1+NU)
C*******************************
C
C                     DIMENSIONING AND DEFINITION OF ARRAYS
C
      COMMON/CONST/LEC,IMP
      DIMENSION CONSTA(5)
C
C     PRINTING OF THE CONSTANTS
C
            WRITE(IMP,20000)
            WRITE(IMP,20001) (CONSTA(I),I=1,5)
      RETURN
20000  FORMAT(1H1,7X,56('*')/8X,'* * *      CHARACTERISTIC
     *        '                CONSTANTS    * * *'/8X,56('*')////)
20001  FORMAT(////3X,'CONSTA(1) =',E12.4/3X,'CONSTA(2) =',E12.4/3X,
     *        'CONSTA(3) =',E12.4/3X,'CONSTA(4) =',E12.4/3X,
     *        'CONSTA(5) =',E12.4)
      END
```

```
      SUBROUTINE IMCAR
C
C     PRINTING OF THE ARRAY 'CAREL' WHICH CONTAINS THE INTEGRATION
C     PROPERTIES OF EACH ELEMENT
C
C******************** DICTIONARY
C
C     CAREL(1,...,10) = RADIAL AND AXIAL CO-ORDINATES OF THE 5 POINTS OF
C                       INTEGRATION
C     CAREL(11)       = RADIAL COMPONENT OF THE NORMAL
C     CAREL(12)       = AXIAL COMPONENT OF THE NORMAL
C     CAREL(13)       = LENGTH OF THE ELEMENT
C     CAREL(14)       = RADIAL CO-ORDINATE OF THE C.O.G. OF THE ELEMENT
C     CAREL(14)       = AXIAL CO-ORDINATE OF THE C.O.G. OF THE ELEMENT
C*******************************
C
C                     DIMENSIONING AND DEFINITION OF ARRAYS
C
      COMMON/CONST/LEC,IMP,MEMO,MEMOB,NUP,NPER,NPEDP,NCON,NUDOSC
     *               ,NPI,NUDOS,NELEM
      COMMON/CIA/IA(1)
      DIMENSION CAREL(1)
      EQUIVALENCE (CAREL,IA)
      N1=LOCA('CAREL',NPER)-15
C
C     PRINTING OF THE INTEGRATION PROPERTIES
C
            WRITE(IMP,20000)
      DO 10 I=1,NELEM
      N1=N1+15
            WRITE(IMP,20001) I
            WRITE(IMP,20002)
            WRITE(IMP,20003) (J,CAREL(N1+2*J-2),CAREL(N1+2*J-1),J=1,5)
            WRITE(IMP,20004) CAREL(N1+10),CAREL(N1+11)
            WRITE(IMP,20005) CAREL(N1+12)
            WRITE(IMP,20006) CAREL(N1+13),CAREL(N1+14)
   10 CONTINUE
      RETURN
20000  FORMAT(1H1,7X,56('*')/8X,'* * *  CHARACTERISTICS
     *           '      OF    THE   ELEMENT    * * *'/8X,56('*')////)
20001  FORMAT(////5X,'ELEMENT',2X,I3/5X,13('=')//)
20002  FORMAT(5X,'INTEGRATION POI',5X,'RADIAL COOR',5X,'AXIAL COOR'
     *              /5X,15('-'),5X,11('-'),5X,10('-'))
```

```
20003  FORMAT(12X,I1,13X,F10.4,5X,F10.4)
20004  FORMAT(//41X,'RADIAL',2X,F10.4//5X,'*COMPONENTS OF THE NORMAL'
     *     //41X,'AXIAL',3X,F10.4//)
20005  FORMAT(5X,'*LENGTH OF THE ELEMENT',5X,F10.4//)
20006  FORMAT(41X,'RADIAL',2X,F10.4//5X,'*CO-ORDINATES OF THE C.D.G.'
     *     ' OF THE ELEMENT'//41X,'AXIAL',3X,F10.4)
      END
```

```
      SUBROUTINE IMRES
C
C     PRINTING OF THE RESULTS ALONG THE BOUNDARY
C
C******************* DICTIONARY
C
C     VEC(NELEM,4)        = VECTOR OF RESULTS
C     NELEM               = NUMBER OF BOUNDARY ELEMENTS
C*******************************
C
C                     DIMENSIONING AND DEFINITION OF ARRAYS
C
      COMMON/CONST/LEC,IMP,MEMO,MEMOB,NBP,NPER,NPEDP,NCON,NUDOSC
     *        ,NPI,NUDOS,NELEM
      COMMON/CIA/IA(1)
      DIMENSION VEC(1)
      EQUIVALENCE (VEC,IA)
      N1=LOCA('VEC',NPER)-2
C
C     PRINTING OF THE RESULTS
C
            WRITE(IMP,20000)
20000       FORMAT(1H1,/X,39('*')//8X,'* * *  DISPLACEMENTS AND TRACTI'
     *             'ONS   * * *'//8X,'* * *    ALONG THE BOUNDARY *'
     *             ' * *'//8X,41('*'))
            WRITE(IMP,20001)
20001       FORMAT(////5X,'ELEMENT',5X,'RADIAL DISP',5X,'AXIAL DISP'
     *       ,5X,'RADIAL TRA',6X,'AXIAL TRA'/5X,7('-'),5X,
     *       11('-'),5X,10('-'),5X,10('-'),5X,10('-')/)
      DO 10 I=1,NELEM
            N1=N1+2
            N3=2*NELEM+N1
            N4=N3+1
            N2=N1+1
            WRITE(IMP,20002)I,VEC(N1),VEC(N2),VEC(N3),
     *       VEC(N4)
  10   CONTINUE
20002       FORMAT(8X,I3,7X,E10.4,5X,E10.4,6X,E10.4,6X,E10.4)
      RETURN
      END
```

References

1 Margulies, M., Intercombination of the Boundary Element and Finite Element Methods, in *Progress in Boundary Element Method I.* C.A. Brebbia (ed.). Pentech Press, 1981

2 Wendland, W., Asymptotic Accuracy and Convergence, in *Progress in Boundary Element Method I.* C.A. Brebbia (ed.). Pentech Press, 1981

3 Doblaré, M., Formulación Tridimensional del Método de los Elementos de Contorno con Interpolación Parabólica. PhD Thesis Universidad Politécnica, Madrid 1981

4 Cruse, T.A., Snow, D.W., and Wilson, P.B., Numerical Solutions in Axisymmetric Elasticity. Computers and Structures **7**, pp. 445–451, 1977

5 Danson, D., A Boundary Element Formulation of Problems in Linear Isotropic Elasticity with Body Forces. 3rd Int. Seminar in Recent Advances in B.E.M. C.A. Brebbia (ed.). Irvine (California), 1981

6 Hartmann, F., Elastostatics, in *Progress in Boundary Element Method I.* C.A. Brebbia (ed.). Pentech Press, 1981

7 Gomez-Lera, S., El Método de los Elementos de Contorno en Problemas Axisimétricos. PhD. Thesis Universidad Politécnica, Madrid 1982
8 Mikhlin, S.G., Multidimensional Singular Integrals and Integral Equations. Pergamon Press, 1965
9 Tottenham, N.N., The Boundary Element Method for Plates and Shells, in *Development in Boundary Element Method.* Banerjee and Butterfield (ed.). Applied Science Publishers Ltd. 1979
10 Dominguez, J. and Alarcon, E., Elastodynamics, in *Progress in Boundary Element Method I.* C.A. Brebbia (ed.). Pentech Press 1981
11 Paris, F., Martin, A., and Alarcon, E., Potential Theory, in *Progress in Boundary Element Method I.* C.A. Brebbia (ed.). Pentech Press, 1981
12 Wrobel, L.C. and Brebbia, C.A., Time Dependent Potential Problems, in *Progress in Boundary Element Method I.* C.A. Brebbia (ed.). Pentech Press,1981
13 Lean, M.H., Friedman, M., and Wexler, A., Application of the BEM in Electrical Eng. Problem, in *Development in Boundary Element Methods.* Banerjee and Butterfield (ed.). Applied Science Publishers Ltd., 1979
14 Telles, J.C.F. and Brebbia, C.A., Plasticity, in *Progress in Boundary Element Method I.* C.A. Brebbia (ed.). Pentech Press, 1981
15 García-Benitez, F., El Método de los Elementos de Contorno en Elastoplasticidad. PhD Thesis Universidad Politécnica, Madrid 1981
16 Zienkiewicz, O.C., The Finite Element Method. McGraw Hill, 1977.
17 Brebbia, C.A., Boundary Element Method for Engineers. Pentech Press, London 1978
18 Martin, A., Rodriguez, I., and Alarcon, E., Mixed Elements, in *Recent Advances in B.E.M.* C.A. Brebbia (ed.). CML Publications, 1980
19 Stephan, E. and Wendland, W., Remarks to Galerkin and Least Squares Methods with Finite Elements for General Elliptic Problems. Lecture Notes Math. 564, pp. 461, 471. Springer-Verlag, 1976
20 Wendland, W., On Galerkin Collocation Methods for Integral Equations of Elliptic Boundary Value Problems, in *Numerical Treatment of Integral Equations.* J. Albrecht and L. Collatz (eds.). Int. Ser. Num. Math. Vol. 53. Birkhäuser, 1980
21 Hsiao, G.C., Kopp, P., and Wendland, W.L., The Synthesis of the Collocation and the Galerkin Method Applied to some Integral Equations of the First Kind, in *New Developments in Boundary Element Methods.* C.A. Brebbia (ed.). CML Publications, pp. 136–422, 1980
22 Lachat, J.C., A Further Development of the Boundary Integral Technique for Elastostatics. PhD Thesis, University of Southampton (U.K.), 1975
23 Danson, D.J., Beasy. A Boundary Element Analysis System. 4th Int. Sem. on Boundary Elem. Meth. in Eng. C.A. Brebbia (ed.). Southampton (U.K.), 1982
24 Doblaré, M. and Alarcon, E., A Three-Dimensional B.E.M. Program. 1st Int. Seminar in Finite Elements. C.A. Brebbia (ed.). CML Publications, Southampton (U.K.), 1981
25 Felippa, C.A., Data Base Management in Scientific Computing I. General Description. Computers and Structures, Vol. 10, pp. 53–61, 1979
26 Felippa, C.A., Data Base Management in Scientific Computing II. Data Structures and Program Architecture. Computers and Structures, Vol. 12, pp. 131–145, 1980.
27 Bettes, J.A., A Data Structure for Finite Element Analysis. Int. J. Numerical Methods in Engineering, Vol. II, 1977
28 Date, C.J., An Introduction to Database Systems. Addison-Wesley, Reading, Massachusetts, 1975
29 Kamel, M.D., McCabe, M.W., and Deshazo, P.G., Optimum Design of Finite Element Software subject to Core Restrictions. Computers and Structure, Vol. 10, pp. 63–80, 1979
30 Katzan, M., Computer Data Management and Data Base Technology, Van Nostrand Reinhold, 1975
31 McCornick, J.M. and Wright, J., Managing Data in Fortran Programs. ASCE, 1979
32 García-Benitez, F., Doblaré, M., Paris, F., Martin, A., and Alarcon, E., PECET, un Programa Tridimensional de Elementos de Contorno en Elasticidad. Technical Rapport Required by RENFE, 1981

33 Burgos, C., Doblaré, M., and Alarcon, E., Automatic mesh generation processor for 3-D parabolic boundary discretization. Adv. Eng. Software, 1983, Vol. 5, No. 3
34 Birkhoff, G. and Gordon, W.J., The draftsman's and related equations. J. Approx. Theory, **1**, pp. 191–208, 1968
35 Gordon, W.J., Blending-function Methods of Bivariate and Multivariate Interpolation and Approximation, SIAM J. Num. Anal. **8**, pp. 158, 1971
36 Gordon, W.J. and Hall, C.A., Construction of Curvilinear Coordinates Systems and Application to Mesh Generation. Int. J. Num. Meth. Eng. **7**, pp. 461, 1973
37 Gordon, W.J. and Hall, C.A., Transfinite Element Methods Blending-function Interpolation over Arbitrary Curved Element Domains, Num. Meth. **21**, pp. 109, 1973
38 Paris, F., El Método de les Elementos de Contorno en Teoría del Potencial y Elasticidad. PhD Thesis, Universidad Politécnica, Madrid 1979
39 Davis, P. and Rabinowitz, P., Abscisses and Weights for Gaussian Quadratures of Higher Order. J. Res. Nat. But. Standards **56**, No. 1, pp. 35–37, 1956
40 Davis, P. and Rabinowitz, P., Additional Abscisses and Weights for Gaussian Quadratures of Higher Order. J. Res. Nat. But. Standards, **60**, No. 6, pp. 613–614, 1960
41 Abramowitz, M. and Stegun, I., Handwork of Mathematical Functions. Applied Mathematical Series, No. 55, National Bur. of Standards, Dover 1965
42 Benitez, F.G. and Alarcon, E., Simulación Numérica por el Método de los Elementos de Contorno. Utilización de los Elementos con interpolación Parabólica, in *1st Symposium sobre Modelado y Simulación.* Comité español de la IFAC, 1980
43 Anza, J.J., El Método de los Elementos de Contorno en Termoelasticidad. PhD Thesis Universidad Politécnica de Madrid, 1982
44 Dominguez, J., Cálculo de Tensiones en las Inmediaciones de Anclajes. Aplicación del Método de los Elementos de Contorno. PhD Thesis University of Sevilla, 1977
45 Telles, J.C.F. and Brebbia, C.A., The Boundary Element Method in Plasticity. 2nd Int. Seminar in Recent Advances in B.E.M. C.A. Brebbia (ed.), pp. 295–317, Southampton (U.K.), 1980
46 Cruse, T.A., Application of the Boundary Integral Equation Method to Three-Dimensional Stress Analysis. Computers and Structures, Vol. 3, pp. 509–527, 1973
47 Wrobel, L.C. and Brebbia, C.A., Axisymmetric Potential Problems. 2nd Int. Seminar on Recent Advances in B.E.M. C.A. Brebbia (ed.), pp. 77–79, Southampton (U.K.), 1980
48 Doblaré, M., Alarcon, E., and Brebbia, C.A., Boundary Element Formulation for Elastoplastic Analysis of Axisymmetric Bodies. Applied Mathematical Modelling, Vol. 6, 1982.
49 Andersen, D.G., Gaussian Quadrature Formulae for $\int_0^1 \ln x f(x)\, dx$. Math. Comp. **19**, 477–481, 1965
50 Krylow, V.I. and Pal'tsen, A.A., Tables for Numerical Integration of Functions with Logarithmic and Power Singularities. Israel Program for Scientific Translations. Jerusalem 1974
51 Critescu, M. and Loubignac, G., Gaussian Quadrature Formulas for Functions with Singularities in $1/r$ over Triangles and Quadrangles in Recent Advances in B.E.M. C.A. Brebbia (ed.). Pentech Press, 1978
52 Pina, N.L.G., Fernandes, J.L.M., and Brebbia, C.A., Some Numerical Integration Formulae over Triangles and Squares with a $1/R$ Singularity. Applied Mathematical Modelling, Vol. 5, pp. 209–211, 1981
53 Davis, P.J. and Rabinowitz, P., Methods of Numerical Integrations. Academic Press, 1975.
54 Sluis, A. van der and Zweerus, J.R., An Appraisal of some Methods for Computing Cauchy Principal Value Integrals. Numerische Integration. G. Hammerlin (ed.). Birkhäuser Verlag, I.S.N.M., Vol. 45, 1979
55 Longman, I.M., On the Numerical Evaluation of Cauchy Principal Values of Integrals. Math. Comp. **12**, pp. 205–207, 1958
56 Hunter, D.B., The Numerical Evaluation of Cauchy Principal Values of Integrals by Gaussian Integration. Numer. Math. **21**, pp. 185–192, 1973
57 Piessens, R., Numerical Evaluation of Cauchy Principal Values of Integrals. B.I.T. **10**, pp. 476–480, 1970.

58 Lebcdev, V.I. and Baburin, O.V., Calculation of the Principal Values, Weights and Nodes of the Gauss Quadrature Formulae of Integrals. U.S.S.R. Comp. Math. and Math. Phys 5, No. 3, pp. 81–92, 1965
59 Hunter, D.B., Some Gauss-Type Formulae for the Evaluation of Principal Values of Integrals. Numer. Math. **19,** pp. 418–424, 1972
60 Elliot, D., On the Convergence of Hunter's Quadature Rule for Cauchy Principal Value Integrals. BIT **19,** pp. 457–462, 1979
61 Paget, D.F. and Elliot, D., An Algorithm for the Numerical Evaluation of Certain Cauchy Principal Values Integrals. Numer. Math. **19,** pp. 373–385, 1972
62 Chawla, MM. and Ramakrishnan, T.R., Modified Gauss-Jacobi Quadrature Formulas for the Numerical Evaluation of Cauchy Type Singular Integrals. BIT **14,** pp. 14–21, 1974
63 Theocaris, P.S. and Kazantzakis, J.G., On the Numerical Evaluation of Two and Three-Dimensional Cauchy Principal Values Integrals. Arch. Mech. **39,** pp. 105–138, 1981
64 Theocaris, P.S., Iokamidis, N.I., and Kazantzakis, J.G., On the Numerical Evaluation of Two-Dimensional Principal Values Integrals. Int. J. Numer. Math. in Eng. **15,** pp. 611–634, 1980
65 Kazantzakis, J.G. and Theocaris, P.S., The Evaluation of Certain Two-Dimensional Singular Integral used in Three-Dimensional Elasticity. Int. I. Solid and Struct. 05, pp. 203–207, 1979
66 Hartmam, R., Quadratur furela von Gauss typ Singulare Integrale und ihre Anwendung zur Cauchy Singularer Integralgleichungen, 17 Technische Hochschule Karl Marx Stadt, Sektion Mathematic, Karl Marx Stadt 1982
67 Davis, P. and Rabinowitz, P., Ignoring the Singularity in Approximate Integration. S.I.A.M. J. Numer. Anal. 2, pp. 367–383, 1965
68 El-Tom, M.E.D., On Ignoring the Singularity in Approximate Integration. S.I.A.M. J. Numer. Anal. 8, pp. 412–427, 1971
69 Lachat, J.C. and Watson, J.O., Effective Numerical Treatment of Boundary Integral Equation: a Formulation for Three Dimensional Elastostatics. Int. J. for Numerical Methods in Eng., Vol. 10, 1971
70 Chaudonneret, M., Sur la Resolution du Probleme de la Discontinuite du vecteur Contrainte dans les Calculus des Structures par la Method des Equations Integrals. C.R. Acad. des Sc. Paris t 284, Serie A, pp. 463–466, 1977
71 Paris, F., SERBA: a B.I.E. program with linear element for 2-D elastostatics analysis. Advances in Engineering Software, Vol. 2, 1980
72 Doblaré, M., Discontinuidades de Tensión en Interfaces Tridimensionales. Tratamiento con Elementos de Contorno. Aplicaciones del Método de los Elementos Finitos en Ingeniería. Barcelona (Spain), 1982
73 Chaudouet, A., Cyclic Symmetry and Sliding Between Structures by the Boundary Integral Equation Method, in 3rd Int. Sem. on Recent Advances in B.E.M. C.A. Brebbia (ed.). Newport Beach (California), 1981
74 Schren, E., Computer Implementation of the Finite-Element Procedure. Numerical and Computer Methods in Structural Mechanics. S.J. Fenves, N. Perrone, A.R. Rubison and W.C. Schnobrich (ed.). Academic Press, pp. 79–121, 1973
75 Lachat, J.C. and Watson, J.D., Progress in the use of Boundary Integral Equations, Illustrates by Examples. Comp. Math. Appl. Eng. **10,** pp. 273–289, 1977
76 Crotty, J.M., A Block Equation Solver for Large Unsymmetric Matrices Arising in the Boundary Integral Equation Method. Int. J. Num. Meth. in Eng., Vol. 18, pp. 997–1018, 1982
77 Dvornik, J., Generalization of the C.G. Method Applied to Linear and Nonlinear Problems. J. Comp. and Struct. **10,** pp. 217–223, 1979
78 Gambolati, G., Fast Solution to Finite Element Flow Equations by Newton Iteration and Modified Conjugate Gradient Method. Int. J. Num. Meth. in Eng. **15,** pp. 661–675, 1980
79 Jenninos, A. and Malik, G.M., The Solution of Sparse Linear Equations by The Conjugate Gradiente Method. Int. J. Num. Meth. in Eng. **12,** pp. 141–158, 1978

Other Bibliography

1 Baker, C., The Numerical Treatment of Integral Equations. Oxford University Press, 1977
2 Brebbia, C.A. (ed.), Recent Advances in B.E.M. Pentech Press, 1978
3 Engels, H., Numerical Quadrature and Cubature. Academic Press, 1980
4 Fengels, W., An Economical Storage Organization for Large and Small Finite Element Systems with Reference to the Application of Iterative Approximation Methods to Non-linear Material Behavior, Vol. 15, pp. 701 – 712, 1980
5 Ivanov, V., The Theory of Approximate Methods and Their Applications to the Numerical Solution of Integral Equations. Noordhoff, 1976
6 Jaswon, M. and Symm, G., Integral Equation Methods in Potential Theory and Elastostatics. Academic Press, 1973
7 Kupradze, V.D. (ed.), Three Dimensional Problems of the Mathematical Theory of Elasticity and Thermoelasticity. North Holland Publishing Co., 1979
8 Muskhelishvili, W.I., Singular Integral Equations, Pergamon Press, 1956
9 Stroud, A.N. and Secrett, D., Gaussian Quadrature Formulae. Prentice Hall, 1966
10 Zabreyko, P.P. et al., Integral Equation. A Reference Text, Noordhoff, 1975

Chapter 5

The Edge Function Method (E.F.M.) for Cracks, Cavities and Curved Boundaries in Elastostatics

by P. M. Quinlan and M. J. A. O'Callaghan

5.1 Introduction

The following paper represents a considerable development of the paper, Quinlan and O'Callaghan (1984), presented at the *Sixth International Conference of B.E.M. in Engineering* and published in the Conference proceedings. The section on curved boundaries has been greatly expanded with appropriate examples.

The Edge-Function Method may be described as a boundary element method that pieces together suitable 'asymptotic' solutions for the several parts of a domain D and satisfies the boundary conditions in an acceptable approximate manner, using collocation or harmonic fitting.

The method was originally developed by Quinlan (1964) for the torsion of prismatic bars of polygonal cross-section. Since then it has been successfully applied to isotropic and aeolotropic thin plates, Quinlan (1968), Grannell (1980); to Elastostatics, Quinlan (1971, 1974); moderately thick plates and shallow shells, Desmukh (1973); vibrations of thin plates, Tai (1973); vibration of shallow shells, O'Callaghan (1975); to 3-D Elasticity, Quinlan (1977); and to 3-D embedded elliptical cracks, Quinlan (1982).

Over the past two years the first author has returned to 2-D crack cavities and stress concentration problems in Elastostatics. A corresponding comprehensive program EQUINP has been developed, whose input requires no knowledge of E.F.M., and is directly comparable to both F.E.M., and B.E.M *when these are backed by a fully automatic mesh generator.*

The E.F.M. can be used to provide special substructures for F.E.M. programs, which would be considerably more powerful and less time consuming than the special elements in current use. It has much in common with B.E.M. in that they are both integral methods that satisfy the equations exactly, and approximate the boundary conditions. However E.F.M., while yet much less widely used than B.E.M., incorporates a far greater range of singular functions in its solution scheme – functions that B.E.M. might be encouraged to examine with a view to their incorporation into its work. With this latter aim in mind an effort has been made to make the present paper a reasonably self-contained exposé of the Edge-Function Method for Elastostatics.

The present paper presents the Complex Displacement Method for 2-D Elastostatics, which can be seen to be superior to the classical Complex Variable approach. The input routine, as given in Appendix A, does not require any knowledge of E.F.M., and in the complicated case of Example A – an irregular pentagon

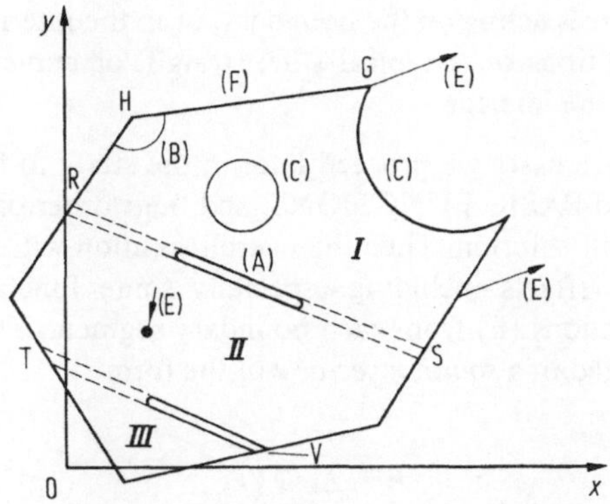

Fig. 1. Two-dimensional region

with a skew edge crack and mixed boundary conditions with parabolic loading on one edge – only eighteen lines of input are required. Other examples presented include concentrated loads, interior straight line cracks, notches, and circular or elliptic cavities.

Solutions for up to four increasing sizes of matrix can be computed if requested. Each solution vector provides an *exact* solution to the specified problem with slightly different boundary values, the difference from the prescribed values being termed the *boundary residuals.* If the residuals are within the limits within which an engineer can specify the boundary values, then E.F.M. offers as *'exact'* a solution as can be obtained to the problem.

Stresses and fracture coefficients are obtained as accurately as the displacements, and are readily computed and require little computer time. When the Solution Vector has been obtained, it then provides *analytical expressions for* both stresses and displacements. This is in marked contrast to standard element methods, and facilitates contour plots of results.

5.2 The Edge Function Method (E.F.M.) (a Qualitative Description)

Consider a two-dimensional domain, as in Figure 1, with edge and interior cracks, sharp vertices, cavities, concentrated forces, and with either stresses or displacements prescribed on its boundary segments. It is useful to examine the various critical regions in the problem by constructing and analysing simplified problems – called BASIC PROBLEMS – that portray corresponding singular features. In particular we wish to study

(A) A semi-infinite straight line crack in an infinite medium.
(B) A typical semi-infinite wedge to portray stress concentrations at a vertex.
(C) A circular hole in an infinite medium – or a circular indentation or mound on part of the boundary.
(D) An elliptic hole – or an elliptic indentation or mound on part of the boundary.

(E) Concentrated forces acting on the boundary, or in the interior.
(F) How edge conditions on a typical side, straight or curved, are imposed and propagated into the interior.

In each of the above cases we proceed later in the study to find the corresponding functions – called BASIC FUNCTIONS, and their order of importance in contributing to the overall solution. Then the overall solution will be taken as a superposition of relevant effects including especially some functions (B) from each vertex and some functions (F) from each boundary segment. The resulting solution scheme can be arranged in a *solution vector* $\boldsymbol{u}$ of the form

$$\boldsymbol{u} = \sum_{k=1}^{k'} c_k u_k, \tag{1}$$

involving some specified number k' arbitrary constants c_k for $k = 1$ to k', where $\boldsymbol{u}$ is the primary displacement component in Eqs. (2).

The constants c_k must then be determined so that the derived stresses and/or displacements satisfy the prescribed boundary conditions to an acceptable level of approximation. The simplest way to do this is by *collocation* in which $k'/2$ collocation points – involving two equations at each point – are selected in a special manner on the boundaries; slightly better results can usually be attained by using *harmonic fitting,* Quinlan (1977), as outlined in Appendix A, to approximate the boundary conditions.

However the most important requirement for a good approximation to the problem is that $\boldsymbol{u}$ should include *all* relevant BASIC FUNCTIONS in its solution scheme.

Simple rules for setting up the solution vector (1) will be given when the BASIC PROBLEMS outlined above have been analysed. But first the COMPLEX DISPLACEMENT METHOD, as developed by the authors, will be outlined as a suitable means for obtaining all the BASIC FUNCTIONS required.

5.3 Complex Displacement Method for Elastostatics

The displacement components (u, v) in an isotropic elastostatics problem, with zero body forces, satisfy the Navier equations

$$(\lambda + 2\mu)\frac{\partial^2 u}{\partial x^2} + \mu\frac{\partial^2 u}{\partial y^2} + (\lambda + \mu)\frac{\partial^2 v}{\partial x\,\partial y} = 0$$

$$(\lambda + \mu)\frac{\partial^2 u}{\partial x\,\partial y} + \mu\frac{\partial^2 v}{\partial x^2} + (\lambda + 2\mu)\frac{\partial^2 v}{\partial y^2} = 0, \tag{2}$$

in any cartesian axes (x, y), where λ and μ are Lame's elastic constants. On eliminating v it follows that u satisfies the biharmonic equation

$$\left(\frac{\partial^2}{\partial x^2} + \frac{\partial^2}{\partial y^2}\right)^2 u = 0. \tag{3}$$

On transforming to complex quantities $(z, \bar{z})$, where

$$z = x + \mathrm{i}\, y; \quad \bar{z} = x - \mathrm{i}\, y$$

$$\frac{\partial}{\partial x} = \frac{\partial}{\partial z} + \frac{\partial}{\partial \bar{z}}; \quad \frac{\partial}{\partial y} = \mathrm{i}\left(\frac{\partial}{\partial z} - \frac{\partial}{\partial \bar{z}}\right), \tag{4}$$

Eq. (3) reduces to

$$\frac{\partial^4 u}{\partial z^2 \partial \bar{z}^2} = 0,$$

which has a solution in the complex form

$$u = F(z) + \bar{z}\, G(z), \tag{5}$$

where $F(z)$ and $G(z)$ are arbitrary functions, which will be called *Complex Potentials; u* being termed the *primary displacement component*.

A coupling equation for v follows on using operateurs (4) to transform the first of Eq. (2) into

$$\frac{\partial^2 u}{\partial z^2} + \frac{\partial^2 u}{\partial \bar{z}^2} + 2T\frac{\partial^2 u}{\partial z\, \partial \bar{z}} + \mathrm{i}\left(\frac{\partial^2}{\partial z^2} - \frac{\partial^2}{\partial \bar{z}^2}\right) v = 0, \tag{6}$$

where $T = \dfrac{\lambda + 3\mu}{\lambda + \mu}$.

Since v also satisfies a biharmonic equation we seek its solution in form (5) for which $\partial^2 v/\partial \bar{z}^2 = 0$. Accordingly Eq. (6) gives, on substituting for u from Eq. (5) and integrating twice with respect to z, the following form for v:

$$v = \mathrm{i}\,[F(z) + \bar{z}\, G(z)] + H(z), \tag{7}$$

where

$$H(z) = 2T\mathrm{i} \int G(z)\, dz. \tag{8}$$

A more balanced solution for Eqs. (2) is obtained by superposing on the above displacements for (u, v), the displacement system $[\mathrm{i}\, H(z)/2, \ -H(z)/2]$, an obvious special case of results (5) and (7), giving:

$$\tilde{u} = F(z) + \bar{z}\, G(z) + \mathrm{i}\, H(z)/2$$
$$\tilde{v} = \mathrm{i}\, F(z) + \mathrm{i}\, \bar{z}\, G(z) + H(z)/2. \tag{9}$$

The tilda ˜ on the displacements is used to emphasise that these are complex displacement components, and the physical displacements follow – when required – as the real parts of $\tilde{u}$ and $\tilde{v}$.

5.4 Displacements and Stresses for Arbitrary Cartesian Axes

Consider axes (x', y') at any point P, either on the boundary or in the interior, of a 2-D region, where the x'-axis is at angle ϕ to $0x$. On resolving displacement

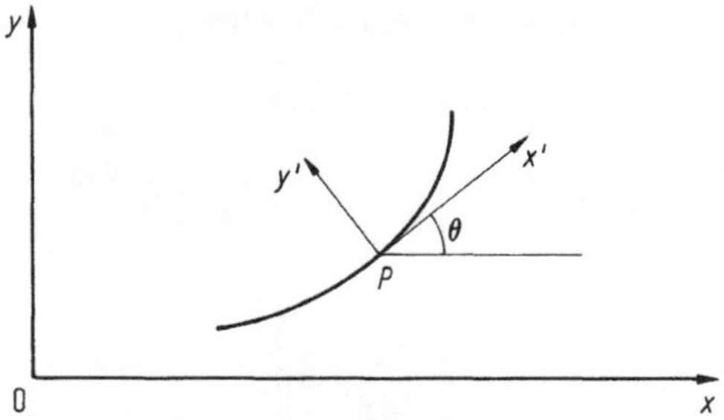

Fig. 2. Arbitrary cartesian axes

components along x' and y' respectively it follows that

$$\tilde{u}' = \tilde{u}\cos\phi + \tilde{v}\sin\phi; \qquad \tilde{v}' = -\tilde{u}\sin\phi + \tilde{v}\cos\phi$$
$$\tilde{u}' = e^{i\phi}[F(z) + \bar{z}\,G(z)] + i\,e^{-i\phi}H(z)/2$$
$$\tilde{v}' = i\,e^{i\phi}[F(z) + \bar{z}\,G(z)] + e^{-i\phi}H(z)/2. \tag{10}$$

The resulting stress components $\tilde{\tau}'_{xy}$, $\tilde{\tau}'_{yy}$ follow readily from the corresponding equations

$$\tilde{\tau}'_{xy} = \mu\left(\frac{\partial\tilde{u}'}{\partial y'} + \frac{\partial\tilde{v}'}{\partial x'}\right)$$
$$\tilde{\tau}'_{yy} = \lambda\left(\frac{\partial\tilde{u}'}{\partial x'} + \frac{\partial\tilde{v}'}{\partial y'}\right) + 2\mu\frac{\partial\tilde{v}'}{\partial y'}, \tag{11}$$

by applying to Eq. (10) the operators [1]

$$\frac{\partial}{\partial x'} = e^{i\phi}\frac{\partial}{\partial z} + e^{-i\phi}\frac{\partial}{\partial\bar{z}},$$
$$\frac{\partial}{\partial y'} = i\,e^{i\phi}\frac{\partial}{\partial z} - i\,e^{-i\phi}\frac{\partial}{\partial\bar{z}}, \tag{12}$$

to obtain the derivatives required in Eqs. (11), leading to

$$\tilde{\tau}'_{xy} = 2\mu\,i\,e^{2i\phi}\left[\frac{dF}{dz} + \bar{z}\frac{dG}{dz}\right]$$
$$\tilde{\tau}'_{yy} = -2\mu\,e^{2i\phi}\left[\frac{dF}{dz} + \bar{z}\frac{dG}{dz}\right] - 4\mu G \tag{13}$$

Observe in the Complex Displacement method that $\tilde{u}$ in solution (5) is equivalent to the complex stress function $X(z)$ used in classical complex variable methods for Elastostatics, as in Sokolnikoff (1956). The general displacement and stress expressions (10) and (13) are considerably simpler than those in current use, and are much more easily programmed.

[1] $\partial V/\partial x' = (\partial V/\partial z)(\partial z/\partial x') + (\partial V/\partial\bar{z})(\partial\bar{z}/\partial x')$, where $z' = (z - z_p)\,e^{-i\phi}$; $\partial z/\partial x' = e^{i\phi}$, $\partial\bar{z}/\partial x' = e^{-i\phi}$; $z' = x' + i\,y'$ and z_p is the location of P relative to 0.

5.5 Basic Problems

We now proceed to study the *BASIC PROBLEMS* arising in Elastostatics, as outlined earlier.

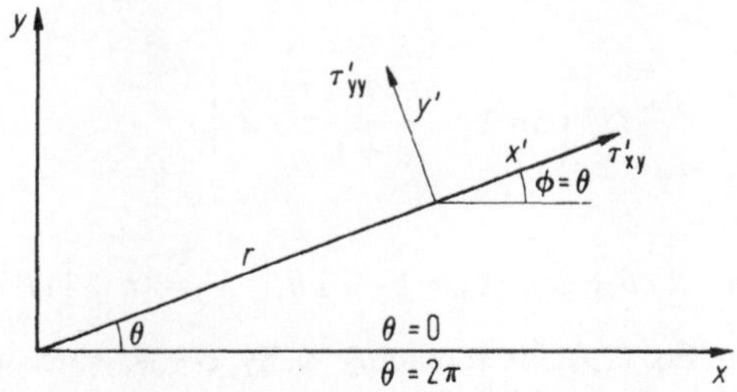

Fig. 3. Semi-infinite Crack

(A) *Semi-infinite Straight Line Crack: Crack Functions*

Consider a crack along the x-axis from $(0, \infty)$, where the appropriate boundary conditions are

$$\tau'_{xy} = \tau'_{yy} = 0 \quad \text{on} \quad \theta = 0 \quad \text{and} \quad \theta = 2\pi, \quad \text{for } r \text{ in } (0, \infty). \tag{14}$$

Accordingly we require functions that are homogeneous[2] in r, which result from taking the complex potentials in (5) as:

$$F(z) = F^n z^{n+1}; \quad G(z) = G^n z^n; \quad z = r\,e^{i\theta}, \tag{15}$$

where the complex constants introduced are

$$F^n = F_1^n + i\,F_2^n; \quad G^n = G_1^n + i\,G_2^n, \tag{16}$$

involving the four real arbitrary constants

$$F_1^n,\ F_2^n,\ G_1^n,\ G_2^n,$$

which can depend on n as is indicated by the superscript.

On substituting from (15) in expressions (13) and taking their real part, it follows that:

$$\begin{aligned}\tau'_{xy} = \operatorname{Re}[\tilde{\tau}'_{xy}] = 2\mu\, r^n [&-(n+1)\{F_1^n \sin X_1 + F_2^n \cos X_1\} \\ &- n\{G_1^n \sin X_2 + G_2^n \cos X_2\}]\end{aligned}$$

$$\begin{aligned}\tau'_{yy} = 2\mu\, r^n [&-(n+1)\{F_1^n \cos X_1 - F_2^n \sin X_1\} \\ &- G_1^n\{n \cos X_2 + 2\cos X_0\} + G_2^n\{n \sin X_2 + 2 \sin X_0\}]\end{aligned} \tag{17}$$

where

$$X_0 = n\,\theta; \quad X_1 = n\,\theta + 2\phi; \quad X_2 = X_1 - 2\theta.$$

---

[2] Alternatively if we replace $(n+1)$ in $F(z)$ by n', the first of Eqs. (20) then becomes $n' r^{n'-1} F_2 + n\, r^n G_2 \equiv 0$; r in $(0, \infty)$, which requires $n = n' - 1$, or $n' = n + 1$, to permit removal of the factor r^n.

Similar expressions follow for u', v' from Eqs. (10), on noting from intgral (8) that

$$H(z) = 2T\,\mathrm{i}\, z^{n+1}/(n+1); \quad n \neq -1, \tag{18}$$

$$u'/r^{n+1} = F_1^n \cos X_3 - F_2^n \sin X_3 + G_1^n \left\{\cos X_4 - \frac{T}{n+1}\cos X_5\right\}$$
$$- G_2^n \left\{\sin X_4 - \frac{T}{n+1}\sin X_5\right\}, \tag{19}$$

where

$$X_3 = (n+1)\,\theta + \phi; \quad X_4 = X_3 - 2\theta; \quad X_5 = (n+1)\,\theta - \phi$$

and v' follows from Eq. (19) on replacing ϕ by $\phi + \frac{\pi}{2}$, since v' is the displacement component at an angle $\phi + \frac{\pi}{2}$ to $0x$.

On applying boundary conditions (14) to Eqs. (17), it follows, on omitting the superscript n and setting $\phi = 0$, that

$$\theta = 0: \quad \left.\begin{aligned} &(n+1)\,F_2 + n\,G_2 = 0 \\ &(n+1)\,F_1 + (n+2)\,G_1 = 0 \end{aligned}\right\} \tag{20}$$

$$\theta = 2\pi: \quad \left.\begin{aligned} &[(n+1)\,F_1 + n\,G_1]\sin 2\pi n = 0 \\ &[(n+1)\,F_2 + (n+2)\,G_2]\sin 2\pi n = 0 \end{aligned}\right\}, \tag{21}$$

on noting in Eqs. (21) that the coefficients of the $\cos 2\pi n$ term in each is zero by virtue of Eqs. (20). Accordingly a non-trivial solution is given by

$$\sin 2\pi n = 0,$$

from which *either* n is an integer, which gives non-singular polynomial solutions, *or* the important singular solutions:

$$n = -\tfrac{1}{2}, \tfrac{1}{2}, \tfrac{3}{2}, \tfrac{5}{2}, \ldots, \tag{22}$$

where the displacements (19) are finite at $r = 0$ provided that $n + 1 > 0$, or $n > -1$.

Note. $n = -\frac{1}{2}$ procedures a stress singularity of order $r^{-0.5}$ at $r = 0$. The derived functions (17) and (19) are called *Crack Functions*, and involve but two arbitrary constants F_1^n and F_2^n, since from Eqs. (20) G_1 and G_2 are given by

$$G_1^n = -F_1^n (n+1)/(n+2); \quad G_2^n = -F_2^n (n+1)/n. \tag{23}$$

Fracture Coefficients

On putting $n = -\frac{1}{2}$ in Eqs. (23) we obtain

$$G_1 = -F_1/3; \quad G_2 = F_2, \tag{24}$$

and the displacements and stresses – in the reference axes in Fig. 3, for which $\phi = 0$ – follow on substituting in Eqs. (19) and (17), as:

$$u = \left[F_1 \left\{ \left(1 + \frac{2T}{3}\right) \cos\frac{\theta}{2} - \frac{1}{3}\cos\frac{3\theta}{2} \right\} + F_2 \left\{ -(1-2T)\sin\frac{\theta}{2} + \sin\frac{3\theta}{2} \right\} \right] \sqrt{r}$$

$$\tau_{yy} = \left[\frac{F_1}{3} \left\{ \cos\frac{\theta}{2} - \cos\frac{5\theta}{2} \right\} + F_2 \left\{ -5\sin\frac{\theta}{2} + \sin\frac{5\theta}{2} \right\} \right] \frac{\mu}{\sqrt{r}} \tag{25}$$

$$\tau_{xy} = \left[\frac{F_1}{3} \left\{ 3\sin\frac{\theta}{2} + \sin\frac{5\theta}{2} \right\} + F_2 \left\{ -\cos\frac{\theta}{2} + \cos\frac{5\theta}{2} \right\} \right] \frac{\mu}{\sqrt{r}},$$

with similar expressions for v and τ_{xx}. These check with those given in books on fracture mechanics e.g. Parker (1981).

The corresponding *fracture coefficients* at each crack tip ($r \to 0$, $\theta = \pi$) are defined as

$$K_1 = \operatorname*{Lt}_{r \to 0} \sqrt{2\pi r}\, \tau_{yy}; \quad K_2 = \operatorname*{Lt}_{r \to 0} \sqrt{2\pi r}\, \tau_{xy} \tag{26}$$

for the crack opening mode (mode 1) and the crack sliding mode (mode 2) respectively, and on substituting from Eqs. (24) and (25) it follows that

$$\begin{aligned} K_1 &= -\mu\sqrt{2\pi}\,(F_2^n + 3G_2^n) = -4\mu\sqrt{2\pi}\,F_2^n \\ K_2 &= \mu\sqrt{2\pi}\,(F_1^n - G_1^n) \quad = \frac{4\mu\sqrt{2\pi}}{3} F_1^n; \quad n = -1/2. \end{aligned} \tag{27}$$

The coefficients F_1^n and F_2^n, $n = -\frac{1}{2}$, are readily located amongst the arbitrary constants c_k in solution vector (1), which is much simpler and more accurate than current methods, Parker (1981), of determining fracture coefficients.

The coefficients K_1 and K_2 are usually normalised by dividing by $P_0\sqrt{c_0}$, where

$$P_0 = \text{normalising load}; \quad c_0 = \text{normalising crack length}. \tag{28}$$

(B) *Stress Concentrations at a Vertex: Vertex Functions*

Consider the vertex Q when, for example, stresses are given on side QR and displacements given on QP. Complex potentials (15), with origin at Q, can be used to

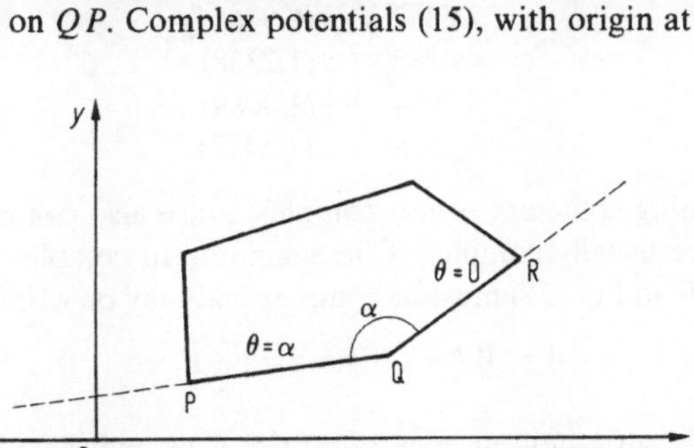

Fig. 4. Semi-infinite sector

model an eigenvalue situation where the stress components are set to zero on $\theta = 0$ and the displacements to zero on $\theta = \alpha$ of the semi-infinite wedge with vertex at Q.

The analysis follows, as in basic problem (A), where analogous to Eqs. (20) and (21) we obtain, on setting $\phi = \theta$

$$\begin{aligned}
\theta = 0{:}\quad & (n+1)\, F_2 + n\, G_2 = 0 \\
& (n+1)\, F_1 + (n+2)\, G_1 = 0 \\
\theta = \alpha{:}\quad & F_1 \cos n'\alpha - F_2 \sin n'\alpha + G_1 (1 - T') \cos n\alpha \\
& \qquad - G_2 (1 - T') \sin n\alpha = 0 \\
& F_1 \sin n'\alpha + F_2 \cos n'\alpha + G_1 (1 + T') \sin n\alpha \\
& \qquad + G_2 (1 + T') \cos n\alpha = 0,
\end{aligned} \tag{29}$$

in which $n' = n + 2$, $T' = T/(n+1)$.

On eliminating F_1, F_2, G_1, and G_2 the resulting 4×4 determinant $\psi(n)$ of the coefficients matrix reduces to

$$\psi(n) \equiv 4\, T \sin^2 (n+1)\, \alpha + 4\, (n+1)^2 \sin^2 \alpha - (1+T)^2 = 0. \tag{30}$$

On solving the first three of Eqs. (29) in determinant form we obtain

$$F_1/U_1 = F_2/U_2 = G_1/U_3 = G_2/U_4,$$

where U_1, U_2, U_3, U_4 are the corresponding cofactors. On setting $F_1/U_1 = W$ it follows that

$$F_1 = U_1 W, \quad F_2 = U_2 W, \quad G_1 = U_3 W, \quad G_2 = U_4 W, \tag{31}$$

where W is an arbitrary constant that may depend on n.

A comprehensive subroutine EMACOS, has been developed to set up and solve the appropriate eigenvalue equation[3] $\psi(n) = 0$, similar to Eq. (30), for any of the nine possible boundary combinations resulting from all three possible conditions – fixed, free or symmetric – on each edge.

The first six roots arranged in order of singularity, computed by EMACOS for $\alpha = 0.1695$ in eigen-equation (30), are

$$\begin{aligned}
-0.31960 &+ \mathrm{i}\,(0) \\
0.60136 &+ \mathrm{i}\,(0.41837) \\
2.5342 &+ \mathrm{i}\,(0.97655) \\
4.4232 &+ \mathrm{i}\,(1.2338) \\
6.2984 &+ \mathrm{i}\,(1.4089) \\
8.1668 &+ \mathrm{i}\,(1.5427),
\end{aligned}$$

and the corresponding cofactors – also complex – are also computed and stored. The eigenvalues are usually complex. Corresponding to complex values for n the arbitrary constant W in Eq. (31) must be complex and may be written:

$$W = W_1^n + \mathrm{j}\, W_2^n; \quad \mathrm{j} = \sqrt{-1}. \tag{32}$$

---

[3] This can be left in matrix form in using EMACOS, if the reduction is too complicated as in the case of anisotropic elasticity. The program uses complex arithmetic to deal with complex values for n.

Hence, as with Crack Functions, functions derived from potentials (15) are called *Vertex Functions*, and involve the two arbitrary constants W_1^n and W_2^n. The condition that the displacements should be finite at $r=0$ requires that $\mathrm{Re}\,(n+1)>0$.

(C) *Circular Boundaries, Mounds, or Indentations; Curved Edge Functions*

This requires a slight modification of case (A) as θ is now the boundary parameter and this requires that the functions $F(z)$ and $G(z)$ be homogeneous in $e^{i\theta}$. Accordingly functions (15) must be replaced by

$$F(z) = F^n z^{n-1}; \quad G(z) = G^n z^n, \tag{33}$$

where the appropriate values for n are:

Boundary	n
Circular Outer Boundary	N
Circular Inner Boundary	$-N$
Circular Mound on Boundary	$2\pi N/\alpha$
Circular Indentation on Boundary	$-2\pi N/\alpha$

N being an integer, and α the central angle subtended by the arc. The order of importance of the functions is $N=0, 1, 2, \ldots$. This can be established by considering, in a manner analogous to the loaded semi-infinite strip for Edge-Functions (59), a curvilinear region bounded by $\theta=0$, $\theta=\alpha$; $r=a$ and $r=b$ where $b\to\infty$, and loaded on the circular edge $r=a$ on which the parameter θ is in the interval $(0,\alpha)$. For each $N>0$, and for each displacement, or stress, component four functions with coefficients F_1^n, F_2^n, G_1^n and G_2^n – called *Curved Edge Functions* – derive from potentials (33), as in Eq. (37) under.

It follows from Definition (8) that

$$\begin{aligned} H(z) &= 2T\mathrm{i}\,G^n z^{n+1}/n+1; \quad && n \neq -1 \\ &= 2T\mathrm{i}\,G^n \log z; && n=-1 \end{aligned} \tag{34}$$

The resulting displacements (u,v) from Eqs. (9) are multivalued in θ when $n=-1$, as in the case of a circular hole. These can be rendered single valued by superposing on them the displacements

$$\bar{u} = \mathrm{Re}[C\log z]; \quad \bar{v} = \mathrm{Re}[\mathrm{i}\,C\log z], \tag{35}$$

got by substituting $F(z)=C\log z$ in Eq. (9), and on setting the resulting multivalued parts to zero we obtain

$$C = -T\bar{G}^n; \quad \bar{G}^n = G_1^n - \mathrm{i}\,G_2^n, \quad n=-1. \tag{36}$$

The resulting displacements and stresses follow from Eqs. (10) and (13) respectively on using Eqs. (33) and (34), which on a circular boundary, for which $\phi=\frac{\pi}{2}+\theta$, reduce to

$$\begin{aligned} u' &= -[F_1^n r^{n-1} + G_1^n r^{n+1}(1+T/n+1)]\sin n\theta \\ &\quad -[F_2^n r^{n-1} + G_2^n r^{n+1}((1+T/n+1)]\cos n\theta \\ \tau'_{xy} &= 2\mu r^{n-2}\{[(n-1)F_1^n + n r^2 G_1^n]\sin n\theta \\ &\quad [(n-1)F_2^n + n r^2 G_2^n]\cos n\theta\}, \quad n\neq -1, \end{aligned} \tag{37}$$

with similar expression for v' and τ'_{yy}. An appropriate modification involving superposing system (35), is required when $n=-1$.

A simple subroutine EPOLC computes the numerical values of the various Curved Edge-Functions[4] with arbitrary constants F_1^n, F_2^n, G_1^n, G_2^n respectively, at any point (x, y) with reference to axis ϕ, given n and the coordinates of the centre and the slope of the corresponding local axis.

In the special case when $N=0$ the first pair of these functions F^0/z correspond to an inner boundary or indentation, and the second pair G^0 correspond to an outer circular boundary or mound.

(D) *Elliptic Boundaries, Mounds or Indentations: Curved Edge Functions*

These follow by using the Joukowski mapping of the region outside the limiting ellipse, with foci at $\pm c$, on the region $|\zeta| \geqq 1$ – where as $z \to \infty$, $\zeta \to \infty$ – by the function

$$z = 0.5\, c\,(\zeta + 1/\zeta) \quad \text{or} \quad \zeta = [z + \sqrt{z^2 - c^2}]/c; \quad \zeta = \varrho\, e^{i\psi}.$$

Points ζ on circles $|\zeta| = \varrho > 1$ map onto a series of confocal ellipses, where ψ is in the interval $(-\pi, \pi)$.

Analogous to polar harmonics (33) we take elliptic harmonics

$$F(z) = F^n \zeta^{n-1}; \quad G(z) = G^n \zeta^n. \tag{38}$$

Exponents n follow as in harmonics (33); $\alpha = \psi_2 - \psi_1$ being the parametric angular extent of the elliptic arc of semi-axes (a, b) which is mapped on a circle of radius ϱ_0, where

$$x = a \cos\psi, \quad y = b \sin\psi \quad \text{and} \quad c^2 = a^2 - b^2,$$

where

$$a = c\,(\varrho_0 + 1/\varrho_0)/2 \quad \text{and} \quad b = c\,(\varrho_0 - 1/\varrho_0)/2. \tag{39}$$

On substituting in Def. (8) for $G(z)$ and dz, where

$$\frac{dz}{d\zeta} = 0.5\, c\,(1 - 1/\zeta^2), \tag{40}$$

we obtain:

$$H(z) = c\, T\, i\, G^n \left[\frac{\zeta^{n+1}}{n+1} - \frac{\zeta^{n-1}}{n-1}\right]; \quad n \neq \pm 1. \tag{41}$$

In the exceptional cases when $n = \pm 1$ we must replace the term $\zeta\%$ in the above by $\log \zeta$ since

$$\int \frac{d\zeta}{\zeta} = \log \zeta.$$

[4] For example, the four curved edge-functions for u' in Eq. (37) are:
$-r^{n-1} \sin n\theta$, $-r^{n-1} \cos n\theta$, $-r^{n+1}(1 + T/_{n+1}) \sin n\theta$, $-r^{n+1}(1 + T/_{n+1}) \cos n\theta$.

The resulting displacements (u, v) from Eqs. (9), when $n = \pm 1$, are multivalued in ψ where $\zeta = \varrho\, e^{i\psi}$. These can be rendered single valued by superposing on them the displacements

$$\bar{u} = \mathrm{Re}[C \log \zeta]; \quad \bar{v} = \mathrm{Re}[i\, C \log \zeta], \tag{42}$$

obtained by substituting $F(z) = C \log \zeta$ in equations (9), and on setting the resulting multivalued parts to zero we obtain:

$$G = n\, c\, T \bar{G}^n/2; \quad n = \pm 1 \tag{43}$$

where

$$\bar{G}^n = G_1^n - i\, G_2^n.$$

The displacements $(\bar{u}', \bar{v}')$ follow as in Eq. (10) on superposing systems (38) and (42). The corresponding stresses follow, as in Eq. (13), on noting that

$$\frac{d}{dz}(\zeta^n) = n\, \zeta^{n-1} \frac{d\zeta}{dz} = \frac{n\, \zeta^n}{\sqrt{z^2 - c^2}}. \tag{44}$$

As in the case of circular boundaries, the order of importance of the functions (38) is $N = 0, 1, 2, \ldots$ This can be established by considering a curvilinear region bounded by $\psi = \psi_1$, $\psi = \psi_2$, $\varrho = \varrho_1$ and $\varrho = \varrho_2$ where $\varrho_2 \to \infty$, and loaded along the curved edge $\varrho = \varrho_1$ on which the parameter ψ is in the interval (ψ_1, ψ_2).

Mound, or a Solid Ellipse

In the case of a mound, or a solid ellipse, the focal cut SS' for ζ, from $z = c$ to $z = -c$, occurs in the material body and corresponds to the limiting ellipse $|\zeta| = 1$, as under in Figure 5.

At any coincident points P_0 and P_0', with respective parameters ψ_0 and $2\pi - \psi_0$, the displacements and stresses must be single valued. On putting $\phi = 0$, Eq. (9) gives, where n' is $n-1$, n, or $n+1$:

$$u = \mathrm{Re}\left[F^n \zeta^{n-1} + \bar{z}\, G^n \zeta^n - \frac{c\,T}{2} G^n \left(\frac{\zeta^{n+1}}{n+1} - \frac{\zeta^{n-1}}{n-1} \right) \right],$$

which involves terms like $\sin(n' \psi_0)$ at P_0 and $\sin\{n'(2\pi - \psi_0)\} = -\sin(n' \psi_0)$ at P_0', that cause a discontinuity in u, unless it is obliterated. Accordingly we must

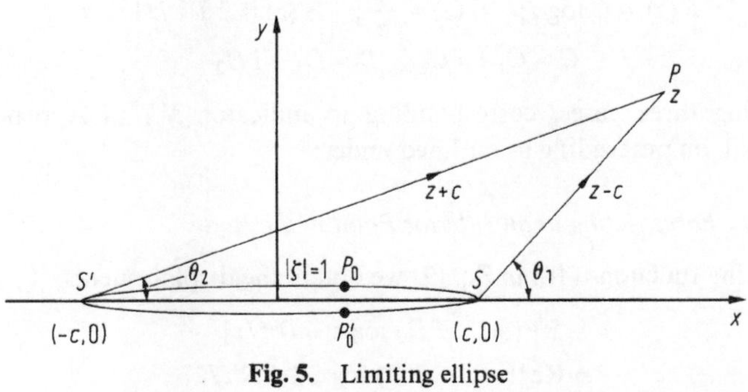

Fig. 5. Limiting ellipse

superpose the requisite potentials, which are easily found to be

$$F^c(z) = F^n \zeta^{-n+1}; \quad G^c(z) = G^n \zeta^{-n}, \tag{45}$$

involving a second program loop in the corresponding point routine EJOUK with replacements

$$n \to -n+2 \text{ in } F(z); \quad n \to -n \text{ in } G(z). \tag{46}$$

It is a simple exercise to show that the displacements and stresses resulting from the superposed potentials $F(z)+F^c(z)$, $G(z)+G^c(z)$ are continuous on SS', on taking the cut for $\sqrt{z^2-c^2}$ as $(-c, c)$ which corresponds to taking $\arg(z+c)$ and $\arg(z-c)$ in $(0, 2\pi)$, and noting from Fig. 5:

$$\begin{aligned} &\theta_1=\pi, \quad \theta_2=0 \text{ at } P_0, \quad \theta_1=\pi, \quad \theta_2=2\pi \text{ at } P_0'; \\ &\arg(\sqrt{z^2-c^2}) = \tfrac{1}{2}(\theta_1+\theta_2) \\ \therefore \quad &\sqrt{z^2-c^2} \text{ at } P_0' = -\sqrt{z^2-c^2} \text{ at } P_0. \end{aligned} \tag{47}$$

Also we can show that the stresses at S are finite, on taking a point $z=c+z_2$, with $\phi=0$, close to S and noting that

$$\operatorname*{Lt}_{z_2\to 0} \frac{\zeta^n-\zeta^{-n}}{\sqrt{z^2-c^2}} = \operatorname*{Lt}_{z_2\to 0} \frac{[\{1+K\sqrt{z_2}+\ldots\}^n - \{1+L\sqrt{z_2}+\ldots\}^{-n}}{2c\sqrt{z_2}} \Rightarrow \frac{n}{c}K,$$

which is finite. Similarly for the stresses at the point S'.

As in basic problem (C) for circular curves, in the $N=0$ case the first pair of functions (38), F^0/ζ, correspond to an inner boundary or indentation and the second pair, G^0, to an outer boundary or elliptic mound.

A subroutine EJOUK does the computation for Curved Edge-Functions derived fom the elliptic harmonics (38) analogous to that done by EPOLC for Curved Edge-Functions derived from polar harmonics (33).

Note. In the case of an elliptic mound that it is necessary to confine ψ to $(-\pi, \pi)$, to ensure that $\cos(n'\psi_0)$ at P_0 is equal to $\cos(n'\psi_0')$ at P_0' for continuity requirements where n' is not an integer. Continuity follows since $\psi_0' = -\psi_0$.

(E) *Concentrated Forces on Boundary or Interior – also Moments*

These follow readily on setting the complex potentials to

$$\begin{aligned} F(z) = C\log z; \quad G(z) = \frac{D}{z}; \quad H(z) = 2T\mathrm{i}\,D\log z; \\ C = C_1 + \mathrm{i}\,C_2; \quad D = D_1 + \mathrm{i}\,D_2. \end{aligned} \tag{48}$$

The following three cases, corresponding to indicator NV in Appendix A, are readily solved, on proceeding as outlined under:

Case $NV=1$: Force Acting at an Interior Point in the Body

On substituting functions (48) in Eq. (9) we obtain the displacements:

$$\begin{aligned} u &= \mathrm{Re}\,[(C-TD)\log z + D\,\bar{z}/z] \\ v &= \mathrm{Re}\,[(\mathrm{i}\,(C+TD)\log z + \mathrm{i}\,D\,\bar{z}/z] \end{aligned} \tag{49}$$

and on setting the coefficient of θ to zero, to render the displacements single valued, it follows that

$$C = -T\bar{D}. \tag{50}$$

On excluding the force point $r = 0$ by a circular cavity, as shown in (a) in Fig. 6 under, it follows from Eqs. (13) on using substitutions (49) and (50) that:

$$\tau'_{xy}/2\mu = \mathrm{Re}\left[-\mathrm{i}\,\mathrm{e}^{2i\theta}\left(\frac{C}{z} - \frac{D\bar{z}}{z^2}\right)\right] = [D_1 \sin\theta - D_2 \cos\theta]\,\frac{(1-T)}{r}$$

$$\tau'_{yy}/2\mu = \mathrm{Re}\left[\mathrm{e}^{2i\theta}\left(\frac{C}{z} - \frac{D\bar{z}}{z^2} - \frac{2D}{z}\right)\right] = [D_1 \cos\theta - D_2 \sin\theta]\,\frac{(3+T)}{r}, \tag{51}$$

and on equating the resultant action of the cavity on the body to $\bar{\boldsymbol{P}} = (P_x, P_y)$ we obtain

$$P_x = -\int_0^{2\pi} (\tau'_{xy} \sin\theta + \tau'_{yy} \cos\theta)\, r\, d\theta$$

$$P_y = \int_0^{2\pi} (\tau'_{xy} \cos\theta - \tau'_{yy} \sin\theta)\, r\, d\theta, \tag{52}$$

from which on substituting from Eq. (51) and integrating it follows that

$$D_1 = P_x/4\pi\mu(1+T); \quad D_2 = P_y/4\pi\mu(1+T). \tag{53}$$

Case NV = 3: Force Acting at a Vertex of a Polygon

Consider a force $\bar{\boldsymbol{P}} = (P_x, P_y)$ acting at a vertex Q, as shown in Fig. 6b under, the stresses resulting from potentials (48) on any radial line through Q, $\phi = \theta$, follow from Eq. (13):

$$\tau'_{xy}/2\mu = -[(C_1 + D_1)\sin\theta + (C_2 - D_2)\cos\theta]/r$$
$$\tau'_{yy}/2\mu = [-(C_1 + D_1)\cos\theta + (C_2 - D_2)\sin\theta]/r. \tag{54}$$

If $r \neq 0$, then the above reduce to zero on every radial line – especially on the sides of the vertex – if

$$C_1 + D_1 = 0; \quad C_2 - D_2 = 0 \quad \text{or} \quad C = -\bar{D}. \tag{55}$$

On excluding the vertex $r = 0$ and proceeding as in integrals (52) the integration being from 0 to α, we obtain on solving for D_1 and D_2:

$$D_1 = (P_x V_2 - P_y V_3)/V_4$$
$$D_2 = -(P_x V_3 - P_y V_1)/V_4, \tag{56}$$

where

$$V_1 = 2\alpha + \sin 2\alpha$$
$$V_2 = 2\alpha - \sin 2\alpha$$
$$V_3 = 1 - \cos 2\alpha$$
$$V_4 = 2\mu(V_1 V_2 - V_3^2), \tag{57}$$

and the axis $\theta = 0$ is along the right side of the angle and the vertex angle is α; the required cut being taken as the bisector of the external vertex angle.

Case $NV = 2$: Force Acting at a Point on a Side

This follows from case $NV = 3$ on putting $\alpha = \pi$, which on substituting in Eqs. (57) and (56) gives:

$$D_1 = P_x/4\pi\mu; \quad D_2 = P_y/4\pi\mu. \tag{58}$$

Moments can be deduced by treating each moment as a force-pair and performing the appropriate limiting, either algebracially or numerically, for displacements and stresses.

An appropriate subroutine EPOLP is available in EQUINP to compute the displacements and stresses at any point due to the above concentrated loads and Moments.

Distributed forces are dealt with by Gaussian integration in EPOLP, which effectively reduces them to equivalent concentrated forces.

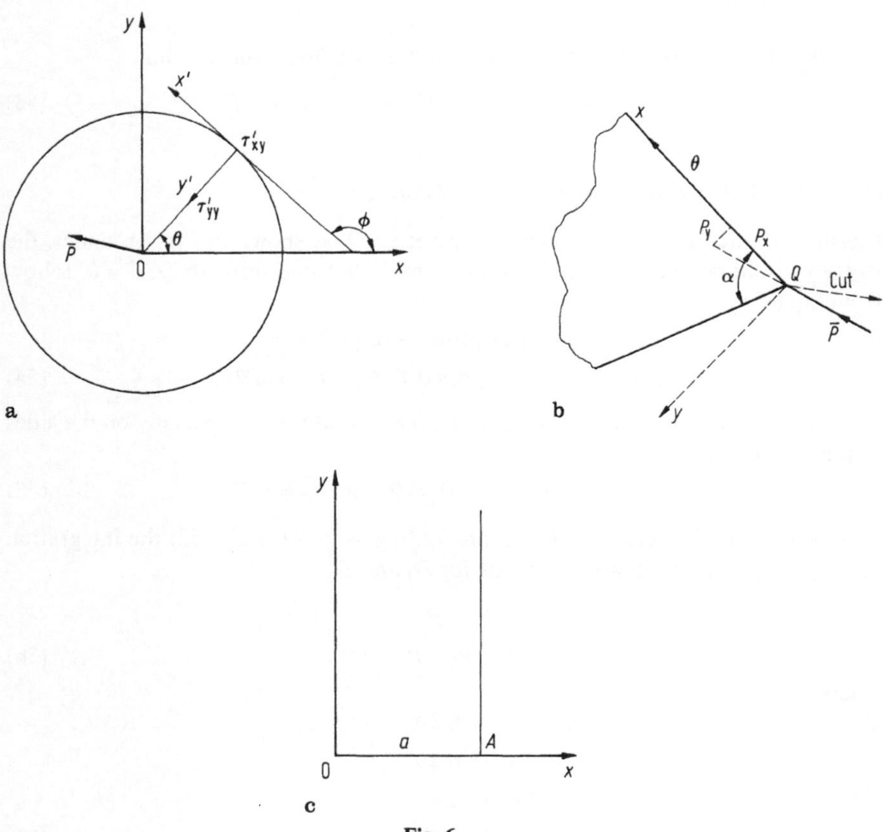

Fig. 6

(F) *Loaded Semi-infinite Strip*: *Edge-Functions*

Consider a loaded semi-infinite strip, as in Fig. 6c, bounded by $x=0$, $x=a$, and $y=0$, where all stresses and displacements tend to zero as $y \to \infty$ and either stresses or displacements are prescribed on the short edge OA, $y=0$.

This basic problem was first examined in Quinlan (1974), the resulting functions, called *Edge-Functions* being obtained by applying separation of the variables to Eq. (3). However they can also be obtained from the complex potentials

$$F(z) = F^n e^{inz}; \quad G(z) = G^n e^{inz}, \tag{59}$$

by replacing solution form (5) by the alternative biharmonic form

$$u = F(z) + 2yG(z); \quad y = (z - \bar{z})/2i, \tag{60}$$

the factor 2 being introduced for convenience. The above can be expressed in form (5) by making the replacements

$$F(z) \to F(z) - izG(z); \quad G(z) \to iG(z), \tag{61}$$

and displacements and stresses then follow from Eqs. (10) and (13).

On setting $\phi = 0$ in Eq. (13) it follows, on using potentials (59) with substitutions (61) and taking real parts, that

$$\begin{aligned}\tau_{xy}/2\mu &= e^{-ny}[-n\{F_1^n \cos nx - F_2^n \sin nx\} \\ &\quad + (1-2ny)[G_1^n \cos nx - G_2^n \sin nx\}],\\ \tau_{yy}/2\mu &= e^{-ny}[n\{F_1^n \sin nx + F_2^n \cos nx\} \\ &\quad + (1+2ny)\{G_1^n \sin nx + G_2^n \cos nx\}].\end{aligned} \tag{62}$$

These expressions are a slight modification of these given in Quinlan (1973), but the difference is unimportant.

If the stresses on $y=0$ are prescribed by

$$\tau_{xy} = \psi_1(x); \quad \tau_{yy} = \psi_2(x); \quad 0 \leqq x \leqq a, \tag{63}$$

they can be matched by taking a superposition in n of stresses (62), with $y=0$, leading to the *boundary identities* on $y=0$ for $0 \leqq x \leqq a$:

$$\begin{aligned}&\sum_{(n)} [(-nF_1^n + G_1^n)\cos nx + (nF_2^n - G_2^n)\sin nx] = \psi_1(x)\\ &\sum_{(n)} [(nF_2^n + G_2^n)\cos nx + (nF_1^n + G_1^n)\sin nx] = \psi_2(x).\end{aligned} \tag{64}$$

The above become Fourier Series identities for x in $(0, a)$ on putting

$$n = \frac{2\pi N}{a}; \quad \sum_{(n)} = \sum_{N=0}^{\infty}. \tag{65}$$

On equating corresponding Fourier coefficients in identities (64), we obtain

$$-nF_1^n + G_1^n = \frac{2}{a}\int_0^a \psi_1(x)\cos nx\,dx \tag{66}$$

$$n F_1^n - G_1^n = \frac{2}{a} \int_0^a \psi_2(x) \sin n x \, dx, \tag{67}$$

with similar expressions for F_2^n and G_2^n.

The essential feature of identities (64) *is that the Edge-Functions for* $N = 1, 2, 3 \ldots$ *match the corresponding harmonics of the applied boundary stresses and propagate their effects into the interior of the region with a decay factor* e^{-ny}. *This is, in effect, an example of St. Venant's principle.*

The higher the boundary harmonic N the faster its decay away from the boundary and the lesser its effects on the interior and on the other boundaries.

If the Edge-Function series in Eq. (64) is truncated at L to $N = 1$ to L, the residuals, or errors, remaining on the boundary are seen to be due to the unmatched boundary harmonics greater than L resulting from the boundary stresses (63). Since the corresponding decay factors, e^{-ny}, decrease exponentially the residuals reduce rapidly to insignificance as L increases. Examples will illustrate this later by showing comparative results for a range of values of L, and in practice L can be quite small.

From the above analysis of a loaded semi-infinite strip model it could be conjectured that the Nth Edge-Functions generated from a side j of length a_j of a polygon, as in Fig. 1, ought play a major role in enabling the stresses, or displacements, derived from solution vector (1) to match the Nth harmonics of the specified boundary stresses, or displacements, on side j. The corresponding origin is taken at the start of side j, with associated constants

$$n_j = 2\pi N / a_j; \quad F_{1,j}^N, F_{2,j}^N, G_{1,j}^N, G_{2,j}^N; \tag{68}$$

there being 4 arbitrary constants for each set N of Edge-Functions from any side j, as designated by an appropriate subscript j.

Note. The two zero harmonics on each side j, as in identities (64) corresponding to boundary stresses (63), *require separate treatment and must be matched by a corresponding number of non-edge-functions,* since the member $N = 0$ in series (65) is not an edge-function.

If the boundary effects on side j are truncated at L_j harmonics, then this would require that L_j sets of Edge-Functions, whether straight or curved, corresponding to $N = 1, 2, \ldots L_j$, be included for each side j, in solution vector (1). Each Edge-Function set provides four functions for inclusion in set u_k in (1).

It is reasonable to assume that L_j is proportional to a_j. If j_0 is the shortest side with corresponding truncation level $L_{j_0} = L$, then, as defined in Appendix B,

$$L_j = L^* \text{nhs}(j), \tag{69}$$

where nhs (j) is the proportionality factor for side j. Strict proportionality is not required in practice, and nhs (j) can inputted as directed in Appendix B.

Sub-Division into Regions, or Sub-Structures (70)

Note that each Edge-Function must be confined to a region where $y \geqq 0$, as required by the corresponding decay factor e^{-ny}. Accordingly regions that have *re-*

entrant angles, or cracks, must be subdivided into convex regions, or sub-structures, as for example, in Fig. 1, which is subdivided into regions I, II and III by cuts as indicated by the dashed lines RS and TV.

Edge-Functions set up on the edges of each cut and confined to their corresponding regions, play a major part in ensuring continuity of displacements and stresses, by collocation or localised least squares, across each cut.

A solution vector, similar to vector (1), must be set up for each region and should include all functions designated as active in that region. Some functions may be designated as active in two or more regions, and accordingly these will have *built-in continuity* across the corresponding cuts.

For computational convenience it is better to set up a *General Solution Vector* of type (1) that contains *all* functions required to model the whole domain, the numbers of the regions in which each function is active being then designated by the indicators JE1, JE2, JV1, JV2, ... as defined in the boundary data card in Appendix B.

In the case of a *composite side,* as for example the one made up by segments 7 and 8 in Fig. 7, the Edge-Function sets relate to the whole side, a_j in constants (68) being the corresponding length.

A subroutine EDGEF does the computation for Edge-Functions derived from harmonics (59), analogous to that done by EPOLC and EJOUK for the corresponding Curved Edge-Functions.

Boundary Equations

The two boundary conditions on any segment j are specified by MBC(j) as defined in Appendix B, and lead to corresponding *boundary identities* as in Eq. (64). These involve specifying two of the four quantities u^j, v^j, τ^j_{xy}, τ^j_{yy} – the displacement and stress components along and perpendicular to side j, as obtained from Eqs. (10) and (13) where $\phi = \phi_j$ is the corresponding slope.

Accordingly if the tangential displacement u^j is specified as $\psi(x'_j)$ on side j, this gives on using solution vector (1), the boundary identity

$$\sum_{k=1}^{k'} c_k u^j_k - \psi(x'_j) = 0; \quad 0 \leqq x'_j \leqq a_j. \tag{71}$$

This may be satisfied, in an approximate manner, using *Collocation,* or *Point Matching,* by satisfying the above identity at m_j points x'_{jm}, selected in some suitable manner, for each boundary condition on each side j, leading to the point-equations:

$$\sum_{k=1}^{k'} c_k u^j_k(x'_{jm}) - \psi(x'_{jm}) = 0. \tag{72}$$

In this manner k' equations are generated to determine the k' unknowns c_k in solution vector (1), which *boundary equations* may be written in the matrix form:

$$\tilde{A}\,\bar{c} = \bar{b}, \tag{73}$$

where $\tilde{A}$ is called the *coefficients matrix,* the elements of which are the point-values $u_k(x'_{jm})$.

Guidance on the selection of collocation points is given at (A-16) in Appendix A, wherein is developed the method of *Harmonic Fitting* now adopted in E.F.M. computer programs. In the latter boundary Eqs. (73) are built up from *Harmonic Equations*, N_H, the discrete version of Eqs. (A-4), got by using the trapezoidal quadrature form, and the associated *Vertex Equations*, N_V, each set being computed for each side j and its two boundary conditions.

Care must be taken to exclude Vertex Equations that are *not* independent. The total number of equations N_T in (73) follows as

$$N_T = \sum_{(j)} N_H + \sum_{(j)} N_V, \tag{74}$$

and a corresponding number of functions, $k' = N_T$, must be included in solution vector (1). The functions are selected:

(i) by taking L_j sets of Edge-Functions from each side j, whether straight or curved, with each set consisting of four functions;
(ii) by making up the balance, to provide a total of k' functions, by taking equal numbers of Vertex Functions (32) and Crack Functions (15), choosen in order of singularity at each vertex and crack tip. A small number, NNPOL – see Appendix B – of the required functions may be provided by Polar Functions (Polynomials) generated from potentials (33) with $n = N$.

Note. The matrix $\tilde{A}$ is pseudo-banded, and can be decomposed into

$$\tilde{A}_1 \bar{c} = \bar{b} - \tilde{A}_2 \bar{c}, \tag{75}$$

where the terms in $\tilde{A}_2$ are of minor significance, and their effects can be obtained by iteration from the *banded matrix* $\tilde{A}_1$. Usually three or four iterations provide an acceptable solution. The time required for each iteration is only a few percent of that required to solve the banded matrix $\tilde{A}_1$. Thus E.F.M. is now firmly within the banded matrix field and is comparable to F.E.M. in this respect.

The decomposition of $\tilde{A}$ is carried out while the matrix is undergoing Gaussian elimination, insignificant terms being, figuratively, shifted to the right hand side in equation (75) and ignored during the first pass.

Computer Program EQUINP

A comprehensive program EQUINP, as outlined in Appendix C, has been developed for 2-D Elastostatics. A series of problems, Figs. 7 to 14, illustrate the scope of the program, amongst which Figs. 13, and 14 deal with curved boundaries.

The domain in Fig. 7 is subdivided into convex regions I and II in accordance with Sect. (70), by producing the crack to meet the opposite side. The boundaries are described in a continuous anticlockwise manner by segments numbered 1 to 10; the coordinates of the corresponding points being given at each point, and the boundary type number MBC being written on each side e.g. (1) denotes that MBC = 1. In the examples taken normal distributed loadings of parabolic form $p_0 t(1-t)$ act on some sides where p_0 is given on the diagram.

The data inputs are described in Appendix B, and, as listed under for Example A, are seen to require no knowledge of E.F.M., and can readily be checked by the reader.

Data Inputs, with underlined headings, as in Appendix B, for *Example A* as given in Figure 7.

Problem Description

Irregular Pentagonal Beam with Edge Crack.
Fixed on one side; Parabolic normal Loading on another.

NJOB	NS	NCUT	NCRACK	NHOLE	NBDRY	JPRT
1	10	0	0	0	1	0

NCL	NCC	NM	L	LS	NNPOL	FSET
0	0	0	4	4	8	1.30

NTYP	NBC	NHS	X	Y	JP1	JP2	JE1	JE2	JV1	JV2	NTOT	I_1	J_1	I_2	J_2
0	2	1	0.0	0.6	1	0	1	0	1	0	0				
0	2	2	0.4	0.0	1	0	1	0	1	0	0				
0	2	1	2.4	0.0	1	0	1	0	1	0	0				
0	2	1			2	0	2	0	0	0	0	7	8	3	5
0	1	2	2.0	1.0	2	0	2	0	2	0	0				
0	2	1	0.0	1.0	2	0	2	0	2	0	0				
0	4	2	0.0	0.6	2	0	2	0	2	0	2				
0	5	2	0.8	0.5	1	2	2	0	0	0	0				
0	−5	2			1	2	1	0	0	0	2	−4			
0	−4	2	0.8	0.5	1	0	1	0	0	0	0				

σ	E	UX	UY	UR
0.30	30×10^6	0	0	0

N_1	N_2	N_3
0	0	0
2	0	0

C_1	C_2	C_3	C_4	C_5	C_6
0.00	2.40	−2.40	0	0	0

Production Cards: (*one* for each set of results)

U_1	V_1	U_2	V_2	Fee	MT	N	JELT
0.4	0.0	2.0	0.3	0.0	4	11	2
0.4	0.0	2.0	0.3	0.0	2	11	2

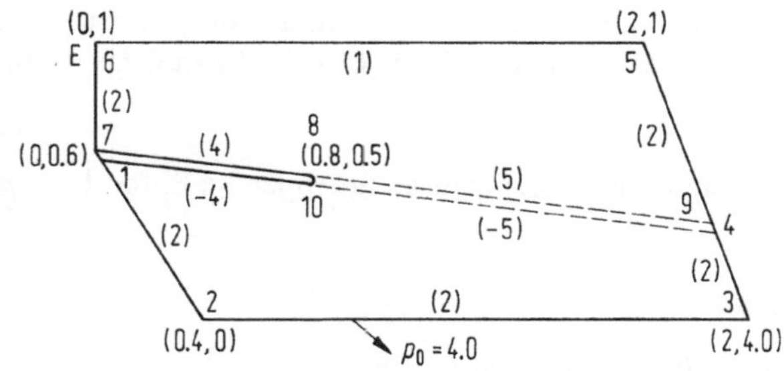

Fig. 7. Skew edge crack (change points j from 1 to 10)

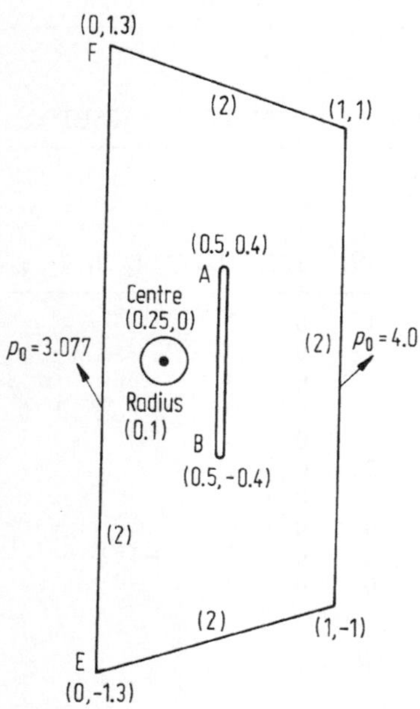

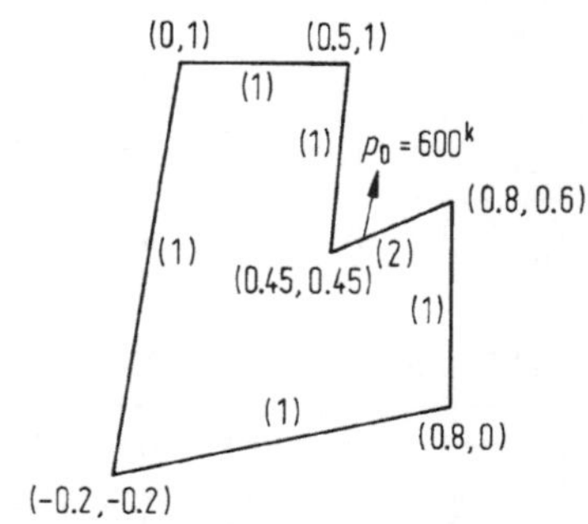

Fig. 9. Re-entrant polygon

Fig. 8. Trapezium with interior crack and hole

Comments: The above data inputs, in accordance with Appendix B, require but eighteen lines to input the complicated problem given in Figure 7. The computed boundary residuals follow in Table 1 for four matrix sizes ranging from 134 to 278, and their continued marked decrease with increasing size of matrix is obvious.

The *maximum stress residuals decrease* from 0.03 to 0.00002, or from 3 to 0.002 per cent of maximum stress, which is seen from Table 2 under to be of order 1.0. While the maximum *displacement* residuals decrease from 0.2 to 0.0002 per cent of the maximum displacement, which Table 3 shows is or order 0.4 D-0.7. The difference between the quantities in Tables 2 and 3 is seen to be consistent with the residuals in Table 1, and a matrix of 182 equations is seen to give all the quantities correct to at least one per cent of the corresponding maximum stress or displacement.

Results (all for four matrix sizes from MAT = 134 to 278)

Table 1. *RMS of Residuals* on side *j* for boundary Conditions MT

MT	Side J	MAT = 134	MAT = 182	MAT = 230	MAT = 278
3	1	0.2728 D-02	0.7017 D-03	0.7511 D-04	0.2389 D-05
3	2	0.3153 D-01	0.9078 D-02	0.4340 D-03	0.8009 D-05
3	3	0.1228 D-02	0.1337 D-02	0.3330 D-04	0.2707 D-05
3	4	0.1627 D-02	0.1564 D-02	0.7866 D-04	0.1377 D-04
1	5	0.1027 D-10	0.1837 D-11	0.5713 D-13	0.3189 D-13
3	6	0.4074 D-04	0.3775 D-05	0.1649 D-05	0.1105 D-05
3	7	0.3455 D-03	0.2978 D-03	0.1591 D-04	0.1855 D-04
1	8	0.6977 D-11	0.1317 D-10	0.3051 D-12	0.9574 D-13
3	9	0.3623 D-02	0.1083 D-02	0.4825 D-04	0.5557 D-05
3	10	0.3817 D-02	0.9940 D-03	0.6287 D-04	0.9171 D-05
4	1	0.1985 D-02	0.1261 D-02	0.9142 D-04	0.1160 D-04
4	2	0.3346 D-01	0.1153 D-01	0.5470 D-03	0.2135 D-04
4	3	0.2247 D-02	0.2251 D-02	0.4943 D-04	0.4896 D-05
4	4	0.2086 D-02	0.1132 D-02	0.6110 D-04	0.1367 D-04
2	5	0.7034 D-10	0.3927 D-11	0.2457 D-12	0.1638 D-13
4	6	0.2209 D-03	0.3271 D-04	0.1170 D-04	0.9171 D-05
4	7	0.2182 D-03	0.3315 D-03	0.1899 D-04	0.1203 D-04
2	8	0.2573 D-10	0.1636 D-10	0.6377 D-12	0.5381 D-13
4	9	0.4530 D-02	0.9058 D-03	0.1380 D-04	0.5316 D-05
4	10	0.3324 D-02	0.7089 D-03	0.1770 D-04	0.3371 D-05

Table 2. *Stress MT = 4* at 9 equidistant points (x, y) on Line from (0.4, 0) to (2, 0.4); with Fee = 0

MT	Side J	MAT = 134	MAT = 182	MAT = 230	MAT = 278
Point (x, y)					
0.40	0.00	0.4271 D-13	0.9326 D-12	0.3439 D-12	0.1234 D-14
0.60	0.05	0.3287 D-00	0.3517 D-00	0.3534 D-00	0.3527 D-00
0.80	0.10	0.6692 D-00	0.6627 D-00	0.6620 D-00	0.6621 D-00
1.00	0.15	0.8964 D-00	0.8886 D-00	0.8904 D-00	0.8904 D-00
1.20	0.20	0.9419 D-00	0.9513 D-00	0.9508 D-00	0.9507 D-00
1.40	0.25	0.9694 D-00	0.9655 D-00	0.9660 D-00	0.9660 D-00
1.60	0.30	0.9460 D-00	0.9446 D-00	0.9430 D-00	0.9430 D-00
1.80	0.35	0.8653 D-00	0.8597 D-00	0.8621 D-00	0.8620 D-00
2.00	0.40	0.6939 D-00	0.6897 D-00	0.6882 D-00	0.6878 D-00

Table 3. *Displacement MT = 2* at 9 equidistant points (x, y) on Line from (0.4, 0) to (2, 0.4); with Fee = 0

MT	Side J	MAT = 134	MAT = 182	MAT = 230	MAT = 278
Point (x, y)					
0.40	0.00	0.3661 D-07	0.3821 D-07	0.3875 D-07	0.3878 D-07
0.60	0.05	0.3237 D-07	0.3344 D-07	0.3378 D-07	0.3379 D-07
0.80	0.10	0.2701 D-07	0.2740 D-07	0.2755 D-07	0.2756 D-07
1.00	0.15	0.2249 D-07	0.2259 D-07	0.2265 D-07	0.2265 D-07
1.20	0.20	0.2034 D-07	0.2041 D-07	0.2042 D-07	0.2042 D-07
1.40	0.25	0.1941 D-07	0.1937 D-07	0.1936 D-07	0.1936 D-07
1.60	0.30	0.1879 D-07	0.1869 D-07	0.1867 D-07	0.1867 D-07
1.80	0.35	0.1869 D-07	0.1862 D-07	0.1855 D-07	0.1855 D-07
2.00	0.40	0.2001 D-07	0.1983 D-07	0.1972 D-07	0.1972 D-07

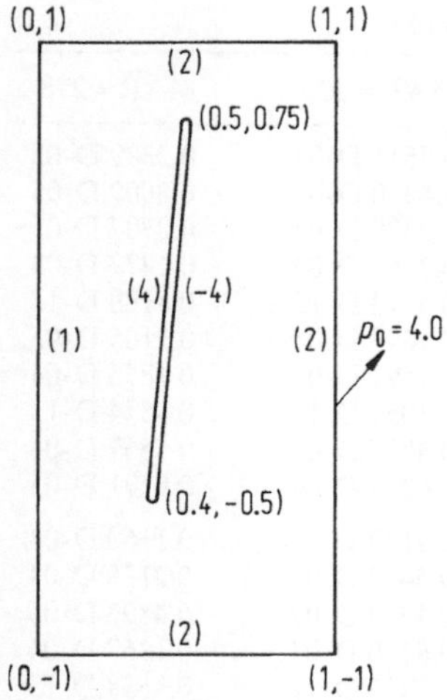

Fig. 10. Interior crack

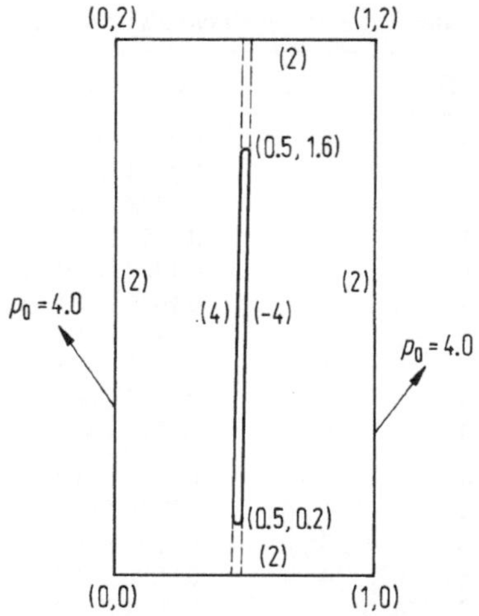

Fig. 11. Interior symmetrical crack

Table 4. Fracture coefficient for crack opening mode (mode I) in Fig. 7 (two cases)

MAT →	134	182	230	278
EF Free	0.3730	0.3813	0.3846	0.3847
EF Fixed	0.3860	0.3859	0.3857	0.3855

Note the slight change of about 2% in the Fracture coefficient when the side EF is fixed rather than free. Fracture coefficients for Crack Sliding Mode (Mode II) follow similarly.

Further illustrative examples are given in Figures 8 to 14. To conserve space only a few illustrative results of particular interest are given under.

Table 5. Fracture coefficients (mode I) and maximum residuals for problem in Fig. 9

Fracture Coeffs.	MAT = 161	MAT = 221	MAT = 281
A	1.1107	1.176	1.164
B	1.077	1.165	1.160
% Resids: HOLE	6.0	0.4	0.09
% Resids: EF	3.0	0.4	0.20

In the case when the circular hole is replaced by an elliptical hole with the same centre and semi-axes, (0.30, 0.10) the fracture coefficients at A and B were 1.183 and 1.181, or they agree to 0.2%, the worst residuals were 1% on the hole when using 281 equations.

Table 6. Singularity coefficients at notch in Fig. 9 for roots n.

Re (n)	Coeffs.	MAT = 124	MAT = 172	MAT = 220	MAT = 268
−0.6937	C_1	−0.2171 D-04	0.1876 D-04	0.1846 D-04	0.1841 D-04
−0.1836	C_2	−0.9636 D-04	−0.9781 D-04	−0.9782 D-04	−0.9787 D-04
0.0298	C_3	−0.1555 D-03	−0.1582 D-04	−0.1591 D-03	−0.1592 D-03
0.3248	C_4	−0.2001 D-03	−0.2983 D-03	−0.3132 D-03	−0.3155 D-03
0.7857	C_5	−0.8148 D-03	−0.7158 D-03	−0.6851 D-03	−0.6775 D-03

The stability of the above results, even for the higher singularities, is impressive. Infinite stresses occur at the notch and result from vertex functions for which $\mathrm{Re}(n) < 0$.

Table 7. Skew interior crack (Fig. 10)

Fracture Coeffs.	MAT = 154	MAT = 210	MAT = 266
A	1.122	1.122	1.122
B	1.160	1.163	1.163
% Max. Resids	2.0	0.3	0.05

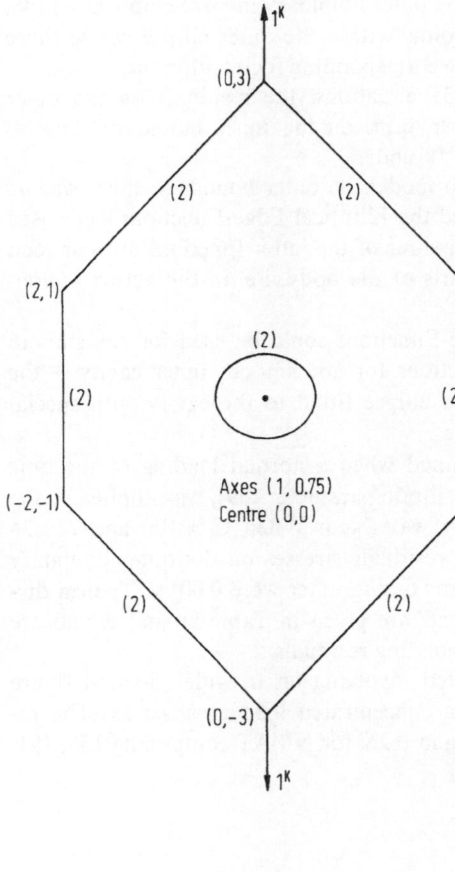

Fig. 12. Hexagon loaded at vertices with elliptic hole

Table 8. Interior crack, using symmetry (Fig. 11)

Fracture Coeffs.	MAT = 157	MAT = 193	MAT = 229	MAT = 265
A	0.8217	0.8218	0.8220	0.8219
B	0.9144	0.9145	0.9146	0.9146
% Max. Resids	0.1	0..05	0.01	0.002

Table 9. Hexagon with an elliptic cavity and concentrated loads at vertices (Fig. 12)

% Max. Resids	MAT = 133	MAT = 167	MAT = 203
on sides	0.022	0.024	0.004
on ellipse	0.0093	0.002	0.00004

Curved Boundaries

Figure 13 represents an ellipse with an elliptical hole, not necessarily confocal, subject to concentrated loads at P and Q, both boundaries being otherwise free from stress. The Curved Edge-Functions for the outer boundary involve elliptical set (46) and accordingly are continuous at all points within the outer ellipse, while those for the inner boundary are singular on the corresponding focii within the cavity.

On using 28 harmonics, involving 231 equations, the residuals on the outer boundary were reduced to 0.01%, and to 0.1% on the inner boundary. Typical stresses on the line EF are given in Table 10 under.

On using Polar Edge-Functions (33) to model the outer boundary, there was no appreciable difference in results provided the Elliptical Edge-Functions were used on the inner boundary; the singular behaviour of the latter functions at their focii produce higher stresses at the nearer parts of the body, i.e. in the vertex regions R and S outside the cavity.

It follows that Polar (or Elliptic) Edge-Functions could be used for any smooth outer boundary, and Elliptic Edge-Functions for any smooth inner cavity – the latter functions being generated from an ellipse fitted to the cavity with special attention to the parts of sharpest curvature.

Equally satisfactory results were obtained when a normal loading of the type $C_1 t(1-t)$, where $t = \psi/2\pi$, ψ being the elliptic parameter (39), was applied to the outer boundary and the inner boundary was fixed. When $C_1 = 100$ and $L = 24$ harmonics, involving 196 equations, the residuals stresses on the outer boundary were 0.01%, and the residual displacements on the inner were 0.001%. Typical displacements at a few sample interior points are given in Table 11 under, and are seen to be fully consistent with the corresponding residuals.

As a final example, Fig. 14 is presented involving an irregular 5-sided figure with a deep elliptic indentation under a concentrated load at a vertex. The residuals on the indentation were 2%, 1%, and 0.2% for NMAT comprising 155, 191, and 227 equations respectively.

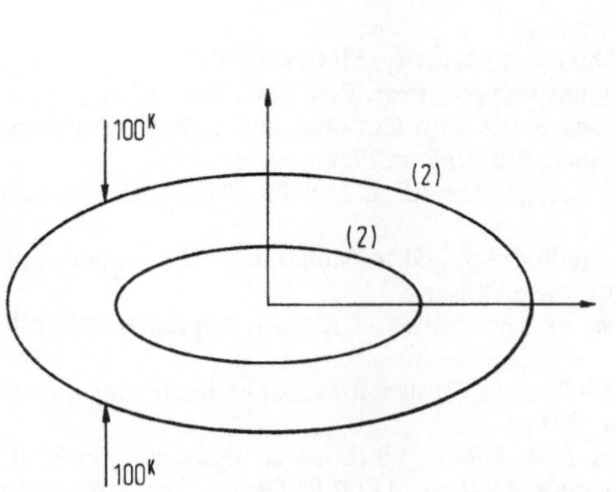

Fig. 13. Elliptic plate axes (1.5, 2.0) with elliptic hole (2.13, 1.6)

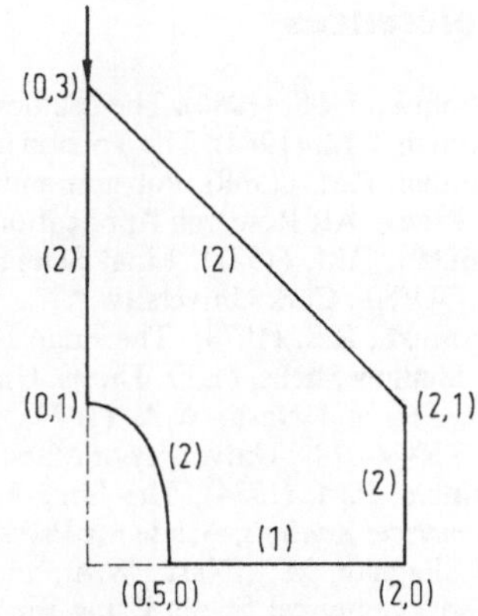

Fig. 14. Pentagon with elliptic indentation

Table 10. Normal stress (MT = 4) at points on EF (Fig. 13)

Point (X, Y)		NMAT = 215	NMAT = 223	NMAT = 231
2.500	0.000	0.10122 D-03	0.10117 D-03	0.10118 D-03
2.426	0.000	0.61979 D-02	0.61051 D-02	0.61046 D-02
2.352	0.000	0.18614 D-02	0.18596 D-02	0.18582 D-02
2.278	0.000	−0.27409 D-02	−0.27429 D-02	−0.27452 D-02
2.204	0.000	−0.78795 D-02	−0.78749 D-02	−0.78762 D-02
2.130	0.000	−0.13845 D-03	−0.13785 D-03	−0.13766 D-03

Table 11. Displacement along Y-axis for some interior points

Point (X, Y)		NMAT = 180	NMAT = 188	NMAT = 196
1.800	1.348	0.11165 D-04	0.11174 D-04	0.11172 D-04
1.400	1.404	0.77198 D-05	0.77203 D-05	0.77222 D-05
1.000	1.460	0.25696 D-05	0.25691 D-05	0.25689 D-05
0.600	1.516	−0.12824 D-05	−0.12824 D-05	−0.12824 D-05
0.200	1.572	−0.15675 D-05	−0.15674 D-05	−0.15674 D-05

Conclusion

The E.F.M. is presented for appraisal on the above varied and difficult examples, wherein the computer time in each is less than twice that required to solve the corresponding matrix by Gaussian elimination. Highly satisfactory checks have been obtained against a wide range of standard fracture problems. A particular advantage of E.F.M. is the rapidity with which point values for functions MT can be obtained from the corresponding analytical expressions, and accordingly principal stress values and directions can easily be computed and hooked up to a computer graphics package for visual presentation.

References

Sokolnikoff, I.S. (1956), The Mathematical Theory of Elasticity, McGraw-Hill

Quinlan, P.M. (1964), The Torsion of an Irregular Polygon, Proc. Roy. Soc., Vol. 282 A

Quinlan, P.M. (1968), Polygon and Swept-Back plates with Cut-Outs and Column Supports, Proc. OAR Research Applications Conference, Washington 1968

Quinlan, P.M. (1971), Final Scientific Report on Grant AF EOAR-69-0049, with Program QUINP. Cork University

Desmukh, R.S. (1973), The Edge-Function Method Applied to Moderately Thick Plates and Shallow Shells, Ph.D. Thesis, University of Massachusetts

Tai, I.H. and Nash, W.A. (1973), Vibrations of Thin Plates – A New Approach, AFOSR TR-74-0789, University of Massachusetts

Quinlan, P.M. (1974), The Edge-Function Method in Elastostatics, paper in *Studies in Numerical Analysis,* Academic Press, London 1974

O'Callaghan, M.J., Nash, W.A., and Quinlan, P.M. (1975), Vibration Analysis of Thin Shallow Spherical Shells Using the Edge-Function Method, AFOSR TR-72-2340, University of Massachusetts

Quinlan, P.M., Fitzgerald, J.E., and Atluri, S.N. (1977), The Edge-Function Method, Proceedings First International Symposium on Innovative Numerical Methods in Engineering Science, Versailles

Quinlan, P.M., Grannell, J.J., Atluri, S.N., and Fitzgerald, J.E. (1979), Boundary Discretisation using the Edge-Function Method, App. Math. Modelling, 3, 1979

Grannell, J.J. and Quinlan, P.M. (1980), The Edge-Function Method for Thin Aeolotropic Plates, Proc. Roy. Irish Academy

Parker, A.P. (1981), The Mechanics of Fracture and Fatigue, E.S.F.N. Spon, London

Quinlan, P.M. Grannell, J.J., Atluri, S.N., and Fitzgerald, J.E. (1982), The Edge-Function Method for Three-Dimensional Stress Analysis, included Embedded Elliptical Cracks, Proceedings of the Fourth International Conference on Boundary Element Methods in Engineering

Quinlan, P.M. and O'Callaghan, M.J.A. (1984), The Edge-Function Method (E.F.M.) for Cracks, Cavities, and Stress Concentrations in Elastostatics, Proceedings of the Sixth International Conference on Boundary Element Methods in Engineering

Appendix A – Harmonic Fitting

Consider the identity, similar to Eq. (71), for each boundary condition on each side j, the subscript j being omitted:

$$S(x') \equiv \sum_{k=1}^{k'} c_k u_k(x') - \psi(x') = 0; \quad 0 \leqq x' \leqq a', \tag{A-1}$$

which, on expanding in a Fourier series in $(0, a')$ and truncating at L harmonics, gives

$$S(x') \approx a_0' + \sum_{N=1}^{L} a_N \cos n x' + \sum_{N=1}^{L} b_N \sin n x', \tag{A-2}$$

where

$$a_N = \frac{2}{a'} \int_0^{a'} S(x') \cos n x' \, dx', \quad a_0' = 1/2 \, a_0$$

$$b_N = \frac{2}{a'} \int_0^{a'} S(x') \sin n x' \, dx'; \quad n = \frac{2 N \pi}{a'} \tag{A-3}$$

Consequently identity (A-1) is achieved, in an approximate manner, by setting the several coefficients in series (A-2) to zero, resulting in $(2L+1)$ *harmonic equations*[5] in the k' unknowns c_k:

$N = 0, L$

$$a_N = 0 \Rightarrow \sum_{k=1}^{k'} c_k \int_0^{a'} u_k(x') \cos n x' \, dx' = \int_0^{a'} \psi(x') \cos n x' \, dx',$$

$N = 1, L$ (A-4)

$$b_N = 0 \Rightarrow \sum_{k=1}^{k'} c_k \int_0^{a'} u_k(x') \sin n x' = \int_0^{a'} \psi(x') \sin n x' \, dx',$$

which are similar to point Eqs. (72). The above process is called *Harmonic Matching*, and was introduced in Quinlan (1968).

The effectiveness of approximation (A-2) depends on the convergence of the Fourier series representation of $S(x')$, which will have convergence of order $1/N^3$ provided that $S(x')$ has continuity of order two at all points $0 \leqq x' \leqq a'$. This is best achieved by setting $S(x')$ and $dS(x')/dx'$ to zero at all points of discontinuity, including the end points of the interval at $x' = 0$ and $x' = a'$. Accordingly we are led to the equations, called *Vertex Equations*, again for each boundary condition on each side j:

$$S(x') = 0; \quad x' = 0 \quad \text{and} \quad x' = a'$$
$$\frac{dS(x')}{dx'} = 0; \quad x' = 0 \quad \text{and} \quad x' = a', \tag{A-5}$$

which has been found to be advantageous to include with harmonic equations (A-4) in setting up boundary equations (73).

The main disadvantage of Eqs. (A-4) is that each of the elements involves a numerical integration with an oscillatory integrand.

The above integrals are replaced by relatively short series in the method of *Harmonic Fitting* introduced in Quinlan (1977), which is based on considering approximation (A-2) as the interpolation problem of fitting a trigonometric series of orthogonal functions to $S(x')$. The principal methods of fitting are:

(i) *Continuous Least-Squares* for fitting a truncated trigonometric series to $S(x')$, which, as is well known, leads again to integrals (A-3) – and hence shows no computational gain, and

(ii) *Discrete Least-Squares.* Consider the approximation (A-2), when written in the form:

$$S(x) \approx \sum_{M=0}^{\bar{M}} d_M T_M(x); \quad \bar{M} = 2L+1, \tag{A-6}$$

where the dash has been dropped from x', and $T_M(x)$ denotes the set of functions for $M = 0, \bar{M}$:

$$T_M(x) = [1, \cos x, \sin x, \cos 2x, \sin 2x, \ldots, \cos n x, \sin n x \ldots], \tag{A-7}$$

[5] These are seen to follow from boundary identity (A-1) on multiplying by $\cos n x'$ and $\sin n x'$ respectively and integrating 0 to a'.

and accordingly define

$$d_0 = a_0'; \quad d_M = a_{N'}, \quad \text{if } M \text{ is even with } N' = M/2$$
$$= b_{N''}, \quad \text{if } M \text{ is odd with } N'' = (M-1)/2. \qquad \text{(A-8)}$$

Let η^2 denote the sum of the weighted residuals, or errors, of the approximation (A-6), at $(\lambda' + 1)$ data points x_λ in the interval $(0, a)$, where

$$\eta^2 = \sum_{\lambda=0}^{\lambda'} w_\lambda \left[S(x_\lambda) - \sum_{M=0}^{\bar{M}} d_M T_M(x_\lambda) \right]^2, \qquad \text{(A-9)}$$

and w_λ denotes the weight factor at data points x_λ. The discrete points x_λ are chosen so that the set of functions $\{T_M(x)\}$ is orthogonal over these points with weight factors w_λ, which implies that

$$\sum_{\lambda=0}^{\lambda'} w_\lambda T_M(x_\lambda) T_{M'}(x_\lambda) = 0, \qquad M \neq M'$$
$$= J_M, \qquad M = M', \qquad \text{(A-10)}$$

where J_M is some constant that they may depend on M.

The discrete least-squares method minimises η^2 with respect to variables d_j for $j = 0, \bar{M}$, which leads to the equations

$$\frac{\partial(\eta^2)}{\partial d_j} = 0; \quad j = 0, \bar{M}, \qquad \text{(A-11)}$$

from which it follows, on using Eqs. (A-10), that

$$d_j = \frac{1}{J_j} \sum_{\lambda=0}^{\lambda'} w_\lambda S(x_\lambda) T_j(x_\lambda). \qquad \text{(A-12)}$$

It can be shown[6], as in Quinlan (1976), that the set (A-7) is orthogonal, in a

[6] On noting the quadrature expression

$$\int_0^a e^{inx} dx = \operatorname*{Lt}_{\lambda' \to \infty} \sum_{\lambda=0}^{\lambda'} w_\lambda e^{inx_\lambda},$$

we are motivated to seek some quadrature formula involving w_λ and x_λ, for finite λ', for which $Q_N = 0$, where

$$Q_N = \sum_{\lambda=0}^{\lambda'} w_\lambda e^{inx_\lambda}; \quad n = \frac{2\pi N}{a}; \quad N \neq 0.$$

Fortunately trapezoidal quadrature, involving w_λ and x_λ as given in (A-13), can be shown to make

$$Q_N = 0, \qquad N \neq 0$$
$$= \lambda', \qquad N = 0;$$

Orthogonality conditions (A-10) then follow on noting that:

$$\cos n x_\lambda = (1/2)[e^{inx_\lambda} + e^{-inx_\lambda}]; \quad \sin n x_\lambda = (1/2i)[e^{inx_\lambda} - e^{-inx_\lambda}]$$
$$\cos^2 n x_\lambda = (1/2)[1 + \cos 2n x_\lambda]; \quad \sin n x_\lambda \cos n x_\lambda = (1/2) \sin 2n x_\lambda.$$

discrete sense over any $(\lambda'+1)$ equidistant points x_λ in $(0, a)$ with weight factors w_λ where

$$
\begin{aligned}
x_\lambda &= \lambda\, a/\lambda'; \quad \lambda = 0, \lambda' \\
w_\lambda &= 1/2; \qquad \lambda = 0, \text{ and } \lambda = \lambda' \\
&= 1; \qquad \text{otherwise} \\
J_j &= \lambda'/2 .
\end{aligned}
\tag{A-13}
$$

It follows from (A-12), on using Def. (A-8), that

$$
a_N = \frac{2}{\lambda'} \sum_{\lambda=0}^{\lambda'} w_\lambda\, S(x_\lambda) \cos m x_\lambda, \quad b_N = \frac{2}{\lambda'} \sum_{\lambda=0}^{\lambda'} w_\lambda\, S(x_\lambda) \sin n x_\lambda, \tag{A-14}
$$

and, on substituting for $S(x_\lambda)$ from (A-1) the *harmonic equations* follow similar to (A-4); the integrals in the latter being replaced by the corresponding series in (A-14). In practice the series (A-14) involve considerably fewer point calculations than would be required to evaluate the integrals (A-4) to an acceptable degree of accuracy.

A comparison of Eqs. (A-3) and (A-14) shows that the series expressions in the latter derive *formally* from integrals (A-3) on "evaluating" them by the *trapezoidal rule* – which involves dividing the interval $(0, a')$ into λ' equal parts with the corresponding trapezoidal weight factors w_λ as given in (A-13).

The above method of generating harmonic equations (A-4) is called *Harmonic Fitting*, involving the formal use of the trapezoidal rule to "evaluate" the integrals. This does *not* imply that the integrals are evaluated to an acceptable degree of accuracy, rather than the trapezoidal rule analogy is used as a convenient means of generating series (A-14).

The following special cases of Harmonic Fitting are of interest:

(a) If the number of data points x_λ used in (A-97) is equal to the number of unknown D_M, then

$$
\lambda' = \bar{M},
$$

and the minimum value of η^2 is obviously zero, resulting from setting the residual at each point x_λ to zero – which is equivalent to collocating at these points.

Accordingly Collocation is seen as a limiting case of harmonic approximation. (A-2), *and hence requires, as in Vertex Equations* (A-5), *the equations*[7]

$$
\frac{dS(x')}{dx'} = 0; \quad x' = 0 \quad \text{and} \quad x' = a' \tag{A-16}
$$

for each side j, to make it effective; the data points x'_j being equally distributed in $(0, a')$. Thus in Collocation special weight is given to the vertices by means of Vertex Equations (A-16).

---

[7] The equations $S(x') = 0$ are already included in the collocation equations.

(b) The number of data points x'_λ on each side must be at least equal to the corresponding number of unknowns d_M to ensure that the system (A-11) is determinate, or

$$\lambda' \geqq \bar{M} = 2L + 1 . \qquad \text{(A-17)}$$

In practice a slight excess of data points – called *Least-Squares Smoothing*, say 20–30% over that required for collocation has proved beneficial, and involves only a similar increase in computer time over collocation in generating the boundary equations (73).

(c) If $\lambda' \to \infty$ then the series (A-14) become the Riemann quadrature expressions for integrals (A-3). Thus Harmonic Matching is seen as the limiting case of Harmonic Fitting, when a large number of data points x'_λ are taken.

(d) The Edge-Functions from side j, involving sets of constants F_1^n, F_2^n, G_1^n and G_2^n corresponding to four of coefficients c_k in Solution Vector (1), reduce, in the boundary identities on the corresponding side j, to Fourier Series analogous to these in identities (64).

Let $\bar{\eta}^2$ denote the sum of the squares of the boundary errors $S(x')$ at points ξ_λ on side j, where

$$\bar{\eta}^2 = \sum_{\lambda=0}^{\lambda'} w_\lambda [S(x_\lambda)]^2 , \qquad \text{(A-18)}$$

analogous to Eqs. (A-9). On noting from equations (A-4) that $a_N = 0$, $b_N = 0$, Eqs. (A-14) show that

$$\sum_{\lambda=0}^{\lambda'} w_\lambda S(x_\lambda) \cos n x_\lambda = 0, \quad \sum_{\lambda=0}^{\lambda'} w_\lambda S(x_\lambda) \sin n x_\lambda = 0 , \qquad \text{(A-19)}$$

and these imply from Eq. (A-18) that

$$\left(\frac{\partial \bar{\eta}^2}{\partial F_1^n}\right) = \left(\frac{\partial \bar{\eta}^2}{\partial F_2^n}\right) = 0 , \qquad \text{(A-20)}$$

and

$$\left(\frac{\partial \bar{\eta}^2}{\partial F_1^n}\right) = 0 \;\Rightarrow\; \left(\frac{\partial \bar{\eta}^2}{\partial G_1^n}\right) = 0, \quad \left(\frac{\partial \bar{\eta}^2}{\partial F_2^n}\right) = 0 \;\Rightarrow\; \left(\frac{\partial \bar{\eta}^2}{\partial G_2^n}\right) = 0 . \qquad \text{(A-21)}$$

Accordingly Hamonic Fitting is equivalent to the L o c a l i s e d L e a s t - S q u a r e s M i n i m i s a t i o n *of* $\bar{\eta}^2$ *for each boundary condition and for each side j, with respect to the corresponding Edge-Function coefficients* F_1^n *and* F_2^n *(for side j) for* $N = 1, L_j$. *Results* (A-21) *show that minimisation of* $\bar{\eta}^2$ *w.r.t.* G_1^n *and* G_2^n *is implied.*

The above equivalence applies likewise to Harmonic Matching, and both must be complemented by the $N = 0$ member of (A-4) – the zero harmonic – and Vertex Equations (A-5).

In contrast to Localised Least-Squares *Global Least-Squares* seeks to minimise the total sum $[\sum \eta]^2$, over all sides and their corresponding boundary conditions, wih respect to all the superposition constants c_k in solution vector (1). This is a time consuming cumputer task requiring the use of a continuous, or integral, least-squares minimisation criterion.

Energy and Variational Methods of determining the constants c_k are heavier in computer demands, than Global Least Squares.

It has been found in practice that the most important requirement in producing a good mathematical model to an elastostatic problem is that the solution vector (1) should include all relevant Basic Functions, and then *Harmonic Fitting* gives results at least as good as can be got from the more computer demanding Energy or Variational Schemes.

Appendix B – Data Inputs for 2-D Elastostatic Program "EQUINP"

Job description 2 cards, or lines of input

Geometry outline *one* card

Read (*) NJOB, NS, NCUT, NCRACK, NHOLE, NBDRY, JPRT.

NJOB = Job no. assigned to problem,

NS = No. of segments required for continuous description of outer boundary including *surface cracks* and their associated cuts

NCUT = No. of pure cuts – required only in large problems

NCRACK = No. of interior straight cracks

NHOLE = No. of interior holes

NBDRY = If set to 1, indicates that some non-zero boundary values are encountered and must be inputted later.

JPRT = If set to 1 eigenvalues and function assignment in solution mix will be printed. If set to 2 in addition to above the special points assigned and solution vectors will be printed.

Load and control indicators one card

Read (2) NCL, NCC, NM, L, LS, NNPOL, FSET.

NCL = No. of loads acting on the body whether point, line, or distributed

NCC = No. of points held fixed in the interior

NM = No. of concentrated moments acting on the body

L = Control on matrix size, corresponds to $2L+1$ interior *collocation* points on smallest segment, *or* to truncation at L harmonics when using *harmonic fitting*

LS = No. of mathematical models (or solutions) required

NNPOL = No. of polar functions (or polynomials) to be included in solution mix, usually 6 or 8. If left blank program sets NNPOL to 8.

FSET = If left blank program does collocation. If set to 1.3, say, uses 30% more calculation points than required for collocation, and sets up to matrix using Harmonic Fitting.

Boundary data one card required for *each* segment

Read (*) NTYP, MBC, NHS, X, Y, JP1, JP2, JE1, JE2, JV1, JV2, NTOT, I_1, J_1, I_2, J_2.

NTYP (segment type)
- 0, straight line
- 1, elliptic (incl. circular) indentation
- 2, elliptic mound
- 3, closed interior curve (ellipse)
- 4, closed exterior curve (ellipse)

MBC (boundary type)
- 1, displacement boundary
- 2, stress boundary
- 3, axis of symmetry
- 4, upper side of straight crack
- − 4, lower side of straight crack
- 5, upper side of straight cut
- − 5, lower side of straight cut
- 6, upper side of curved crack
- − 6, lower side of curved crack

NHS	Rough proportionality factor for calculation points on side; set to 1 for smallest side a; set to 2 for a side b, $2a < b < 4a$; and to 3 for $\beta > 4a$. *If left blank EQUINP selects suitable value for NHS.*
X, Y	Coordinates of point at beginning of segment
JP1, JP2	In what region, or regions are points on this segment? – Enter their numbers
JE1, JE2	In what regions are the Edge-Functions (straight or curved), associated with this segment, active or included in solution mix?
JV1, JV2	In what regions are the vertex functions, associated with this vertex active?
NTOT	Total number of segments in *composite line or curve;* all such segments have common Edge-Functions.

Note. If more convenient can specify, the point x, y as the intersection of the lines joining vertices (I_1, J_1) and (I_2, J_2). If this vertex is the same as vertex JJ, can specify it by putting $I_1 = -$ JJ.

Continuous non-linear curve: one additional card for each curve

Read (*) A, B, XC, YC, ZC

(A, B)	= Semi-axes of ellipse (inc. circle)
(XC, YC)	= Centre of ellipse
ZC	= Inclination of major axis as a multiple of $\frac{\pi}{2}$.

Note. If the curve is other than an ellipse, then insert provision in program for its decription.

Material properties and body forces: one card

Read (*) σ, E, UX, UY, UR

σ	= Poisson's Ratio
E	= Young's Modulus

(UX, UY) = Components of Body Force
UR = Magnitude of Rotary Inertia.

Fixed points NCC, loads NCL; moments NM

NCC = *One* card is required for each fixed point to give location and fixing specification.
NCL = *One* card for each loading to give location, load-components, and load type[8]. In the case of a distributed loading *another* card is required to give the load distribution.
NM = *One* card to specify centre, magnitude, and type of moment.

Non-zero boundary conditions: one card for each condition

If NBDRY is set to 1 in Geometry Outline card, then two cards are required to give segment numbers that have non-zero conditions – one card for each cycle of boundary conditions:

Read (*) $N_1, N_2, N_3, N_4, \ldots,$
Read (*) $N_1, N_2, N_3, N_4, \ldots,$

One card is required to give *each* specified non-zero distribution, provision being made for functions of the type

$$C_0 + C_1 t + C_2 t^2 + C_3 t^3 + C_4 t^4 + C_5 t^5,$$

where points on the segment are designated by $t\,(0, 1)$. The corresponding input is

Read (*) $C_0, C_1, C_2, C_3, C_4, C_5$.

Production: one card for each function required on specified line AB, which must be confined to a single element.

Read (*) U_1, V_1, U_2, V_2, FEE, MT, N, JELT

$(U_1, V_1), (U_2, V_2)$ = gives coordinates of A and B
FEE = Function axis as multiple of $\frac{\pi}{2}$
MT:
1, displacement component u'
2, displacement component v'
3, stress component τ'_{xy}
4, stress component τ'_{yy}
5, principal stresses
N = No. of equidistant points on AB at which function MT is required
JELT = No. of element in which AB occurs.

Note. If JELT set to -1 production is taken for points on a similar ellipse to ellipse centred at JEE given by input V_1 where U_1 inputs the magnification factor λ for the corresponding semi-axes.

[8] Load types are NV = 1, 2, or 3 as derived from Complex Potentials (48).

Appendix C – Program EQUINP

Outline of Main Program

The program sets up the coefficients matrix by rows, or equations, where the solution vector consists of function sets comprising of 1, 2, or 4 similar functions that are calculated together.

The *basic calculation* is a point-value determination, at a point on the boundary for a specified boundary condition, of the functions in each function set.

Two separate types of equations are generated; one cycle sets up the *Harmonic Equations*, and another sets up the *Vertex Equations*. The harmonic equations are set up by harmonic fitting, involving a weighted combination of the point values, at equidistant points, on each side and for each boundary condition.

Program SOLCOR is called to solve the coefficients matrix and provides solutions for a specified LS different sizes of matrix, in ascending order. All the reduced matrices are derived from the matrix as set up, by omitting certain rows and columns.

The LS solutions obtained provide LS *Mathematical Models* of the given *Physical Problem*. This is equivalent to being able to obtain comparative solutions for coarser meshes from a fine mesh computation, without remeshing and without using significantly more computer time – a facility not available in either FEM or BEM.

The matrix size is controlled by the number L, which corresponds to $(2L+1)$ fixed points on the standard side when using collocation, *or* to truncation of the boundary residuals at L_j harmonics on side j when using harmonic fitting.

The boundary effects produced by each solution vector obtained in SOLCOR are then checked, at a number of points on each side, against the corresponding prescribed boundary values, yielding the residual errors for each of the corresponding two boundary conditions. The root mean square of the residuals on each side and for each boundary condition are computed, and provide a suitable means of accessing the accuracy of the solutions obtained, and act as acceptance tests for the Mathematical Models obtained.

Finally the program proceeds to production, at a specified number of points on a specified line AB as described in the Data Inputs, for displacements, stresses, or principal stresses and principal axes of stress. Since each solution vector leads to an analytical formula for each of the above quantities, production proceeds very rapidly, and the resulting array of point values can, if required, be hooked up to a graphics package for visual presentation of results.

Subroutines

1 ESDATA	Inputs Data, as described in Appendix B, and does some processing – Converts all distributed loads into equivalent concentrated loads.

2 EMACOF — Decides on the special independent point equations – called Vertex Equations – that are required at each vertex j to give proper weight to vertex effets. The data for the resulting point equations is stored for use.

3 EMACOG — Assigns functions to solution scheme and sets up the required information for the calculation of point-values of these functions. Also sets up claculation points for all curves, other than straight lines.

4 ELTFN — Decides whether function being processed in function set NP is operative at calculation point being computed on side j, i.e. is the function included in solution scheme that applies in the element in which the calculation point lies?

5 EMACOS — Extracts eigenvalues for vertex functions required in solution scheme, and sets up the corresponding cofactors.

6 EDGEF — Evaluates the Edge-Function set, denoted by NFN = 1, for separation constant n, and specified origin and side.

7 EPOLC — Point values for Polar Functions, NFN = 5, generated from a specified circle for given n ... similar to EDGEF.

8 CRACK — Point values for Crack Functions, NFN = 7, for a specified crack tip and given n ... similar to EPOLC.

9 EPOLW — Point values for Vertex Function, NFN = 4, for a speficied vertex and a given root n.

10 EPOLP — Point values for particular functions, indicated by NFN = 3, including body forces and concentrated loads and/or moments, and non-zero boundary functions specified on the side in equation. These constitute the r.h.s. of the coefficients matrix.

11 EJOUK — Point values for Joukoski Polar functions, NFN = 6, generated from a specified ellipse.

12 SOLCOR — Subroutine to solve the matrix equations for L harmonics, and also the reduced matrices corresponding to $L-1$, $L-2$, harmonics, giving LS solution sets in all. It checks the equations for redundancy, and eliminates any equations found redundant. Prints out solutions vectors when indicator JPRINT is set to 2.

13 EELPS — NCase = 1 find corner slopes and ϕ-parameters for any elliptic (inc. circular) indentation, or mound.
NCase = 2 computes, and stores, the required calculation points, and slopes, for elliptic, or part elliptic, boundaries.

14 EGAUSS — Given weight-factors $w(k)$, and division points $d(k)$, for any specified N divisions. If Ncode set to 3 program uses Gaussian integration, otherwise uses equal subdivisions (equivalent to trapezoidal rule) as required for harmonic fitting.

15 QPOLAR — A $\tan^{-1}$ subroutine that gives the angle in $(0, 2\Pi)$.

16 QCUT — Adjusts angle values to lie in range $(\text{CUT}, 2\Pi + \text{CUT})$, where the value of CUT is specified.

Chapter 6

Theoretical and Practical Aspects of Multigrid Methods in Boundary Element Calculations

by H. Schippers

Summary

In this paper multigrid methods are advocated for the fast solution of the large nonsparse systems of equations that occur in boundary-element methods. Multigrid methods combine relaxation schemes and coarse-grid corrections. Ample attention is given to the decomposition of the system matrix in order to obtain a relaxation scheme that reduces the high-frequency components of the iteration error. It is shown that the decomposition should take the edges of the boundary into account, because they have a strong influence on the smoothing property of the relaxation scheme. The practical aspects of the multigrid method are concerned with the use of the method in boundary element calculations. The choice of the coarse-grid operators, the interactions between the grids and the implementation of the algorithm are discussed. The theoretical investigations show that the multigrid method converges more rapidly as the number of boundary elements increases. This is illustrated for two plane problems: (1) potential flow around an aerofoil and (2) interior fundamental problem of elasticity.

6.1 Introduction

In recent years, the boundary-element method has become an acceptable numerical technique for solving boundary-value problems in various fields of applied mechanics. The essential feature of the boundary-element method is the transformation of the governing differential equations into a set of integral equations on the boundary of the domain. Consequently, only the boundary and the variables on the boundary have to be discretized. The dimensionality of the problem is thereby effectively reduced by one: a three-dimensional (volume) problem becomes a two-dimensional (surface) one, while a two-dimensional (plane) problem involves only one-dimensional (line) integrations. This is the main advantage of the boundary-element method versus domain-type methods such as finite-difference and finite-element methods, which approximate the solution of the differential equation at many interior points. An immediate consequence of this is that the discretized system of equations, generated by the boundary-element method, is considerably smaller than that generated by an equivalent finite-difference or finite-element approximation. Nevertheless, in many applications the dimension of the discretized system that occurs in the boundary-element method may be large, so that iterative techniques are needed for their solution.

Most iterative processes that have been found in the literature are based on techniques that are also applied to finite-difference or finite-element equations, although the structure of the system matrix is quite different. In the boundary-element method, the system matrix is fully distributed and its coefficients strongly depend on the geometry of the original problem. This is in contrast with the finite-difference of finite-element approach, which yields a sparse matrix, of which the coefficients are related to the differential equation. For the solution of these sparse matrices fast solvers have been developed recently. Most of these solvers appear to be robust and, in particular when the governing differential equation is linear (e.g. the Laplacian Δ or the bi-Laplacian Δ^2), they converge rapidly, independent of the geometry of the original problem. It is this class of differential equations for which the boundary-element method is appropriate. For this method, however, no fast solvers are available, but it is decisive to develop them from considerations of competition. Therefore, in the present paper multigrid methods are advocated for the solution of boundary-element systems. The fast convergence of the multigrid method is established for two plane problems: (1) calculation of potential flow around aerofoils and (2) interior fundamental problem of elasticity.

Multigrid methods are iterative schemes that work with a sequence of computational grids of increasing refinement. The solutions of the different but related problems on these grids interact with each other to obtain an approximation to the continuous solution of the boundary integral equation. On each grid in a multigrid process a boundary-element system of equations is approximately solved by some relaxation scheme (for example: Jacobi or successive over relaxation) which reduces the oscillating error vectors with short wavelength, compared with the scale of the grid. The components with long wavelength are reduced by a coarse-grid correction. Such multigrid methods have recently received increased attention for solving partial differential equations [5, 12, 33, 37]. In the present paper, it is shown that they can also be used advantageously for the solution of the non-sparse systems of equations that occur in boundary-element methods for two-dimensional problems. Ample attention is given to the decomposition of the system matrix, in order to obtain a relaxation scheme that reduces the high-frequency errors. When the relaxation scheme satisfies this smoothing property, it can be proven that the multigrid method converges more rapidly as the number of boundary elements increases. For non-smooth boundaries, the decomposition should take the singularities of the geometry into account, because they strongly affect the influence coefficients of the system matrix. This is illustrated for the calculation of potential flow around an aerofoil with a corner singularity near the trailing edge. For first kind equations, the decomposition is related to the logarithmic kernel function. It hinges upon the decomposition of the integral operator into a positive definite pseudo-differential operator and a compact operator, as proposed in [34]. This is explained for the interior second fundamental problem of elasticity. It is not an easy task to extend this type of decomposition to three-dimensional problems, because then the kernel function does not have a logarithmic singularity. Further research is necessary to develop relaxation schemes for boundary integral equations of the first kind of three-dimensional problems.

6.2 Boundary Integral Equations

A large class of elliptic boundary value problems can be formulated in terms of integral equations on the boundary of a given domain, in general, when a fundamental solution of the differential equation is explicitly available. Such boundary integral equations are found in many applications: potential flow problems [15, 22], acoustics [10, 21], elasticity [20, 28], electrostatics [3, 9], diffraction problems [4]. In general, these equations are of the form

$$A\,\mu + B\,\omega = g, \quad \Lambda\,\mu = b, \tag{2.1}$$

where A is a $p \times p$ matrix of linear boundary integral operators mapping the p-vector-valued functions μ on the boundary S into p-vector-valued functions on S; B is a given $p \times q$ matrix of smooth functions on S and Λ is a given $q \times p$ matrix of linear functionals. The p-vector-valued function g and $b \in \mathbb{R}^q$ are given, μ and $\omega \in \mathbb{R}^q$ are the unknowns.

As simple examples, the following applications are considered.

Example 1. Calculation of incompressible potential flow around aerofoils.

For incompressible flow, a velocity potential ϕ exists satisfying Laplace's equation

$$\Delta\phi = 0, \tag{2.2}$$

with boundary conditions

$$\frac{\partial\phi}{\partial n_e} = 0 \quad \text{along the aerofoil } S, \tag{2.3}$$

where $\partial/\partial n_e$ denotes differentiation in the direction of the outward normal to S and

$$\phi(\zeta) \to \phi_\infty(\zeta) \quad \text{for} \quad |\zeta| \to \infty, \tag{2.4}$$

with ϕ_∞ the velocity potential due to a uniform onset flow. If the flow is non-circulatory, we have $\phi_\infty(\zeta) = \boldsymbol{U} \cdot \zeta$, with $\boldsymbol{U}$ the velocity vector of the undisturbed flow. We represent the velocity potenital ϕ as

$$\phi(\zeta) = \phi_\infty(\zeta) + \phi_d(\zeta), \tag{2.5}$$

with ϕ_d the double-layer potential given by

$$\phi_d(\zeta) = -\frac{1}{2\pi} \int_S \mu(z) \frac{\partial}{\partial n_z} (\log|z-\zeta|)\, dS_z, \tag{2.6}$$

where n_z is the outward normal to the boundary S at the point z. The doublet distribution μ is such that ϕ satifies the boundary condition

$$\phi^-(\zeta) = 0, \tag{2.8}$$

where ϕ^- denotes the limit from the inner side to S. Using the Plemelj-Privalov formulae (see [23]), we obtain the following integral equation:

$$\mu(\zeta) - \frac{1}{\pi} \int_S \mu(z) \frac{\partial}{\partial n_z} (\log|z-\zeta|)\, dS_z = -2\phi_\infty(\zeta), \quad \zeta \in S. \tag{2.9}$$

When the boundary S is assumed to be sufficiently smooth, it can be proven that the solution of interior Dirichlet problem (2.8) also satisfies Neumann problem (2.2)–(2.4) for the exterior of the boundary S. In this case the boundary integral operator A is defined as a Fredholm integral operator of the second kind (i.e. $A = I + K$, with I the identity operator and K the boundary integral operator); the unknown ω is not present in (2.1).

Example 2. Interior second fundamental problem of elasticity.

For plane isotropic homogeneous elasticity, a stress function $U(\zeta_1, \zeta_2)$ exists such that

$$\tau_{11} = \frac{\partial^2 U}{\partial \zeta_2^2}, \quad \tau_{12} = -\frac{\partial^2 U}{\partial \zeta_1\, \partial \zeta_2}, \quad \tau_{22} = \frac{\partial^2 U}{\partial \zeta_1^2}, \tag{2.10}$$

where τ_{jk} denote the components of the stress tensor. From the Saint-Venant compatibility conditions for the strains, we obtain the biharmonic equation for U,

$$\Delta^2 U = 0. \tag{2.11}$$

The interior second boundary-value problem consists of (2.11) together with given boundary stress (T_1, T_2),

$$T_1 = (\tau_{11}, \tau_{12}) \cdot n_e = \frac{d}{ds} U_{\zeta_2}, \qquad T_2 = (\tau_{21}, \tau_{22}) \cdot n_e = -\frac{d}{ds} U_{\zeta_1},$$

where n_e is the outward normal to the boundary S. These conditions can be rewritten as

$$U_{\zeta_1} - \omega_1 = -\int_0^s T_2(\sigma)\, d\sigma =: g_1, \qquad U_{\zeta_2} - \omega_2 = \int_0^s T_1(\sigma)\, d\sigma =: g_2. \tag{2.12}$$

Here ω_1, ω_2 are arbitrary integration constants, which are related to rigid motions. Following the approach of Hsiao, Kopp and Wendland [17], we express the stress function U in terms of a single-layer potential, i.e.

$$U(\zeta) = \int_S \sum_{\alpha=1}^{2} \frac{\partial F(\zeta, z)}{\partial z_1^{2-\alpha}\, \partial z_2^{\alpha-1}} \mu_\alpha(z)\, dS_z, \tag{2.13}$$

where

$$F(\zeta, z) = -\tfrac{1}{2}|\zeta - z|^2 \log|\zeta - z| \tag{2.14}$$

is the fundamental solution of the biharmonic Eq. (2.11). Applying the boundary conditions (2.12), we obtain the following integral equations of the first kind,

$$\int_S \log|\zeta - z|\, \mu(z)\, dS_z + \int_S L(\zeta, z)\, \mu(z)\, dS_z + \begin{pmatrix} 1, 0, & \zeta_2 \\ 0, 1, & -\zeta_1 \end{pmatrix} \omega = g(\zeta), \quad \zeta \in S, \tag{2.15}$$

with the kernel-function L given by

$$L_{\alpha\beta}(\zeta, z) = \frac{(\zeta_\alpha - z_\alpha)(\zeta_\beta - z_\beta)}{|\zeta - z|^2} + \tfrac{1}{2}\,\delta_{\alpha\beta}. \tag{2.16}$$

In (2.15) both $\omega = (\omega_1, \omega_2, \omega_3)^T$ and $\mu = (\mu_1, \mu_2)^T$ are unknown. The additional requirements for uniqueness are given by Hsiao and Mac Camy [18] in the form

$$\int_S \mu(z)\, dS_z = (b_1, b_2)^T, \qquad \int_S \mu \cdot t\, dS_z = b_3, \tag{2.17}$$

with t the unit tangent to S and $(b_1, b_2, b_3)^T$ any given constant vector. In this example, the operators A and B of (2.1) are defined by the integral equations (2.15). The linear functionals Λ are related to the additional requirements (2.17).

In classical theory the integral equations (2.9) and (2.15) have very different properties. In recent papers [34, 35] Wendland has shown, however, that they also have a very strong, common property if they are considered as pseudo-differential operators. When the boundary integral operator A of (2.1) is strongly elliptic, which is the case in most problems of practical interest, Wendland [34] shows that there exists a decomposition

$$A = D + C. \tag{2.18}$$

where D is a positive definite pseudo-differential operator and C is a compact operator. This decomposition is found to be essential in the development of multigrid methods in boundary element calculations, because the operators D and C are explicitly used in the relaxation scheme. This is described in Sect. 4. For example 1 it is obvious to choose D as the identity operator $D\mu = \mu$. This can be permitted if the aerofoil S is a smooth contour and does not contain corners, because then the integral operator K of (2.9),

$$K\mu(\zeta) := -\frac{1}{\pi} \int_S \mu(z) \frac{\partial}{\partial n_z} (\log|z - \zeta|)\, dS_z, \tag{2.19}$$

can be proven to be compact on the Sobolev space H^0. In general, however, the aerofoil contains a corner at the trailing edge. Then it is necessary to take this singularity into account and to redefine the operator D. It is assumed that the aerofoil S is a piecewise continuous C^2 curve with non-zero trailing-edge angle (i.e. spines are excluded). The part of the aerofoil which consists of the intersection of S with a disk with radius δ and centre at the trailing edge is denoted by S_δ. This is illustrated in Fig. 1. Then there exists $\delta > 0$ such that the operator

$$K^{(1)}\mu(\zeta) := -\frac{1}{\pi} \int_{S_\delta} \mu(z) \frac{\partial}{\partial n_z} (\log|z - \zeta|)\, dS_z \tag{2.20}$$

defines a contraction in H^0 (cf. [8]). It follows that the operator $D = I + K^{(1)}$ has a bounded inverse. Furthermore, it turns out that the remaining operator C, defined by $K - K^{(1)}$, is compact on H^0.

For Example 2, the definition of the pseudo-differential operator D is connected with the logarithmic singularity of the kernel-function. Let the boundary S be

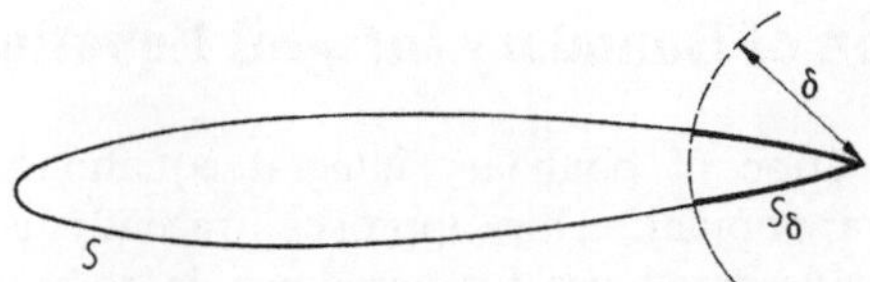

Fig. 1. Definition of the trailing-edge boundary S_δ

given by some parametrization $z(t)$, which is assumed to be 1-periodic, i.e. $z(t+1) = z(t)$, and to satisfy

$$\left|\frac{dz}{dt}\right| \geqq \varkappa > 0 . \tag{2.21}$$

Then, Eqs. (2.15) can be rewritten as

$$\int_0^1 \log|z(\tau) - z(t)|\,\bar{\mu}(t)\,dt + \int_0^1 L(z(t), z(t))\,\bar{\mu}(t)\,dt + \begin{pmatrix} 1, 0, & z_2(\tau) \\ 0\ 1, & -z_1(\tau) \end{pmatrix} \omega = g(z(\zeta)),$$
$$\tau \in [0, 1]\,, \tag{2.22}$$

where $\bar{\mu}$ is the new unknown density, defined by

$$\bar{\mu}(t) := \left|\frac{dz}{dt}\right| \mu(z(t)) .$$

In [16] the following splitting is proposed for the principal part of (2.22),

$$\int_0^1 \log|z(\tau) - z(t)|\,\bar{\mu}(t)\,dt \tag{2.23}$$
$$= \int_{|t-\tau| \leqq 1/2} \log|\tau - t|\,\bar{\mu}(t)\,dt + \int_{|t-\tau| \leqq 1/2} \log\left|\frac{z(\tau) - z(t)}{\tau - t}\right| \bar{\mu}(t)dt .$$

The first integral of this decomposition has the advantage to be independent of the special choice of the curve S and to represent the principal part of (2.22). Here, it also defines the pseudo-differential operator D,

$$D\bar{\mu}(\tau) := \int_{|t-\tau| \leqq 1/2} \log|\tau - t|\,\bar{\mu}(t)\,dt . \tag{2.24}$$

The remaining part of the decomposition (2.23) contains a smooth kernelfunction for all t and τ. This follows from the extention of $\log|(z(\tau) - z(t))/(\tau - t)|$ by $\log|dz/dt|$ for $t = \tau$. We obtain the compact operator C by joining the integral operators with smooth kernels of (2.22) and (2.23), i.e.

$$C\bar{\mu}(\tau) := \int_{|t-\tau| \leqq 1/2} \log\left|\frac{z(\tau) - z(t)}{\tau - t}\right| \bar{\mu}(t)\,dt + \int_0^1 L(z(\tau), z(t))\,\bar{\mu}(t)\,dt . \tag{2.25}$$

6.3 Approximation of Boundary Integral Equations

For the numerical solution of boundary integral equations, boundary-element methods have become popular. This approach, recently advocated anew by Brebbia [6, 7], is in widespread use for aerodynamic computations since 1962, when Hess and Smith [15] introduced the so-called panel-methods for the calculation of non-lifting potential flow about three-dimensional bodies. The asymptotic convergence properties have been studied by Wendland [35, 36] and Nedelec [25].

Boundary-element methods are based on a partition of the boundary S into N segments S_i, such that $S=\bigcup_{i=1}^{N} S_i$ and $S_i \cap S_j = \emptyset$, $i \neq j$. On this partition, the unknown function μ of (2.1) is approximated by finite elements which are usually called boundary elements. In engineering analysis, piecewise constant elements are often applied, which are defined as

$$u_i(\zeta) = \begin{cases} 1, & \zeta \in S_i, \\ 0, & \zeta \notin S_i. \end{cases} \tag{3.1}$$

Then, the function μ is approximated by

$$\mu_N(\zeta) = \sum_{i=1}^{N} \gamma_i u_i(\zeta), \tag{3.2}$$

where the coefficients γ_i are determined by application of a Galerkin procedure to (2.1). This yields the following system of linear equations for γ_i, ω_N,

$$\sum_{j=1}^{N} (A u_j, u_i) \gamma_j + (B \omega_N, u_i) = (g, u_i), \quad \sum_{j=1}^{N} (A u_j) \gamma_j = b, \quad \text{for } i = 1, \ldots, N. \tag{3.3}$$

Here, (,) denotes the inner product of the Sobolev space $H^0(S)$. In (3.3) they are of the following type,

$$(f, u_i) := \int_S f(\zeta) u_i(\zeta) dS_\zeta = \int_{S_i} f(\zeta) dS_\zeta .$$

If f behaves smoothly on S_i, it is sufficient to approximate the latter integral by the rectangular rule:

$$\int_{S_i} f(\zeta) dS_\zeta \cong f(\zeta_i) \Delta S_i , \tag{3.4}$$

where ζ_i is the midpoint and ΔS_i the area of segment S_i. Then, the Galerkin system (3.3) corresponds with the system of equations that would have been obtained by applying a first-order collocation or panel method. This is illustrated for the examples of Sect. 2.

For the calculation of potential flow around an aerofoil, we have to solve the second kind integral equation (2.9). We replace the unknown doublet distribution by the approximation (3.1)–(3.2), which yields

$$\sum_{j=1}^{N} \gamma_j u_j(\zeta) - \frac{1}{\pi} \sum_{j=1}^{N} \gamma_j \int_{S_j} \frac{\partial}{\partial n_z} (\log |z-\zeta|) dS_z = -2\phi_\infty(\zeta). \tag{3.5}$$

The integrals in this formula can be evaluated analytically. To this end we remark that $\log|z-\zeta|$ and $\theta(z-\zeta)$ are conjugate harmonic functions of z, keeping ζ fixed. Here $\theta(z-\zeta)$ denotes the angle vector $z-\zeta$ makes with some fixed direction. Since they satisfy the Cauchy-Riemann equation

$$-\frac{\partial}{\partial n_z}(\log|z-\zeta|)=\frac{\partial}{\partial S_z}\theta(z-\zeta),$$

the integral in (3.5) becomes

$$-\int_{S_j}\frac{\partial}{\partial n_z}(\log|z-\zeta|)\,dS_z=\theta(z_j-\zeta)-\theta(z_{j-1}-\zeta)=:\Psi_j(\zeta). \tag{3.6}$$

Here, $\Psi_j(\zeta)$ is the angle subtended at the point ζ by the segment S_j, having begin and end points at z_{j-1} and z_j. If ζ lies on the segment S_j, formula (3.6) requires a slight modification, since the angle $\theta(z-\zeta)$ jumps by π as z passes through the point ζ. This jump does not contribute to the integral, since it has been accounted by the Plemelj-Privalov formulae in deriving the integral equation (2.9). Therefore, for $\zeta\in S_j$ we have

$$-\int_{S_j}\frac{\partial}{\partial n_z}\log|z-\zeta|\,dS_z=\begin{cases}\Psi_j(\zeta)+\pi, & \text{if } \Psi_j(\zeta)<0,\\ \Psi_j(\zeta)-\pi, & \text{if } \Psi_j(\zeta)>0.\end{cases}$$

Applying the Galerkin procedure to (3.5), we obtain

$$\gamma_i\,\Delta S_i-\frac{1}{\pi}\sum_{j=1}^{N}\gamma_i\int_{S_i}\left\{\int_{S_j}\frac{\partial}{\partial n_z}(\log|z-\zeta|)\,dS_z\right\}dS_\zeta=-2\int_{S_i}\phi_\infty(\zeta)\,dS_\zeta.$$

Next the integrals over segment S_i are approximated by (3.4). This approach yields

$$\gamma_i-\frac{1}{\pi}\sum_{j=1}^{N}\gamma_j\int_{S_i}\frac{\partial}{\partial n_z}(\log|z-\zeta_i|)\,dS_z=-2\,\phi_\infty(\zeta_i), \tag{3.7}$$

for $i=1,\dots,N$, which are exactly the equations of the collocation method.

For the example of elasticity, the computational method of [16] is applied to Eq. (2.22). The mesh covering the interval [0, 1] is uniform with mesh size $h=1/N$. The basis functions are defined as

$$u_j(t)=\begin{cases}1, & (j-1)\,h\le t\le j\,h,\\ 0, & \text{elsewhere},\end{cases}$$

for $j=1,\dots,N$. The Galerkin procedure leads to the following algebraic system:

$$\sum_{j=1}^{N}\gamma_j\{D\,u_j,u_i)+(C\,u_j,u_i)\}+h\,B\,\omega_N=(g,u_i),\quad \text{for } i=1,\dots,N,$$

$$\sum_{j=1}^{N}h\,\gamma_j=(b_1,b_2)^T,\qquad \sum_{j=1}^{N}h\,\gamma_j\cdot\frac{dz}{dt}(h\,(j-1/2))=b_3, \tag{3.8}$$

where

$$(D\,u_j, u_i) = \int_{\tau=0}^{1} \int_{|t-\tau| \leq 1/2} \log|\tau - t|\, u_j(t)\, u_i(\tau)\, dt\, d\tau$$

and

$$(C\,u_j, u_i) = \int_{\tau=0}^{1} \int_{|t-\tau| \leq 1/2} \log\left|\frac{z(\tau)-z(t)}{\tau - t}\right| u_j(t)\, u_i(\tau)\, dt\, d\tau$$

$$+ \int_{\tau=0}^{1} \int_{t=0}^{1} L(z(\tau), z(t))\, u_j(t)\, u_i(\tau)\, dt\, d\tau\,.$$

The essence of the method of [16] is to evaluate the Galerkin weights $(D\,u_j, u_i)$ exactly upto the desired accuracy and to approximate the weights $(C\,u_j, u_i)$ and (g, u_i) by the rectangular rule. This approach yields

$$(D\,u_j, u_i) = h^2 [\log h + W_{\varrho(i-j)}]\,, \tag{3.9}$$

$$(C\,u_j, u_i) \cong h^2 \left\{\log\left|\frac{\zeta_i - \zeta_j}{h\,\varrho(i-j)}\right| + L(\zeta_i, \zeta_j)\right\}, \tag{3.10}$$

$$(g, u_i) \cong h\, g(\zeta_i)\,, \tag{3.11}$$

where

$$\zeta_j = z(h(j-\tfrac{1}{2})) \quad \text{and} \quad \varrho(k) = \min\{|k|, |k+N|, |k-N|\}.$$

For $i = j$, the kernel function in (3.10) is replaced by its smooth extension

$$\log\left|\frac{\zeta_j - \zeta_j}{h\,\varrho(j-j)}\right| := \log\left|\frac{dz}{dt}(h(j-\tfrac{1}{2}))\right|. \tag{3.12}$$

The weights W_ϱ are defined by

$$W_\varrho = \int_{\tau=\varrho}^{\varrho+1} \int_{t=0}^{1} \log|\tau - t|\, dt\, d\tau\,. \tag{3.13}$$

For more details of this numerical scheme we refer to [16].

The examples discussed in this section show that the boundary element methods of the first order, often used in engineering practice, can be treated as a modified Galerkin method. This observation facilitates the study of multigrid convergence, because the asymptotic convergence properties of Galerkin methods are well established.

6.4 Practical Aspects of Multigrid Methods

The numerical treatment of boundary integral equations provides a finite system of linear equations, of which the coefficient matrix, the influence matrix, is fully distributed. Usually, this system of equations is solved by direct methods. When the dimension is large, however, iterative techniques are needed. A fast converging iterative scheme is supplied by the multigrid method. These methods, originally formulated for practical differential equations, combine relaxation schemes for reducing high-frequency errors and coarse-grid correction for diminishing low-frequency errors. In recent papers [30, 32], we have not only shown the applicability of such a method to boundary integral equations but also we have derived theoretical results concerning the convergence of the method. In this section we will discuss some practical aspects.

For the description of multigrid methods, we use the orthogonal projection operator T_N of the Sobolev space H^0 into the finite-dimensional space of piecewise constant elements; that is

$$T_N\colon\ H^0 \to H_N \subset H^0. \tag{4.1}$$

Then, the Galerkin system of Eqs. (3.3) can be written in operator notation as

$$T_N A T_N \mu_N + T_N B\,\omega_N = T_N g\,, \qquad \Lambda T_N \mu_N = b\,. \tag{4.2}$$

In essence a multigrid process consists of two steps: a relaxation algorithm for reducing the high-frequency errors, and a coarse-grid correction for diminishing the low-frequency errors.

For the development of the relaxation alogrithm, we take advantage of the decomposition of A into $D + C$. This leads to the following relaxation process

$$\begin{aligned} T_N D T_N \mu_N^{i+1/2} + T_N B\,\omega_N^{i+1/2} &= T_N g - T_N C T_N \mu_N^{i}\,, \\ \Lambda T_N \mu_N^{i+1/2} &= b\,. \end{aligned} \tag{4.3}$$

Inspection of (4.3) reveals that it is a prerequisite that the operator $T_N D T_N$ can be easily inverted. More exactly, the inversion of $T_N D T_N$ should cost much less operations than the direct inversion of the operator $T_N A T_N$ would require, which asks $2/3\,N^3$ operations if Gaussian elimination is applied. In order to obtain a rapidly converging multigrid process, the above relaxation process should reduce the high-frequency errors, so that the iteration error $\mu_N - \mu_N^{i+1/2}$ can be represented on a coarser grid. When (4.3) is subtracted from (4.2), we are left with the following equations for the iteration error,

$$\begin{aligned} T_N D T_N (\mu_N - \mu_N^{i+1/2}) + T_N B\,(\omega_N - \omega_N^{i+1/2}) &= -\,T_N C T_N (\mu_N - \mu_N^{i})\,, \\ \Lambda T_N (\mu_N - \mu_N^{i+1/2}) &= 0\,. \end{aligned} \tag{4.4}$$

From (4.4) it follows that the iteration error $\mu_N - \mu_N^{i+1/2}$ behaves smoothly if the operator $(T_N D T_N)^{-1}\, T_N C T_N$ has some smoothing property. This is illustrated in Sect. 5.

After the relaxation process (4.3) has been applied, the residue becomes

$$\begin{aligned} r_{1,N}^{i+1/2} &:= T_N g - T_N A T_N \mu_N^{i+1/2} - T_N B\,\omega_N^{i+1/2}, \\ r_{2,N}^{i+1/2} &:= b - \Lambda T_N \mu_N^{i+1/2}\,. \end{aligned} \tag{4.5}$$

When (4.2) is used, these equations can be rewritten as

$$T_N A T_N (\mu_N - \mu_N^{i+1/2}) + T_N B (\omega_N - \omega_N^{i+1/2}) = r_{1,N}^{i+1/2},$$
$$\Lambda T_N (\mu_N - \mu_N^{i+1/2}) = r_{2,N}^{i+1/2}, \tag{4.6}$$

which yields expressions for the exact correction. Equation (4.6) includes, however, the inversion of $T_N A T_N$, which should be avoided. Since the correction can be represented on a coarser grid it is obvious to solve, instead of Eq. (4.6),

$$T_M A T_M \nu_M + T_M B \chi_M = T_M r_{1,N}^{i+1/2}, \qquad \Lambda T_M \nu_M = r_{2,N}^{i+1/2}, \tag{4.7}$$

with $\nu_M \in X_M$ and $M < N$. Note that the residue $r_{1,N}^{i+1/2} \in H_N$ is projected to the subspace H_M. Comparing (4.6) and (4.7), we conclude that ν_M is an approximation for $\mu_N - \mu_N^{i+1/2}$, so that we get the next iterate from

$$\mu_N^{i+1} = \mu_N^{i+1/2} + T_N \nu_M, \qquad \omega_N^{i+1} = \omega_N^{i+1/2} + \chi_M, \tag{4.8}$$

where ν_M is prolongated to H_N. Next, we repeat the relaxation process (4.3). The iterative scheme given by (4.3), (4.5), (4.7) and (4.8) defines a so-called two-level algorithm. It is emphasized that (4.7) is of the same type as (4.2), but that the dimension is lowered. In a multigrid method one usually chooses $M = N/2$ and approximates the solution of (4.7) by two iterations of the above two-level algorithm, except when it has to be solved on the lowest level. In that case the linear system is solved by Gaussian elemination.

It will be clear that the multigrid method is based on a sequence of finite-dimensional subspaces H_p with dimension $N_p = 2^p N_0$ and a sequence of projection operators T_p. In the context of multigrid iteration, the subscript p is called level. We need the following assumptions for $\{H_p\}$,

$$H_0 \subset H_1 \subset \ldots \subset H_l \subset H^0,$$

which is satisfied by the class of piecewise constant functions if the partition of the boundary on level p includes the partition of level $p-1$. This is illustrated below.

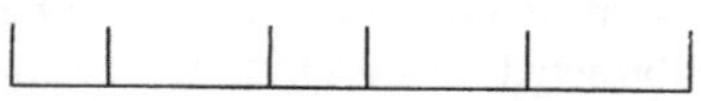

partition of level $p-1$

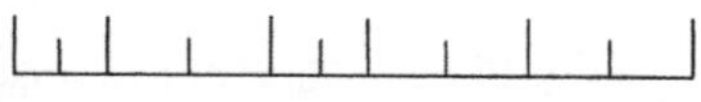

partition of level p.

For curved boundaries this can be realized by the use of some parametrization $z(t)$, $t \in [0, 1]$. Then, a sequence of partitions is introduced on the interval $[0, 1]$ and the corresponding boundary element segmentations of S are obtained by the mapping $z(t)$. For three-dimensional configurations, a similar approach can be followed.

To discuss the multigrid algorithm in more detail, we define

$$D' = \begin{pmatrix} D & B \\ \Lambda & 0 \end{pmatrix}, \quad C' = \begin{pmatrix} C & 0 \\ 0 & 0 \end{pmatrix}, \quad A' = D' + C',$$

the augmented function f by (μ, ω) and the projection operators T'_p from $H^0 \times \mathbb{R}^q$ onto $H_p \times \mathbb{R}^q$ by $T'_p f = (T_p \mu, \omega)$. Then, the two-level algorithm can be abbreviated by

$$f_p^{i+1/2} = f_p^i + (T'_p D' T'_p)^{-1} r_p^i, \tag{4.9}$$

$$f_p^{i+1} = f_p^{i+1/2} + T'_p (T'_{p-1} A' T'_{p-1})^{-1} T'_{p-1} r_p^{i+1/2}, \tag{4.10}$$

where (4.9) describes the relaxation process and (4.10) the coarse-grid correction.

The multigrid algorithm is a recursive procedure with respect to the level number. It can be written in terms of (4.9)–(4.10) by replacing the inverse of the coarse-grid operator by an approximate inverse Q_{p-1}. Then, the coarse-grid correction reads

$$f_p^{i+1} = f_p^{i+1/2} + T'_p Q_{p-1} T'_{p-1} r_p^{i+1/2}. \tag{4.11}$$

In Sect. 6, it will be shown that

$$|||Q_p||| \leq (1 + \tilde{\eta}_p^2)\, |||(T'_p A' T')^{-1}|||, \tag{4.12}$$

where $\tilde{\eta}_p \to 0$ as $p \to \infty$. Here $|||\cdot|||$ denotes the operator norm on the finite-dimensional space H_p. From (4.12) it follows that Q_p approximates $(T'_p A' T'_p)^{-1}$ more accurately as $p \to \infty$. Consequently, the multigrid method converges more rapidly as the number of boundary elements increases.

The recursive procedure in a multigrid method can be easily handled by computer languages like Algol and Pascal. A complete Algol-68 program with multigrid solvers has been given in [31] for the automatic solution of Fredholm equations of the second kind. For the description of the algorithm in Fortran, we refer to [27]. If the total number of levels is relatively small, it is easy to program the multigrid method in Fortran by defining separate two-grid subroutines for each level. This is illustrated in the Fortran subroutine given in Text 1, where the following computational steps are distinguished:

1. $r_p := g_p - T'_p A' f_p$,
2. $r_p := (T'_p D' T'_p)^{-1} r_p$,
3. $f_p := f_p + r_p$,
4. $r_p := g_p - T'_p A' f_p$,
5. $r_{p-1} := T'_{p-1} r_p$,
6. $v_{p-1} := 0$,
7. $v_{p-1} :=$ "result of two applications of this algorithm on level $p-1$",
8. $r_p := T'_p v_{p-1}$,
9. $f_p := f_p + r_p$.

Here steps (1)–(3) define the relaxation process (4.9); the other steps correspond with the coarse-grid correction (4.10). For level $p = 2$, this multigrid algorithm can then be implemented in Fortran as follows.

```
      SUBROUTINE MULGRI2 (SIGMA, F2, G2)
      DIMENSION F2(1), G2(1), R2(128), V1(64), R1(64)
      INTEGER SIGMA
      N2 = N(2)
      N1 = .5*N2
      DO 100 IT = 1, SIGMA
        CALL RESIDUE(F2, G2, R2, N2)    }
        CALL DINV(R2, N2)               } Relaxation process
        CALL ADDVEC(F2, R2, F2, N2)     }
        CALL RESIDUE(F2, G2, R2, N2)    }
        CALL RESTRIC(R2, R1, N2)        }
        CALL ZERO(V1, N1)               } Coarse-grid correction
        CALL MULGRI1(2, V1, R1)         }
        CALL PROLONG(V1, R2, N2)        }
100     CALL ADDVEC(F2, R2, F2, N2)     }
      RETURN
      END
```

Text 1. Implementation of the multigrid algorithm for level 2.

It is emphasized that the subroutines called by MULGRI2 are independent of the level number, except the subroutine MULGRI1. The actual implementation of the subroutines RESTRIC and PROLONG (representing T_{p-1} and T_p, respectively) depends on the choice of $\{H_p\}$ and the projection operators $\{T_p\}$. For the piecewise constant functions and the L_2 projections, they are specified by weighted injection,

$$\mu_{p-1}[i] := \tfrac{1}{2}(\mu_p[2i-1] + \mu_p[2i]), \quad i = 1(1)\, N_{p-1},$$

and piecewise constant interpolation,

$$\mu_p[2i-1] = \mu_p[2i] := \mu_{p-1}[i], \quad i = 1(1)\, N_{p-1}.$$

The multigrid algorithm involves the computation of the influence matrices on the coarser grids. In this paper they are obtained from a computation of the influence coefficients at each level by (Au_i, u_j) for all $u_i, u_j \in H_p$ and for $p = 0, 1, \ldots, l$. Asymptotically, this requires the computation of $4/3N^2$ influence coefficients. In many applications, however, the computing time for the influence matrix is rather high (up to 80% of the total time for solving one boundary-value problem), so that the above method is rather expensive. According to [27], a cheaper method is obtained by defining the coarse-level matrices by the restriction and prolongation operators. On the highest level, given by $p = l$, the operator $A_l = T_l A T_l$ is defined in the classical way. Then, at the lower levels, the coarse-grid operators are obtained from $A_{p-1} = T_{p-1} A_p T_{p-1}$, for $p = l, l-1, \ldots, 1$. For piecewise constant functions and L_2 projections the influence coefficients take the form,

$$A_{p-1}[i,j] := \tfrac{1}{2}(A_p[2i-1, 2j-1] + A_p[2i-1, 2j] + A_p[2i, 2j-1] + A_p[2i, 2j]),$$

for $i, j = 1, \ldots, N_{p-1}$ and $p = l, l-1, \ldots, 1$.

The relaxation scheme (4.9) requires the inversion of the operator $(T_p' D' T_p')$, which is performed in the subroutine DINV (see Text 1). This subroutine depends on the problem to be solved.

For the example of calculating the potential flow around an aerofoil, the matrix corresponding with $(T_p' D' T_p')$ has the following structure,

$$\boldsymbol{D}_p = \begin{bmatrix} \boldsymbol{A}_{11} & \boldsymbol{0} & \boldsymbol{A}_{13} \\ \boldsymbol{A}_{21} & \boldsymbol{I} & \boldsymbol{A}_{23} \\ \boldsymbol{A}_{31} & \boldsymbol{0} & \boldsymbol{A}_{33} \end{bmatrix},$$

where $\boldsymbol{A}_{11}$ and $\boldsymbol{A}_{33}$ are square matrices of dimension N^*. Here N^* denotes the number of elements at the upper and lower part of the trailing edge on S_δ. The matrix $\boldsymbol{D}_p$ can be solved directly by block elemination.

For the second example, the one concerning elasticity, the operator $(T_p' D' T_p')$ is replaced by a $(2N_p+3) \times (2N_p+3)$-matrix $\boldsymbol{D}_p$ which can be partitioned into the form

$$\boldsymbol{D}_p = \begin{bmatrix} \tilde{\boldsymbol{D}} & \boldsymbol{0} & \boldsymbol{B}_1 \\ \boldsymbol{0} & \tilde{\boldsymbol{D}} & \boldsymbol{B}_2 \\ \boldsymbol{\Lambda}_1 & \boldsymbol{\Lambda}_2 & \boldsymbol{0} \end{bmatrix},$$

which can be directly inverted by block elimination. Here $\tilde{\boldsymbol{D}}$ is a $N_p \times N_p$ Toeplitz-matrix, of which the elements are given by a N_p-periodic vector $\boldsymbol{d}$. The sub-matrices $\boldsymbol{B}_k$ and $\boldsymbol{\Lambda}_k$ $(k=1, 2)$ have dimensions $N_p \times 3$ and $3 \times N_p$, respectively.

The matrix $\tilde{\boldsymbol{D}}$ is inverted using the convolution theorem and Fast Fourier Transformation (FFT). The computational costs due to this inversion, being $O(N \log N)$, are negligible with respect to the operation counts for a relaxation process requiring N^2 multiplications. The equation $\tilde{\boldsymbol{D}}\boldsymbol{f}=\boldsymbol{g}$ has the form of a convolution product,

$$\sum_{j=0}^{N-1} d_{k-j} f_j = g_k, \quad k=0, \ldots, N-1.$$

Let $\hat{f}_l$ be the Fourier transform of $\{f_j\}$. Then, the convolution theorem yields:

$$\hat{d}_l \hat{f}_l = \hat{g}_l, \quad l=0, \ldots, (N-1)/2,$$

so that

$$f_j = \frac{1}{\sqrt{N}} \sum_{l=0}^{(N-1)/2} (\hat{g}_l/\hat{d}_l)\, \mathrm{e}^{\mathrm{i}\, 2\pi j l/N}$$

This is illustrated in the subroutine DTINV given in Text 2.

The subroutine DTINV, given in Text 2, determines $\tilde{\boldsymbol{D}}^{-1}\boldsymbol{g}$ ($\boldsymbol{g}$ is a vector of length N) employing the FFT-subroutines C06FAF and C06FBF from the NAG-library [24]. The other subroutines have the following functions: TOEPVEC computes the vector $\boldsymbol{d}$ belonging to the Toeplitz matrix $\tilde{\boldsymbol{D}}$; DIVVEC divides $\hat{g}_l$ by $\hat{d}_l$ for $l=0, \ldots, (N-1)/2$; C06GBF forms the complex conjugate of the vector $\boldsymbol{g}$, and SCALE multiplies $\boldsymbol{g}$ by a scalar.

```
SUBROUTINE DTINV(G, N)
DIMENSION G(1), D(256), WORK(999)
CALL TOEPVEC(D, N)
IFAIL = 0
CALL C06FAF(D, N, WORK, IFAIL)
CALL C06FAF(G, N, WORK, IFAIL)
CALL DIVVEC(G, D, G, N)
CALL C06GBF(G, N, IFAIL)
CALL C06FBF(G, N, WORK, IFAIL)
SQN = 1./SQRT(1.*N)
CALL SCALE(G, N, SQN)
RETURN
END
```

Text 2. Computation of $\tilde{\boldsymbol{D}}^{-1}\boldsymbol{g}$ using the convolution theorem and FFT-subroutines; $\tilde{\boldsymbol{D}}$ is a $N \times N$ Toeplitz-matrix, the elements of which are given by the N-periodic vector $\boldsymbol{d}$.

6.5 Smoothing Property of the Relaxation Process

The basic parts of the multigrid method described in Sect. 4 are the relaxation process (4.3) and the coarse-grid correction (4.7). An efficient multigrid process is obtained if the relaxation process reduces the high-frequency components of the iteration error. This is satisfied if the operator $D^{-1}C$ has the smoothing property to map the space of high frequencies into the space of low frequencies.

When the terminology of low and high frequencies is used, the use of Fourier analysis for the formulation of the smoothing property becomes obvious. To this end we introduce some notations. Let μ be any distribution on the boundary S with the Fourier representation

$$\mu(t) = \int_{-\infty}^{+\infty} \hat{\mu}(\omega)\, e^{2i\pi\omega t}\, d\omega ,$$

with $\hat{\mu}(\omega)$ the Fourier transform defined by

$$\hat{\mu}(\omega) = \frac{1}{2\pi} \int_0^1 \mu(t)\, e^{-2i\pi\omega t}\, dt .$$

Let $H^\alpha(S)$ be the Sobolev space of distributions equipped with the norm

$$\|\mu\|_\alpha := \left\{ \int_{-\infty}^{+\infty} (1+\omega^2)^\alpha\, |\hat{\mu}(\omega)|^2\, d\omega \right\}^{1/2} .$$

Then, for α small the space H^α contains high-frequency functions, having a non-zero spectrum for large values of ω. On the other hand, for large values of α the space H^α only can contain low-frequency functions, which have a non-zero spectrum for small values of ω.

Now the smoothing property is formulated by the requirement that the operator $D^{-1} C$ should be a bounded mapping from $H^0(S)$ into $H^\alpha(S)$ with $\alpha > 0$, i.e.

$$|||D^{-1} C|||_{H^0 \to H^\alpha} := \sup_{\mu \in H^0} \frac{\| D^{-1} C \mu \|_\alpha}{\| \mu \|_0} < \infty. \tag{5.1}$$

This requirement implies that the operator $D^{-1} C$ is compact on H^0, because H^α is compactly imbedded in H^0. Next, this smoothing property is established for the examples of Sect. 2.

For the first example, where a potential flow is calculated around an aerofoil, it can be verified that the operator $D = I + K^{(1)}$ has a bounded inverse on the space $H^\alpha(S)$. Therefore it suffices to investigate the smoothing property of the operator C. From results given in [29] it follows that C is a bounded mapping from $H^0(S)$ into $H^\alpha(S)$ provided the aerofoil S/S_δ is sufficiently smooth. This is illustrated by considering a high-frequency function μ_N (3.2) with coefficients γ_i that are alternately $+1$ and -1, i.e.

$$\gamma_i = (-1)^i, \quad i = 1(1)128.$$

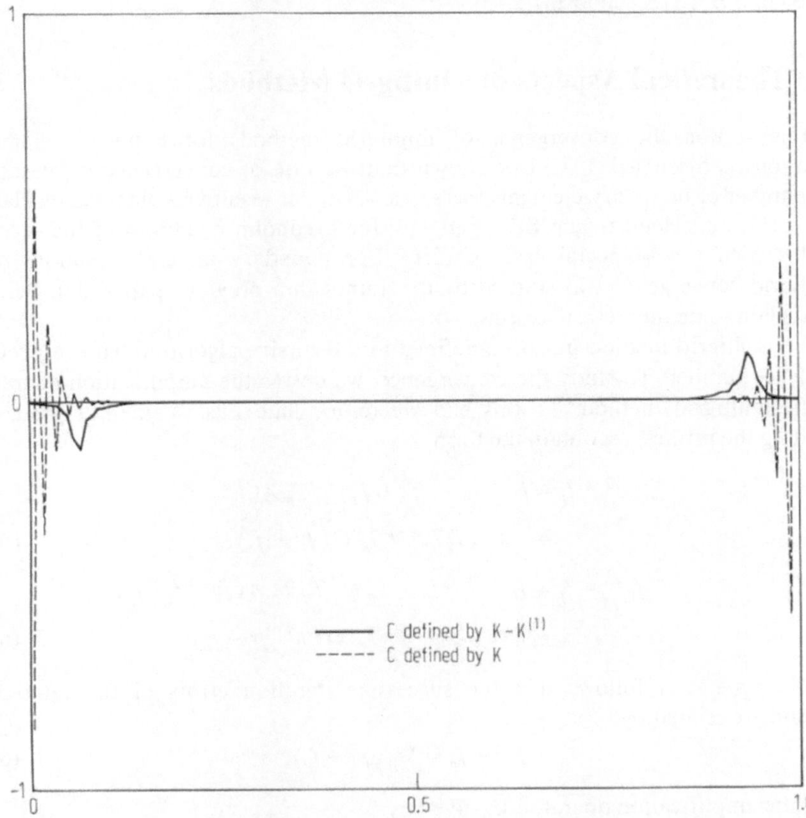

Fig. 2. Integral transform of a high-frequency function μ, which illustrates the smoothing property of the boundary integral operator C occurring in potential-flow calculations around a symmetric aerofoil

Next, the operator C is applied to this oscillating function. We recall that C has been defined as $C = K - K^{(1)}$, where $K^{(1)}$ accounts for the contour integration over S_δ (a small part near the trailing edge) as given by (2.20). From Fig. 2 it is observed that $C\mu_N$ is a smooth function; consequently the operator C satisfies the smoothing property being a bounded mapping from the Sobolev space $H^0(S)$ into $H^\alpha(S)$. The value of α depends on the smoothness of the aerofoil S/S_δ and the size of the cut-off region S_δ. In the extreme case of $\delta = 0$, the operator C equals K. Then, $C\mu_N$ still contains high-frequency components as can be observed from Fig. 2. In this case the operator C is only bounded on $H^0(S)$ and it does not satisfy the smoothing property. Fig. 2 clearly illustrates the necessity to split off the operator $K^{(1)}$ in order to obtain a smoothing integral operator.

For the second example, the one concerning the problem of elasticity, the smoothing property of $D^{-1}C$ follows from theorem 9 of [19]. In this paper it has been shown that the logarithmic integral operator D is a bijective continuous mapping of $H^\alpha(S)$ into $H^{\alpha+1}(S)$ and that the operator C maps $H^{\alpha-1}(S)$ into $H^{\alpha+1}$ continuously, for any real α. It follows that $D^{-1}C$ is a bounded mapping from $H^0(S)$ into $H^1(S)$ by choosing $\alpha = 1$.

6.6 Theoretical Aspects of Multigrid Methods

In this section the convergence of multigrid methods for boundary element equations is discussed. It will be shown that the rate of convergence increases as the number of boundary elements increases. A similar result was obtained by Hackbusch [11] and Hemker and Schippers [14] for Fredholm equations of the second kind, which are a special case of (2.1). For boundary integral equations, the multigrid convergence was studied by the author in a previous paper [32]. Below, we explain some theoretical aspects.

The multigrid method has been defined as a recursive algorithm with respect to the level number. To study the convergence, we derive the amplification operator of the multigrid method. To this end we reformulate Eqs. (4.9), (4.11) and, by omitting the primes, we obtain the form

$$\begin{aligned} f_p^{i+1/2} - f_p &= f_p^i - f_p - (T_p D T_p)^{-1} T_p A (f_p^i - f_p) \\ &= -(T_p D T_p)^{-1} T_p C (f_p^i - f_p), \end{aligned} \tag{6.1}$$

$$\begin{aligned} f_p^{i+1} - f_p &= f_p^{i+1/2} - f_p - T_p Q_q T_q T_p A (f_p^{i+1/2} - f_p) \\ &= (I - T_p Q_q T_q T_p A)(f_p^{i+1/2} - f_p), \end{aligned} \tag{6.2}$$

with $q = p - 1$. It follows that the successive iteration errors of the multigrid method are related by

$$f_p^{i+1} - f_p = \tilde{M}_p (f_p^i - f_p), \tag{6.3}$$

with the amplification operator $\tilde{M}_p$ given by

$$\begin{aligned} \tilde{M}_p &= -(I - T_p Q_q T_q T_p A)(T_p D T_p)^{-1} T_p C T_p \\ &= I - B_p T_p A T_p, \end{aligned} \tag{6.4}$$

where B_p is defined recursively:

$$B_0 = (T_0 A T_0)^{-1},$$

$$B_p = \{I - T_p Q_q T_q T_p C\}(T_p D T_p)^{-1}, \quad p = 1, 2 \ldots$$

with

$$Q_p = \sum_{m=0}^{1} \tilde{M}_p^m B_p . \tag{6.6}$$

Note that (6.4)–(6.6) describe the mathematical definition of the approximate inverse Q_p. From (6.4), (6.6) it follows that

$$Q_p = (I - \tilde{M}_p^2)(T_p A T_p)^{-1}, \tag{6.7}$$

and, consequently,

$$|||Q_p||| \leqq (1 + \tilde{\eta}_p^2)\, |||(T_p A T_p)^{-1}|||, \tag{6.8}$$

with the reduction factor of the multigrid method defined by

$$\tilde{\eta}_p = |||\tilde{M}_p||| .$$

Here, $|||\cdot|||$ denotes the operator norm on the finite-dimensional space H_p. Next it is shown that $\tilde{\eta}_p \to 0$ as $p \to \infty$, which follows from a comparison of the two-grid and the multigrid algorithm. The amplification operator and the reduction factor of the two-grid algorithm (4.9)–(4.10) are obtained, analogously to (6.1)–(6.4),

$$M_p = -(I - T_p (T_q A T_q)^{-1} T_q T_p A)(T_p D T_p)^{-1} T_p C T_p \tag{6.9}$$

and

$$\eta_p = |||M_p||| .$$

Substitution of (6.7), (6.9) into (6.4) leads to the form

$$\tilde{M}_p = M_p - T_p \tilde{M}_q^2 \{M_p + (T_p D T_p)^{-1} T_p C T_p\},$$

so that

$$\tilde{\eta}_p \leqq \eta_p + c\, \tilde{\eta}_q^2 .$$

In [14] it has been proven that the contribution of the second term may be neglected if the number of boundary elements on the lowest level is large enough. Therefore, it suffices to analyse the convergence of the two-grid algorithm (4.9)–(4.10). When the operator G_p is introduced as

$$G_p = T_p (T_{p-1} A T_{p-1})^{-1} T_{p-1} T_p A, \tag{6.10}$$

we can abbreviate the amplification operator M_p by

$$M_p = -(I - G_p)(T_p D T_p)^{-1} T_p C T_p .$$

Next, we can estimate the reduction factor as follows

$$\begin{aligned} \eta_p &= |||(I - G_p)(T_p D T_p)^{-1} T_p C T_p||| \\ &\leqq |||(I - G_p) D^{-1} C||| + |||(I - G_p)(D^{-1} C - (T_p D T_p)^{-1} T_p C T_p)||| \\ &\leqq |||I - G_p|||_{H^\alpha \to H_p} |||D^{-1} C|||_{H_p \to H^\alpha} + c\, |||D^{-1} C - (T_p D T_p)^{-1} T_p C T_p)||| . \end{aligned}$$

For boundary-element methods, the projection operators T_p are bounded. The smoothing property of $D^{-1}C$ and the definition of G_p by (6.10) imply that the first term of this estimate tends to zero as $p \to \infty$. Moreover, the smoothing property of $D^{-1}C$ guarantees that the difference $D^{-1}C - (T_p DT_p)^{-1} T_p C$ is collectively compact. According to the theory of Anselone [2], it converges uniformly to zero, so that also the second term vanishes for $p \to \infty$. Consequently, the reduction factor $\eta_p \to 0$ and therefore the multigrid method converges more rapidly as the number of boundary elements increases.

In the case that piecewise constant elements are employed, a sharper estimate follows from the propositions

$$|||I - G_p|||_{H^\alpha \to H_p} \leqq c\, h_p^{\min(1,\alpha)}$$

and

$$|||D^{-1}C - (T_p DT_p)^{-1} T_p CT_p||| \leqq c\, h_p^{\min(1,\alpha)},$$

for some constant c. Here h_p is a measure of the mesh-size. This implies that

$$\eta_p \leqq c\, h_p^{\min(1,\alpha)}. \tag{6.11}$$

For smooth boundaries ($\alpha > 1$) this estimate shows that the multigrid method converges twice as fast as the number of boundary elements is doubled. This is illustrated in the following section.

6.7 Numerical Results

In this section the fast convergence of the multigrid method is shown for the examples discussed in this paper.

The first example, the one concerning the potential flow around an aerofoil, illustrates the necessity to equip the multigrid method with a relaxation scheme that reduces the high-frequency errors. We calculate the non-circulatory flow around a symmetric Karman-Trefftz aerofoil with a relative thickness of 5%, using the first-order collocation method discussed in Sect. 3. In our experiments, the trailing-edge angle is $\pi/10$. In Table 1 the number of iterations to obtain a residual less than 10^{-6} is given, when the multigrid method (4.9), (4.11) is applied with 32 boundary elements on the coarsest grid. The relaxation scheme is based on a partition of the system matrix into blocks, according to a decomposition of the integral operator, such that the trailing-edge is taken into account. The partitioning

Table 1. Number of multigrid iterations necessary to obtain a residual less then 10^{-6}, with 32 boundary elements on the coarsest grid. N^* = number of boundary elements near the trailing edge.

N^*	$N = 64$	$N = 128$	$N = 256$
0	15	13	11
$N/16$	6	4	2
$N/8$	4	2	2

Table 2. Reduction factors of the multigrid method with 20 boundary elements on the coarsest grid

Number of elements	40	80	160
Reduction factor	0.035	0.014	0.007

yields the matrix D_p, given by (4.13), which contains a free parameter N^*, denoting the number of elements near the trailing edge. This parameter strongly affects the rate of convergence of the multigrid method.

In the extreme case ($N^* = 0$), the multigrid method converges slowly. This can be expected, because the relaxation scheme does not account for the trailing-edge angle. In this case, the matrix D_p reduces to the identity matrix and the operator C equals K, which does not have the smoothing property (5.1) as has been shown in Sect. 5.

In the other cases ($N^* > 0$), the multigrid method converges more rapidly as the number of boundary elements increases. For $N = 256$ only two iterations are necessary to obtain a result of the order of the truncation error.

The second example, where the problem of elasticity is considered, illustrates the applicability to first kind boundary integrals equations. In this case, the relaxation scheme is based on a decomposition of the system matrix into a Toeplitz matrix and a remaining matrix, which satisfy the smoothing property (5.1) if the boundary is sufficiently smooth. If the boundary has some corners, these regions, after having been cut-off should be treated analogously to the trailing-edge angle in the first example. This would imply a modification of the relaxation scheme. Here we only present numerical results for the smooth boundary of an ellips with a relative thickness of 20%, using the Galerkin-collocation method given by (3.8)–(3.13). In Table 2 we give the observed reduction factors (defined by the ratio of the successive iteration errors) of the multigrid method (4.9), (4.11) with 20 boundary elements on the coarsest grid.

From this table, we observe that the multigrid method converges very fast. The reduction factor decreases by a factor 2 if the number of boundary elements increases by a factor 2. Consequently, the multigrid method converges twice as fast if the number of boundary elements is doubled. This is in agreement with the theoretical estimate (6.11).

At this stage, we would give a note of warning: the multigrid method diverges if the number of boundary elements on the coarsest grid is too small. It appears that the fast convergence, indicated in Table 2, is completely lost if the multigrid method is applied with 10 elements on the coarsest grid. Therefore, one should not despair if the multigrid method diverges or converges only slowly, but one should double the number of boundary elements on the coarsest grid, after verification of the smoothing property of the relaxation scheme.

6.8 Conclusions and Recommendations

From the numerical results given in the previous section, two important conclusions have come forward. The first conclusion is that the multigrid method converges very fast. Asymptotically, the method needs only two iterations to obtain a result of the order of the truncation error. This involves an amount of computational work of $6N^2$ operation counts, which is comparable with 6 iterations of the relaxation scheme at the finest grid. The second conclusion is that the fast convergence may be lost if the following two basic assumptions are not satisfied: the relaxation scheme should reduce the high-frequency errors and the number of boundary elements on the coarsest grid should be sufficiently large. Hence, one should first verify the smoothing property of the relaxation scheme that accounts for the boundary singularities, and next one should refine the coarsest grid.

So far we have only discussed the applicability of multigrid methods to two-dimensional (plane) problems. For three-dimensional (volume) problems the use of multigrid methods is restricted at present to second kind boundary integral equations. The applicability to Laplacian problems was investigated by Wolff [38], Nowak [26] and Alund [1]. In the context of three-dimensional Stokes' flow the method has been studied by Hebeker [13]. For first kind equations the use of multigrid methods is hampered because no appropriate decomposition of the system matrix is available. Further research is necessary to improve the current situation. In this context, it is recommendable to investigate the general relaxation schemes that have been developed by Oskam [27] for plane problems. Once relaxation schemes will have been developed for first kind equations, the extension to general boundary-element systems becomes feasible. Then, a multigrid fast solver would become available for the general situation. This would be an important progress for the boundary-element method and would constitute a logical extension of the multigrid method described in this paper.

References

1 Alund, A., Iterative methods to compute the singularity distribution in three-dimensional panel methods, Report L-0-1 R 97, SAAB-Scania, Linköping, Sweden, 1984

2 Anselone, P.M., Collectively compact operator approximation theory, Englewood Cliffs, New Jersey, Prentice-Hall, 1971

3 Barnard, A.C.L., Duck, I.M., Lynn, M.S., and Timlake, W.P., The application of electromagnetic theory to electrocardiology, II. Numerical solution of the integral equations. Biophys. J. **7**, 463–491, 1967

4 Berkhoff, J.C.W., Diffraction of water waves. In: Colloquium Numerical Treatment of Integral Equations (H.J.J. te Riele, ed), MC-Syllabus 41, Mathematisch Centrum, Amsterdam 1979, pp. 241–258

5 Brandt, A., Multi-level adaptive solutions to boundary-valve problems. Mathematics of Computations **31**, 333–390, 1977

6 Brebbia, C. A., The boundary element method for engineers. Pentech Press, London 1978

7 Brebbia, C.A. and Walker, S., Boundary element techniques in Engineering. Butterworth, London 1980

8 Costabel, M. and Stephan, E., Curvature terms in the asymptotic expansions for solutions of boundary integral equations on curved polygons. Preprint Nr. 673, Technical University Darmstadt, Dept. Mathematics, 1982

9 Edwards, T.W. and Van Bladel, K., Electrostatic dipole moment of a dielectric cube. Applied Scientific Research **9**, 151–155, 1961

10 Filippi, P., Layer potentials and acoustic diffraction. J. Sound and Vibration **54,** 473–500, 1977
11 Hackbusch, W., Die schnelle Auflösung der Fredholmschen Integralgleichung zweiter Art. Beiträge zur Numerischen Mathematik **9,** 47–62, 1981
12 Hackbusch, W., Multigrid convergence theory, Multigrid Methods, Proceedings, Köln 1981 (W. Hackbusch and Trottenberg, U., eds.). Lecture Notes in Mathematics, 960, pp. 177–219, Springer-Verlag, Berlin, Heidelberg, New York 1982
13 Hebeker, F.K., On a multigrid method to solve the integral equations of 3-D Stokes' flow, Efficient Solutions of Elliptic Systems. Proceedings, Kiel 1984 (W. Hackbusch, ed.). Vieweg Verlag, Braunschweig/Wiesbaden 1984
14 Hemker, P.W. and Schippers, H., Multiple grid methods for the solution of Fredhom integral equations of the second kind. Math. Comp. **36,** 215–232, 1981
15 Hess, J.L. and Smith, A.M.O., Calculation of potential flow about arbitrary bodies. Progress in Aero Science **8,** 1–138, 1967
16 Hsiao, G.C., Kopp, P., and Wendland, W.L., A Galerkin collocation method for some integral equations of the first kind. Computing **25,** 89–130, 1980
17 Hsiao, G.C., Kopp, P., and Wendland, W.L., Some applications of a Galerkin-collocation method for boundary integral equations of the first kind. Preprint Nr. 768, Technical University Darmstadt, Dept. Mathematics, 1983
18 Hsiao, G.C. and Maccamy, R.C., Solution of boundary value problems by integral equations of the first kind. SIAM Review **15,** 687–705, 1973
19 Hsiao, G.C. and Wendland, W.L., A finite element method for some integral equations of the first kind. J. Math. Anal. Appl. **58,** 449–481, 1977
20 Jaswon, M. A. and Symm, G.T., Integral equation methods in potential theory and elastostatics. Academic Press, London 1977
21 Jones, D.S., Integral equations for the exterior acoustic problem. Quart. J. Mech. Appl. Math. **27,** 129–142, 1974
22 Martensen, E., Berechnung der Druckverteilung an Gitterprofilen in ebener Potentialströmung mit einer Fredholmschen Integralgleichung. Arch. Rat. Mech. Anal. **3,** 235–270, 1959
23 Muschelischwili, N.I., Singuläre Integralgleichungen. Akademie-Verlag, Berlin 1965
24 NAG Fortran Library Manual mark 10, Numerical Algorithms Group. Oxford 1983
25 Nedelec, J.C., Curved finite element methods for the solution of singular integral equations on surfaces in R^3. Comp. Math. Appl. Mech. Eng. **8,** 61–80, 1976
26 Nowak, Z.P., Use of the multigrid method for Laplacian problems in three dimensions, Multigrid Methods, Proceedings, Köln 1981 (W. Hackbusch and U. Trottenberg, eds.). Lecture Notes in Mathematics, 960, pp. 576–598, Springer-Verlag, Berlin, Heidelberg, New York 1982
27 Oskam, B. and Fray, J.M.J., General relaxation schemes in multigrid algorithms for higher order singularity methods. J. Comp. Physics **48,** 423–440, 1982
28 Rizzo, F.J., An integral equation to boundary value problems of classical elastostatics. Q. Appl. Math. **25,** 83–95, 1967
29 Schippers, H., On the regularity of the principal value of the double-layer potential. J. Engineering Math. **16,** 59–76, 1982
30 Schippers, H., Application of multigrid methods for integral equations to two problems from fluid dynamics. J. Comp. Physics **48,** 441–461, 1982
31 Schippers, H., Multiple grid methods for equations of the second kind with applications in fluid mechanics. Ph.D. Thesis, Mathematisch Centrum, Amsterdam 1982, published as Mathematical Centre Tracts 163
32 Schippers, H., Multigrid methods for boundary integral equations, NLR MP 82059 U, National Aerospace Laboratory, Amsterdam 1982, Numerische Mathematik
33 Trottenberg, U., Multigrid methods: fundamental algorithms, model problem analysis and applications, Multigrid Methods, Proceedings, Köln 1981 (W. Hackbusch and U. Trottenberg, eds.). Lecture Notes in Mathematics, 960, pp. 1–176, Springer-Verlag, Berlin, Heidelberg, New York 1982
34 Wendland, W.L., On Galerkin collocation methods for integral equations of elliptic boundary value problems. In: Numerical Treatment of Integral Equations (J. Albrecht and L. Collatz, eds.). Intern. Ser. Numer. Math., 53, pp. 244–275, Birkhäuser, Basel 1980

35 Wendland, W.L., I. Asymptotic convergence of boundary element methods. II. Integral equation methods for mixed boundary value problems. Preprint Nr. 611, Technical University Darmstadt, Dept. Mathematics, 1981

36 Wendland, W.L., Boundary element methods and their asymptotic convergence, Lecture Notes of the CISM Summer-School on *Theoretical Acoustics and Numerical Techniques*, International Centre for Mechanical Sciences, Udine (Italy), 1982 (P. Filippi, ed.), to appear in: Lecture Notes in Physics, Springer-Verlag, Berlin, Heidelberg, New York 1983

37 Wesseling, P., A robust and efficient multigrid method, Multigrid Methods, Proceedings, Köln 1981 (W. Hackbusch and U. Trottenberg, eds.), Lecture Notes in Mathematics, 960, pp. 614–630, Springer-Verlag, Berlin, Heidelberg, New York 1982

38 Wolff, H., Multiple grid method for the calculation of potential flow around 3-D bodies, Preprint NW 119/82, Dept. of Numerical Mathematics, Mathematical Centre, Amsterdam 1982

Chapter 7

Complex Variable Boundary Elements in Computational Mechanics

by T. V. Hromadka

7.1 Introduction

A new and exciting numerical approach to solving two-dimensional potential problems is obtained by use of the Cauchy integral equation for analytic functions. The resulting integral equation is readily solvable by computer, and produces a pair of two-dimensional conjugate harmonic functions which satisfy the Laplace equation over the problem domain.

Of special interest, however, are the approximation error evaluation techniques afforded by the Complex Variable Boundary Element Method (CVBEM). One especially useful technique develops an approximate boundary where the CVBEM solution satisfies the local boundary conditions continuously. Error analysis and reduction then proceeds by the addition of nodal points to the problem boundary where discrepancy between the approximate and problem boundaries is seen to be large.

In this chapter, the CVBEM will be developed in detail with special attention paid to linear and constant basis functions specified on the problem boundary. Generalization to higher order basis functions is also included. Approximation error evaluation techniques are reviewed, with a focus upon the approximate boundary method.

7.2 A Complex Variable Boundary Element Approximation Model

Let Ω be a simply connected domain with an associated simple closed contour boundary Γ. Assume that Γ is a polygonal line composed of V straight line segments and vertices (Fig. 1). Let $\omega(z) = \phi(z) + \mathrm{i}\,\psi(z)$ be a complex variable function which is analytic on $\Omega \cup \Gamma$ where ϕ and ψ are the state variable (or potential) and stream functions, respectively. Then the real variable functions composing $\omega(z)$ are related by the Cauchy-Riemann equations on $\Omega \cup \Gamma$:

$$\frac{\partial \phi}{\partial x} = \frac{\partial \psi}{\partial y}, \quad \frac{\partial \phi}{\partial y} = -\frac{\partial \psi}{\partial x}. \tag{1}$$

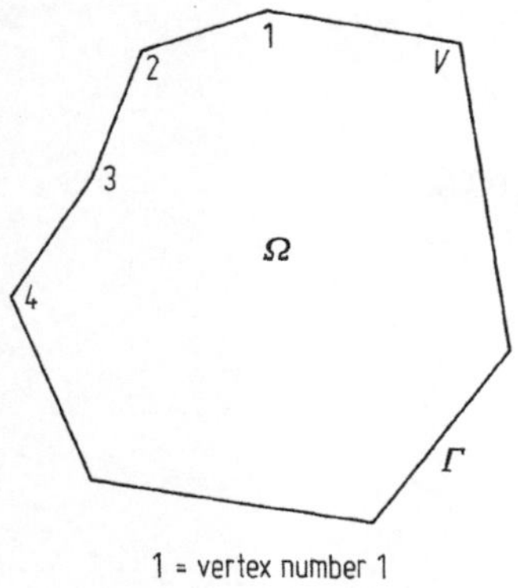

Fig. 1. Domain $\Omega \cup \Gamma$

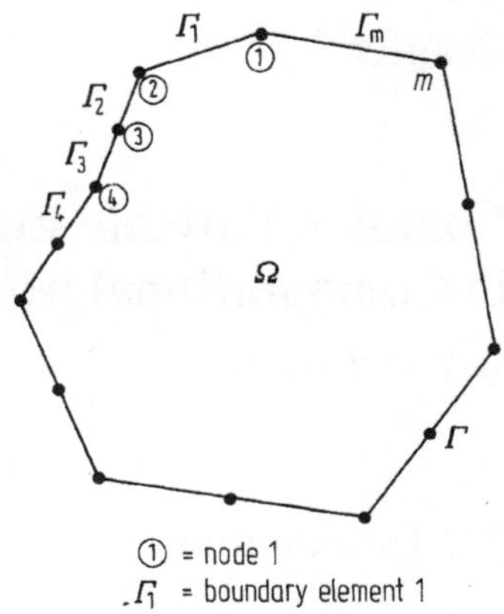

Fig. 2. Nodal point distribution on Γ

Consequently, ϕ and ψ are harmonic functions for $z \in \Omega \cup \Gamma$ such that

$$\frac{\partial^2 \phi}{\partial x^2} + \frac{\partial^2 \phi}{\partial y^2} = 0, \quad \frac{\partial^2 \psi}{\partial x^2} + \frac{\partial^2 \psi}{\partial y^2} = 0. \tag{2}$$

Define on Γ a partition of nodal points $\{z_j, j = 1, 2, \ldots, m\}$ such that at least at each boundary vertex there is a nodal point, i.e. $m \geqq V$. The nodal points are numbered (sequentially from 1) in a counterclockwise direction around Γ (Fig. 2).

At each nodal point z_j, define the symbols $\bar{\phi}_j$ and $\bar{\psi}_j$ to represent specified nodal values where each value is a real number. Similarly, let ϕ_j and ψ_j be notation for the values of $\phi(z_j)$ and $\psi(z_j)$ respectively, $j = 1, 2, \ldots, m$.

Define 2-node boundary elements Γ_j on Γ by

$$\Gamma = \bigcup_{j=1}^{m} \Gamma_j, \tag{3}$$

where

$$\Gamma_j = \{z \in \Gamma\colon z = z_j(1-s) + z_{j+1}s, 0 \leqq s \leqq 1\} \quad \text{and} \quad \Gamma_j \cap \Gamma_{j+1} = z_{j+1}.$$

The numbering of boundary elements follows the nodal point numbering scheme shown in Fig. 2.

A continuous global trial function $G_1(z)$ is defined on Γ by

$$G_1(z) = \sum_{j=1}^{m} N_j(z)\,\bar{\phi}_j + \mathrm{i} \sum_{j=1}^{m} N_j(z)\,\bar{\psi}_j \tag{4}$$

where a typical basis function is defined for nodal point j by

$$N_j(z) = \begin{cases} (z - z_{j-1})/(z_j - z_{j-1}), & z \in \Gamma_{j-1} \\ (z_{j+1} - z)/(z_{j+1} - z_j), & z \in \Gamma_j \\ 0, & z \notin \Gamma_{j-1} \cup \Gamma_{j+1} \end{cases} \tag{5}$$

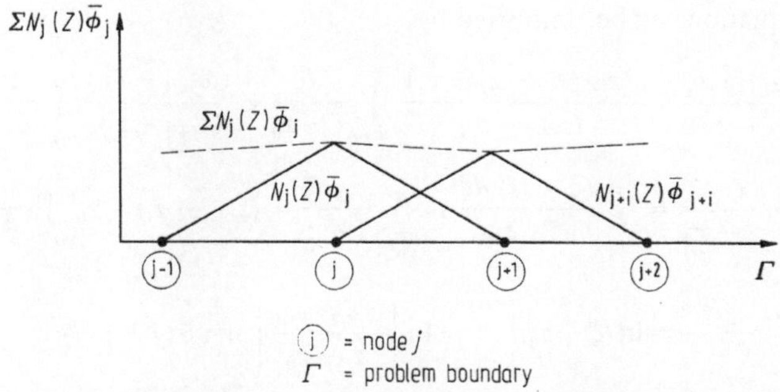

Fig. 3. The linear basis function relationship to the global trial function

Figure 3 shows the linear basis function and its relation to the global trial function $G_1(z)$. The global trial function is seen to be continuous on Γ and $G_1(z_j) = \bar{\phi}_j + i\,\bar{\psi}_j$, $j = 1, 2, \ldots, m$.

An approximation function (an H_1 approximation function) can be developed by defining

$$\hat{\omega}(z) = \frac{1}{2\pi i} \int_{\Gamma} \frac{G_1(\zeta)\,d\zeta}{\zeta - z}, \quad z \in \Omega \tag{6}$$

where z is in the interior of Ω and not on the boundary Γ, and contour integration is in the positive sense.

The $\hat{\omega}(z)$ approximation function will be shown in Sect. 2 to be analytic in Ω and therefore has the property that its real $\hat{\phi}(z)$ and imaginary $\hat{\psi}(z)$ components satisfy the two-dimensional Laplace equation in Ω. The numerical modeling strategy is to try to determine a $\hat{\omega}(z)$ such that $\hat{\omega}(z)$ is arbitrarily close to $\omega(z)$ values for all boundary points $z \in \Gamma$. To develop such an approximator, $\hat{\omega}(z)$ is written in terms of boundary elements Γ_j by

$$\hat{\omega}(z_0) = \frac{1}{2\pi i} \int_{\bigcup_{j=1}^{m} \Gamma_j} \frac{G_1(\zeta)\,d\zeta}{\zeta - z_0}$$

$$= \frac{1}{2\pi i} \sum_{j=1}^{m} \int_{\Gamma_j} \frac{G_1(\zeta)\,d\zeta}{\zeta - z_0}; \quad z_0 \in \Omega, \quad z_0 \notin \Gamma \tag{7}$$

On each Γ_j, $G_1(z)$ is simplified to

$$G_1(z) = N_j\,\bar{\omega}_j + N_{j+1}\,\bar{\omega}_{j+1} = (N_j\,\bar{\phi}_j + N_{j+1}\,\bar{\phi}_{j+1}) + i\,(N_j\,\bar{\psi}_j + N_{j+1}\,\bar{\psi}_{j+1}), \quad z \in \Gamma_j \tag{8}$$

where $\bar{\omega}_j = \bar{\phi}_j + i\,\bar{\psi}_j$, and N_j is used for a shorthand notation for the $N_j(z)$ function. Using (8) the contribution to $\hat{\omega}(z_0)$ from each boundary element Γ_j can be calculated by

$$\int_{\Gamma_j} \frac{G_1(\zeta)\,d\zeta}{\zeta - z_0} = \int_{\Gamma_j} \frac{[(z_{j+1} - \zeta)\,\bar{\omega}_j + (\zeta - z_j)\,\bar{\omega}_{j+1}]/(z_{j+1} - z_j)\,d\zeta}{\zeta - z_0}; \; z_0 \in \Omega,\; z_0 \notin \Gamma_j \tag{9}$$

The above equation can be simplified by

$$\int_{\Gamma_j} \frac{G_1(\zeta)\, d\zeta}{\zeta - z_0} = \frac{(z_{j+1}\,\bar{\omega}_j - z_j\,\bar{\omega}_{j+1})}{(z_{j+1} - z_j)} \int_{\Gamma_j} \frac{d\zeta}{\zeta - z_0} + \frac{(\bar{\omega}_{j+1} - \bar{\omega}_j)}{(z_{j+1} - z_j)} \int_{\Gamma_j} \frac{\zeta\, d\zeta}{\zeta - z_0}$$

$$\int_{\Gamma_j} \frac{\zeta\, d\zeta}{\zeta - z_0} = \int_{\Gamma_j} \frac{(\zeta - z_0)\, d\zeta}{\zeta - z_0} + \int_{\Gamma_j} \frac{z_0\, d\zeta}{\zeta - z_0} = (z_{j+1} - z_j) + z_0 \int_{\Gamma_j} \frac{d\zeta}{\zeta - z_0}$$

$$\int_{\Gamma_j} \frac{d\zeta}{\zeta - z_0} = \ln(\zeta - z_0)\Big|_{z_j}^{z_{j+1}} = \ln\left|\frac{z_{j+1} - z_0}{z_j - z_0}\right| + \mathrm{i}\,\theta(j+1, j)$$

where $\theta(j+1, j)$ is the central angle between the straight line segments joining points z_j and z_{j+1} to central point $z_0 \in \Omega$ (Fig. 4).

Using the above integrations

$$\int_{\Gamma_j} \frac{G_1(\zeta)\, d\zeta}{\zeta - z_0} = \bar{\omega}_{j+1} - \bar{\omega}_j + \bar{\omega}_{j+1} \frac{(z_0 - z_j)}{(z_{j+1} - z_j)} h_j - \bar{\omega}_j \frac{(z_0 - z_{j+1})}{(z_{j+1} - z_j)} h_j \tag{10}$$

where $h_j = \ln(|z_{j+1} - z_0| / |z_j - z_0|) + \mathrm{i}\,\theta(j+1, j)$. The complex value of $\hat{\omega}(z_0)$ is determined as the sum of each Γ_j contribution by

$$2\pi\,\mathrm{i}\,\hat{\omega}(z_0) = \sum_{j=1}^{m} (\bar{\omega}_{j+1} - \bar{\omega}_j) + \sum_{j=1}^{m} [\bar{\omega}_{j+1}(z_0 - z_j) - \bar{\omega}_j(z_0 - z_{j+1})]\, h_j/(z_{j+1} - z_j) \tag{11}$$

where in (11) it is understood that $\bar{\omega}_{m+1} \equiv \bar{\omega}_1$ and $z_{m+1} \equiv z_1$. The first summation term cancels leaving (for linear basis functions)

$$2\pi\,\mathrm{i}\,\hat{\omega}(z_0) = \sum_{j=1}^{m} [\bar{\omega}_{j+1}(z_0 - z_j) - \bar{\omega}_j(z_0 - z_{j+1})]\, h_j/(z_{j+1} - z_j). \tag{12}$$

The above relationship can be written as a complex function

$$\begin{aligned} \hat{\omega}(z_0) &= \hat{\phi}(z_0) + \mathrm{i}\,\hat{\psi}(z_0) \qquad (13) \\ &= \hat{\phi}(z_0, \bar{\phi}_1, \bar{\phi}_2, \ldots, \bar{\phi}_m, \bar{\psi}_1, \bar{\psi}_2, \ldots, \bar{\psi}_m) + \mathrm{i}\,\hat{\psi}(z_0, \bar{\phi}_1, \bar{\phi}_2, \ldots, \bar{\phi}_m, \bar{\psi}_1, \bar{\psi}_2, \ldots, \bar{\psi}_m), \end{aligned}$$

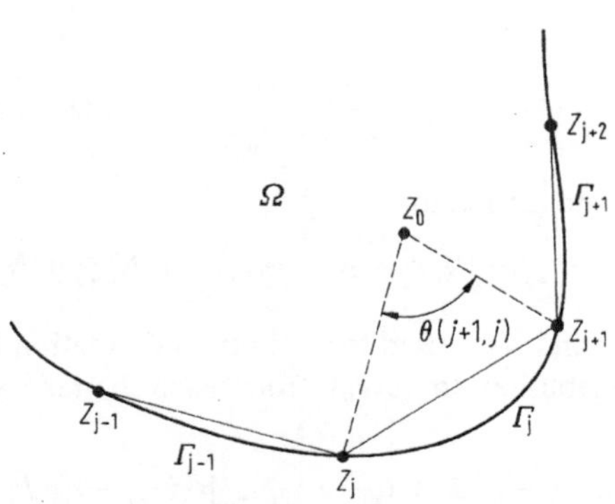

Fig. 4. Nodal point geometry

where z_0 is any point in the interior of Ω, and $\hat{\phi}$ and $\hat{\psi}$ are real valued functions representing the real and imaginary components of the complex function $\omega(z)$. Should values of $\bar{\omega}_j = \bar{\phi}_j + \mathrm{i}\,\bar{\psi}_j$ be known at each z_j, $j = 1, 2, \ldots, m$, then (13) defines a complex valued function which is analytic in Ω, and $\hat{\phi}(x, y)$ and $\hat{\psi}(x, y)$ both satisfy the Laplace equation in Ω. If $\hat{\omega}(z) = \omega(z)$ on Γ, then $\hat{\omega}(z) = \omega(z)$ in Ω and $\hat{\omega}(z)$ is the solution to the boundary value problem.

The usual problem in engineering applications is that only one of the specified nodal value pair $(\bar{\phi}_j, \bar{\psi}_j)$ is known at each z_j and, consequently, part of the modeling task is to evaluate unknown nodal values. A method of developing such an approximation function is to evaluate $\hat{\omega}(z)$ arbitrarily close to each nodal point and in turn, generate an implicit expression of the unknown nodal variable as a function of all the unknown variables. The result is m equations for m unknown nodal values which can be solved by the usual matrix techniques.

The evaluated nodal values are then used along with the original set of known nodal values to complete the definition of the $\hat{\omega}(z)$ approximation function on $\Omega \cup \Gamma$.

For example, suppose that m nodal points are defined on Γ, and $\bar{\psi}_j$ is known for $j = 1, 2, \ldots, k$ where $(0 < k < m)$ and $\bar{\phi}_j$ is known for the remaining nodes $j = (k + 1), \ldots, m$. Two methods of generating the unknown nodal values are provided by the following matrix systems:

Case I

$$
\begin{aligned}
\bar{\psi}_1 &\equiv \hat{\psi}(z_1^-) = \hat{\psi}(z_1^-, \bar{\phi}_1, \ldots, \bar{\phi}_m, \bar{\psi}_1, \ldots, \bar{\psi}_m)\\
&\vdots\\
\bar{\psi}_k &\equiv \hat{\psi}(z_k^-) = \hat{\psi}(z_k^-, \bar{\phi}_1, \ldots, \bar{\phi}_m, \bar{\psi}_1, \ldots, \bar{\psi}_m)\\
\bar{\phi}_{k+1} &\equiv \hat{\phi}(z_{k+1}^-) = \hat{\phi}(z_{k+1}^-, \bar{\phi}_1, \ldots, \bar{\phi}_m, \bar{\psi}_1, \ldots, \bar{\psi}_m)\\
&\vdots\\
\bar{\phi}_m &\equiv \hat{\phi}(z_m^-) = \hat{\phi}(z_m^-, \bar{\phi}_1, \ldots, \bar{\phi}_m, \bar{\psi}_1, \ldots, \bar{\psi}_m).
\end{aligned}
\tag{14}
$$

Case II

$$
\begin{aligned}
\bar{\phi}_1 &\equiv \hat{\phi}(z_1^-) = \hat{\phi}(z_1^-, \bar{\phi}_1, \ldots, \bar{\phi}_m, \bar{\psi}_1, \ldots, \bar{\psi}_m)\\
\bar{\phi}_2 &\equiv \hat{\phi}(z_2^-) = \hat{\phi}(z_2^-, \bar{\phi}_1, \ldots, \bar{\phi}_m, \bar{\psi}_1, \ldots, \bar{\psi}_m)\\
&\vdots\\
\bar{\phi}_k &\equiv \hat{\phi}(z_k^-) = \hat{\phi}(z_k^-, \bar{\phi}_1, \ldots, \bar{\phi}_m, \bar{\psi}_1, \ldots, \bar{\psi}_m)\\
\bar{\psi}_{k+1} &\equiv \hat{\psi}(z_{k+1}^-) = \hat{\psi}(z_{k+1}^-, \bar{\phi}_1, \ldots, \bar{\phi}_m, \bar{\psi}_1, \ldots, \bar{\psi}_m)\\
&\vdots\\
\bar{\psi}_m &\equiv \hat{\psi}(z_m^-) = \hat{\psi}(z_m^-, \bar{\phi}_1, \ldots, \bar{\phi}_m, \bar{\psi}_1, \ldots, \bar{\psi}_m)
\end{aligned}
\tag{15}
$$

where z_j^- indicates a point in Ω which is arbitrary close to boundary node coordinate z_j. Solution of both matrix systems will result in different $\hat{\omega}(z)$ function definitions. The case I matrix system produces values for the unknown nodal values such that

$$
\hat{\omega}(z_j) = \hat{\phi}(z_j) + \mathrm{i}\,\hat{\psi}(z_j) = \begin{cases} \hat{\phi}(z_j) + \mathrm{i}\,\bar{\psi}_j, & j = 1, 2, \ldots, k \\ \bar{\phi}_j + \mathrm{i}\,\hat{\psi}(z_j), & j = k + 1, \ldots, m \end{cases}
\tag{16}
$$

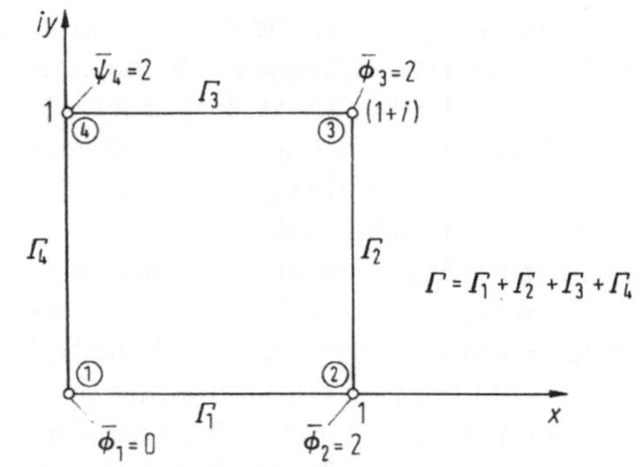

Fig. 5. Problem definition for example. ①, node number 1; o, nodal point on Γ; Γ_1, boundary element number 1

whereas the case II system results in nodal values which do not necessarily agree with the known specified nodal values.

Example 1. A simple example problem will be used to construct a CVBEM approximation function $\hat{\omega}(z)$. Figure 5 shows the assumed problem geometry and the known speficied nodal point values on Γ. The true solution to the boundary value problem is $\omega(z) = 2z$. The objective is to determine the unknown nodal values of $\{\bar{\psi}_1, \bar{\psi}_2, \bar{\psi}_3, \bar{\phi}_4\}$ and develop an approximation function $\hat{\omega}(z)$ on $\Omega \cup \Gamma$.

For node 1, (12) can be used to develop the nodal equation for a point $z_{\bar{1}}$ close to z_1 by

$$2\pi\,\mathrm{i}\,\hat{\omega}(z_{\bar{1}}) = \sum_{j=1}^{4} [\bar{\omega}_{j+1}(z_{\bar{1}} - z_j) - \bar{\omega}_j(z_{\bar{1}} - z_{j+1})]\, h_j/(z_{j+1} - z_j).$$

Letting $z_{\bar{1}}$ be the complex value $\varepsilon_x + \mathrm{i}\,\varepsilon_y$ (where ε_x and ε_y are both positive real numbers) and expanding the nodal equation for $z_{\bar{1}}$ gives

$$\begin{aligned} 2\pi\,\mathrm{i}\,\hat{\omega}(z_{\bar{1}}) = {} & [\bar{\omega}_2(\varepsilon_x + \mathrm{i}\,\varepsilon_y) - \bar{\omega}_1(\varepsilon_x + \mathrm{i}\,\varepsilon_y - 1)]\, h_1/(1) \\ & + [\bar{\omega}_3(\varepsilon_x + \mathrm{i}\,\varepsilon_y - 1) - \bar{\omega}_2(\varepsilon_x + \mathrm{i}\,\varepsilon_y - 1 - \mathrm{i})]\, h_2/(\mathrm{i}) \\ & + [\bar{\omega}_4(\varepsilon_x + \mathrm{i}\,\varepsilon_y - 1 - \mathrm{i}) - \bar{\omega}_3(\varepsilon_x + \mathrm{i}\,\varepsilon_y - \mathrm{i})]\, h_3/(-1) \\ & + [\bar{\omega}_1(\varepsilon_x + \mathrm{i}\,\varepsilon_y - \mathrm{i}) - \bar{\omega}_4(\varepsilon_x + \mathrm{i}\,\varepsilon_y)]\, h_4/(-\mathrm{i}). \end{aligned}$$

From Fig. 6,

$$h_1 = \ln\left|\frac{1 - \varepsilon_x - \mathrm{i}\,\varepsilon_y}{-\varepsilon_x - \mathrm{i}\,\varepsilon_y}\right| + \mathrm{i}\,\theta(2, 1)$$

$$h_2 = \ln\left|\frac{1 + \mathrm{i} - \varepsilon_x - \mathrm{i}\,\varepsilon_y}{1 - \varepsilon_x - \mathrm{i}\,\varepsilon_y}\right| + \mathrm{i}\,\theta(3, 2)$$

$$h_3 = \ln\left|\frac{\mathrm{i} - \varepsilon_x - \mathrm{i}\,\varepsilon_y}{1 + \mathrm{i} - \varepsilon_x - \mathrm{i}\,\varepsilon_y}\right| + \mathrm{i}\,\theta(4, 3)$$

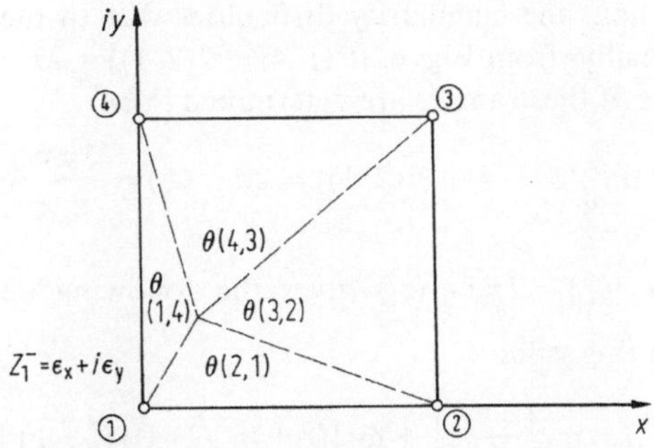

Fig. 6. Geometric values for nodal point ①

$$h_4 = \ln\left|\frac{-\varepsilon_x - i\,\varepsilon_y}{i - \varepsilon_x - i\,\varepsilon_y}\right| + i\,\theta(1,4).$$

All of the above terms are continuous functions for $z_{\bar{1}}$ in the interior of Ω; that is, for $\varepsilon_x > 0$ and $\varepsilon_y > 0$. As $z_{\bar{1}}$ approaches z_1, however, several terms can be evaluated as a limit. These limiting values are as follows:

$$\lim_{z_{\bar{1}} \to z_1} 2\pi\,i\,\hat{\omega}(z_{\bar{1}}) = \lim_{\substack{\varepsilon_x \to 0\\ \varepsilon_y \to 0}} 2\pi\,i\,\hat{\omega}(\varepsilon_x + i\,\varepsilon_y)$$

$$= [0 - \bar{\omega}_1(-1)]\left(\lim_{\substack{\varepsilon_x \to 0\\ \varepsilon_y \to 0}} \ln\left|\frac{1 - \varepsilon_x - i\,\varepsilon_y}{-\varepsilon_x - i\,\varepsilon_y}\right| + i\,\theta(2,1)\right)\Big/(1)$$

$$+ [\bar{\omega}_3(-1) - \bar{\omega}_2(-1-i)]\left(\ln\left|\frac{1+i}{1}\right| + \frac{\pi}{4}i\right)\Big/(i)$$

$$+ [\bar{\omega}_4(-1-i) - \bar{\omega}_3(-i)]\left(\ln\left|\frac{i}{1+i}\right| + \frac{\pi}{4}i\right)\Big/(-1)$$

$$+ [\bar{\omega}_1(-i) - 0]\left(\lim_{\substack{\varepsilon_x \to 0\\ \varepsilon_y \to 0}} \ln\left|\frac{-\varepsilon_x - i\,\varepsilon_y}{i - \varepsilon_x - i\,\varepsilon_y}\right| + i\,\theta(1,4)\right)\Big/(-i)$$

Simplifying terms,

$$\lim_{\substack{\varepsilon_x \to 0\\ \varepsilon_y \to 0}} 2\pi\,i\,\hat{\omega}(\varepsilon_x + i\,\varepsilon_y) = \bar{\omega}_1\left(\lim_{\substack{\varepsilon_x \to 0\\ \varepsilon_y \to 0}} \ln|1 - \varepsilon_x - i\,\varepsilon_y| - \lim_{\substack{\varepsilon_x \to 0\\ \varepsilon_y \to 0}} \ln|-\varepsilon_x - i\,\varepsilon_y| + i\,\theta(2,1)\right)$$

$$+ [-\bar{\omega}_3 + \bar{\omega}_2(1+i)](-i\ln\sqrt{2} + \tfrac{\pi}{4})$$

$$+ [\bar{\omega}_4(1+i) - \bar{\omega}_3\,i](-\ln\sqrt{2} + \tfrac{\pi}{4}i)$$

$$+ \bar{\omega}_1\left(\lim_{\substack{\varepsilon_x \to 0\\ \varepsilon_y \to 0}} \ln|-\varepsilon_x - i\,\varepsilon_y| - \lim_{\substack{\varepsilon_x \to 0\\ \varepsilon_y \to 0}} \ln|i - \varepsilon_x - i\,\varepsilon_y| + i\,\theta(1,4)\right).$$

In the above expression, the singularity difficulties due to the term $\ln|\varepsilon_x + i\,\varepsilon_y|$ sum to zero. Additionally, from Fig. 6, $(\theta(1,4)+\theta(2,1)) = 2\pi - (\theta(4,3)+\theta(3,2))$ and the limiting value of these angles are determined to be

$$\lim_{\substack{\varepsilon_x \to 0 \\ \varepsilon_y \to 0}} (\theta(1,4)+\theta(2,1)) = 2\pi - (\tfrac{\pi}{2}) = \frac{3\pi}{2}.$$

Defining $\lim_{z_{\bar{1}} \to z_1} 2\pi\,i\,\hat{\omega}(z_{\bar{1}}) = 2\pi\,i\,\hat{\omega}(z_1)$ gives the following new nodal equation for node z_1 as the limiting value

$$2\pi\,i\,\hat{\omega}(z_1) = \bar{\omega}_1\left(\frac{3\pi}{2}\,i\right) + \bar{\omega}_2\,[(\tfrac{\pi}{4} + \ln\sqrt{2}) + i\,(\tfrac{\pi}{4} - \ln\sqrt{2})]$$
$$+ \bar{\omega}_3\,[i\ln 2] + \bar{\omega}_4\,[(-\tfrac{\pi}{4} - \ln\sqrt{2}) + i\,(\tfrac{\pi}{4} - \ln\sqrt{2})].$$

Separating the nodal equation into real and imaginary components determines two nodal value equations,

$$-2\pi\,\hat{\psi}(z_1) = \left(-\frac{3\pi}{2}\right)\bar{\psi}_1 + (\tfrac{\pi}{4} + \ln\sqrt{2})\,\bar{\phi}_2 - (\tfrac{\pi}{4} - \ln\sqrt{2})\,\bar{\psi}_2$$
$$+ (-\ln 2)\,\bar{\psi}_3 - (-\tfrac{\pi}{4} - \ln\sqrt{2})\,\bar{\phi}_4 - (\tfrac{\pi}{4} - \ln\sqrt{2})\,\bar{\psi}_4$$

$$2\pi\,\hat{\phi}(z_1) = \frac{3\pi}{2}\,\bar{\phi}_1 + (\tfrac{\pi}{4} + \ln\sqrt{2})\,\bar{\psi}_2 + (\tfrac{\pi}{4} - \ln\sqrt{2})\,\bar{\phi}_2$$
$$+ (\ln 2)\,\bar{\phi}_3 + (\tfrac{\pi}{4} - \ln\sqrt{2})\,\bar{\psi}_4 + (-\tfrac{\pi}{4} - \ln\sqrt{2})\,\bar{\phi}_4.$$

Similar pairs of nodal equations can be determined for the remaining nodal values at z_2, z_3 and z_4 by rotating $\Omega \cup \Gamma$ so that each node in turn occupies the already studied $z_1 = 0 + 0\,i$ location

For this example problem, define constant coefficients

$$\gamma_1 = \tfrac{3}{4},\quad \gamma_2 = \tfrac{1}{8} + \frac{\ln\sqrt{2}}{2\pi},\quad \gamma_3 = \tfrac{1}{8} - \frac{\ln\sqrt{2}}{2\pi},\quad \gamma_4 = \frac{\ln 2}{2\pi}.$$

Then the following nodal value equations are determined:

$$\hat{\psi}(z_1) = \gamma_1\,\bar{\psi}_1 - \gamma_2\,\bar{\phi}_2 + \gamma_3\,\bar{\psi}_2 + \gamma_4\,\bar{\psi}_3 + \gamma_2\,\bar{\phi}_4 + \gamma_3\,\bar{\psi}_4$$

$$\hat{\phi}(z_1) = \gamma_1\,\bar{\phi}_1 + \gamma_2\,\bar{\psi}_2 + \gamma_3\,\bar{\phi}_2 + \gamma_4\,\bar{\phi}_3 - \gamma_2\,\bar{\psi}_4 + \gamma_3\,\bar{\phi}_4$$

Note that in the $\psi(z_1)$ equation the coefficients of the $\bar{\phi}_2$ and $\bar{\phi}_4$ terms sum to zero and the coefficients of $\bar{\psi}_1$, $\bar{\psi}_2$, $\bar{\psi}_3$, and $\bar{\psi}_4$ sum to one. A similar result holds for the $\hat{\phi}(z_1)$ equation. The remaining nodal equations are determined as

$$\hat{\psi}(z_2) = \gamma_1\,\bar{\psi}_2 - \gamma_2\,\bar{\phi}_3 + \gamma_3\,\bar{\psi}_3 + \gamma_4\,\bar{\psi}_4 + \gamma_2\,\bar{\phi}_1 + \gamma_3\,\bar{\psi}_1$$

$$\hat{\phi}(z_2) = \gamma_1\,\bar{\phi}_2 + \gamma_2\,\bar{\psi}_3 + \gamma_3\,\bar{\phi}_3 + \gamma_4\,\bar{\phi}_4 - \gamma_2\,\bar{\psi}_1 + \gamma_3\,\bar{\phi}_1$$

$$\hat{\psi}(z_3) = \gamma_1\,\bar{\psi}_3 - \gamma_2\,\bar{\phi}_4 + \gamma_3\,\bar{\psi}_4 + \gamma_4\,\bar{\psi}_1 + \gamma_2\,\bar{\phi}_2 + \gamma_3\,\bar{\psi}_2$$

$$\hat{\phi}(z_3) = \gamma_1\,\bar{\phi}_3 + \gamma_2\,\bar{\psi}_4 + \gamma_3\,\bar{\phi}_4 + \gamma_4\,\bar{\phi}_1 - \gamma_2\,\bar{\psi}_2 + \gamma_3\,\bar{\phi}_2$$

$$\hat{\psi}(z_4) = \gamma_1\,\bar{\psi}_4 - \gamma_2\,\bar{\phi}_1 + \gamma_3\,\bar{\psi}_1 + \gamma_4\,\bar{\psi}_2 + \gamma_2\,\bar{\phi}_3 + \gamma_3\,\bar{\psi}_3$$

$$\hat{\phi}(z_4) = \gamma_1\,\bar{\phi}_4 + \gamma_2\,\bar{\psi}_1 + \gamma_3\,\bar{\phi}_1 + \gamma_4\,\bar{\phi}_2 - \gamma_2\,\bar{\psi}_3 + \gamma_3\,\bar{\phi}_3.$$

To solve for the unknown nodal values of $\bar{\psi}_1$, $\bar{\psi}_2$, $\bar{\psi}_3$, and $\bar{\psi}_4$, two methods are possible. Determine a case I matrix system such as given in (14) or determine a case II matrix system such as given in (15).

The case I matrix system sets the known nodal values equal to the approximation function values by forcing

Case I

$$\begin{Bmatrix} \bar{\phi}_1 \\ \bar{\phi}_2 \\ \bar{\phi}_3 \\ \bar{\psi}_4 \end{Bmatrix} \equiv \begin{Bmatrix} \hat{\phi}(z_1) \\ \hat{\phi}(z_2) \\ \hat{\phi}(z_3) \\ \hat{\psi}(z_4) \end{Bmatrix}$$

which gives

$$\begin{Bmatrix} \bar{\phi}_1 \\ \bar{\phi}_2 \\ \bar{\phi}_3 \\ \bar{\psi}_4 \end{Bmatrix} = \begin{bmatrix} 0 & \gamma_2 & 0 & \gamma_3 \\ -\gamma_2 & 0 & \gamma_2 & \gamma_4 \\ 0 & -\gamma_2 & 0 & \gamma_3 \\ \gamma_3 & \gamma_4 & \gamma_3 & 0 \end{bmatrix} \begin{Bmatrix} \bar{\psi}_1 \\ \bar{\psi}_2 \\ \bar{\psi}_3 \\ \bar{\phi}_4 \end{Bmatrix} + \begin{Bmatrix} \gamma_1 \bar{\phi}_1 + \gamma_3 \bar{\phi}_2 + \gamma_4 \bar{\phi}_3 - \gamma_2 \bar{\psi}_4 \\ \gamma_1 \bar{\phi}_2 + \gamma_3 \bar{\phi}_3 + \gamma_3 \bar{\phi}_1 \\ \gamma_1 \bar{\phi}_3 + \gamma_2 \bar{\psi}_4 + \gamma_4 \bar{\phi}_1 + \gamma_3 \bar{\phi}_2 \\ -\gamma_2 \bar{\phi}_1 + \gamma_2 \bar{\phi}_3 + \gamma_1 \bar{\psi}_4 \end{Bmatrix}$$

or simply

$$\begin{bmatrix} 0 & \gamma_2 & 0 & \gamma_3 \\ -\gamma_2 & 0 & \gamma_2 & \gamma_4 \\ 0 & -\gamma_2 & 0 & \gamma_3 \\ \gamma_3 & \gamma_4 & \gamma_3 & 0 \end{bmatrix} \begin{Bmatrix} \bar{\psi}_1 \\ \bar{\psi}_2 \\ \bar{\psi}_3 \\ \bar{\phi}_4 \end{Bmatrix} = \begin{Bmatrix} 0 \\ \frac{1}{4} + (\ln \sqrt{2})/\pi \\ 0 \\ \frac{1}{4} - (\ln \sqrt{2})/\pi \end{Bmatrix}$$

which has the solution nodal values $\{\bar{\psi}_1, \bar{\psi}_2, \bar{\psi}_3, \bar{\phi}_4\} = \{0, 0, 2, 0\}$.

Similarly, the case II matrix system sets the unknown nodal values equal to the approximation function values by forcing

Case II

$$\begin{Bmatrix} \bar{\psi}_1 \\ \bar{\psi}_2 \\ \bar{\psi}_3 \\ \bar{\phi}_4 \end{Bmatrix} \equiv \begin{Bmatrix} \hat{\psi}(z_1) \\ \hat{\psi}(z_2) \\ \hat{\psi}(z_3) \\ \hat{\phi}(z_4) \end{Bmatrix}$$

which gives

$$\begin{Bmatrix} \bar{\psi}_1 \\ \bar{\psi}_2 \\ \bar{\psi}_3 \\ \bar{\phi}_4 \end{Bmatrix} = \begin{bmatrix} \gamma_1 & \gamma_3 & \gamma_4 & \gamma_2 \\ \gamma_3 & \gamma_1 & \gamma_3 & 0 \\ \gamma_4 & \gamma_3 & \gamma_1 & -\gamma_2 \\ \gamma_2 & 0 & -\gamma_3 & \gamma_1 \end{bmatrix} \begin{Bmatrix} \bar{\psi}_1 \\ \bar{\psi}_2 \\ \bar{\psi}_3 \\ \bar{\phi}_4 \end{Bmatrix} + \begin{bmatrix} -\gamma_2 \bar{\phi}_2 + \gamma_3 \bar{\psi}_4 \\ -\gamma_2 \bar{\phi}_3 + \gamma_4 \bar{\psi}_4 + \gamma_2 \bar{\phi}_1 \\ \gamma_3 \bar{\psi}_4 + \gamma_2 \bar{\phi}_2 \\ \gamma_3 \bar{\phi}_1 + \gamma_4 \bar{\phi}_2 + \gamma_3 \bar{\phi}_3 \end{bmatrix}$$

or simply

$$\begin{bmatrix} (\gamma_1 - 1) & \gamma_3 & \gamma_4 & \gamma_2 \\ \gamma_3 & (\gamma_1 - 1) & \gamma_3 & 0 \\ \gamma_4 & \gamma_3 & (\gamma_1 - 1) & -\gamma_2 \\ \gamma_2 & 0 & -\gamma_3 & (\gamma_1 - 1) \end{bmatrix} \begin{Bmatrix} \bar{\psi}_1 \\ \bar{\psi}_2 \\ \bar{\psi}_3 \\ \bar{\phi}_4 \end{Bmatrix} = \begin{Bmatrix} (\ln 2)/\pi \\ \frac{1}{4} - (\ln 2)/(2\pi) \\ -\frac{1}{2} \\ -\frac{1}{4} - (\ln 2)/(2\pi) \end{Bmatrix}$$

which has the solution nodal values of

$$\{\bar{\psi}_1, \bar{\psi}_2, \bar{\psi}_3, \bar{\phi}_4\} = \{0, 0, 2, 0\}.$$

Because the assumed basis functions for $\phi(\zeta)$ and $\psi(\zeta)$ (with $\zeta \in \Gamma$) are of the same polynomial order as the solution to the boundary value problem, then the approximation function $\hat{\omega}(z)$ must be the exact solution, and $\hat{\omega}(z) = \omega(z)$ for $z \in \Omega \cup \Gamma$.

From this example, the limiting value of $\hat{\omega}(z_j^-)$ as z_j^- approaches node z_j with z_j^- in the interior of Ω is useful in determining nodal equations. Therefore, a general relationship is needed to evaluate this limiting value.

Consider a linear trial function between successive nodes (z_1, z_2) and (z_2, z_3). Then for $z_0 \in \Omega$ and $\zeta_0 \notin \Gamma$

$$2\pi \mathrm{i}\, \hat{\omega}(z_0) = \int_{z_1}^{z_2} \frac{G_1(\zeta)\, d\zeta}{\zeta - z_0} + \int_{z_2}^{z_3} \frac{G_1(\zeta)\, d\zeta}{\zeta - z_0} + \sum_{z_i \neq z_1, z_2} \int_{z_i}^{z_{i+1}} \frac{G_1(\zeta)\, d\zeta}{\zeta - z_0} \tag{17}$$

Define

$$I = 2\pi \mathrm{i}\, \hat{\omega}(z_0) - \int_{z_1}^{z_2} \frac{G_1(\zeta)\, d\zeta}{\zeta - z_0} - \int_{z_2}^{z_3} \frac{G_1(\zeta)\, d\zeta}{\zeta - z_0}. \tag{18}$$

For the linear basis function assumptions, using (10)

$$\begin{aligned} I = 2\pi \mathrm{i}\, \hat{\omega}(z_0) &- \left[\bar{\omega}_2 - \bar{\omega}_1 + \bar{\omega}_1 \left(\frac{z_2 - z_0}{z_2 - z_1}\right) \ln \left(\frac{z_2 - z_0}{z_1 - z_0}\right) - \bar{\omega}_2 \left(\frac{z_1 - z_0}{z_2 - z_1}\right) \ln \left(\frac{z_2 - z_0}{z_1 - z_0}\right)\right] \\ &- \left[\bar{\omega}_3 - \bar{\omega}_2 + \bar{\omega}_2 \left(\frac{z_3 - z_0}{z_3 - z_2}\right) \ln \left(\frac{z_3 - z_0}{z_2 - z_0}\right) - \bar{\omega}_3 \left(\frac{z_2 - z_0}{z_3 - z_2}\right) \ln \left(\frac{z_3 - z_0}{z_2 - z_0}\right)\right]. \end{aligned} \tag{19}$$

In the limit as z_0 approaches z_2 ($z_0 \in \Omega$, $z_2 \in \Gamma$),

$$\lim_{z_0 \to z_2} I = 2\pi \mathrm{i}\, \hat{\omega}(z_2) - \left[(\bar{\omega}_3 - \bar{\omega}_1) + \bar{\omega}_2 \ln \left(\frac{z_3 - z_2}{z_1 - z_2}\right)\right]. \tag{20}$$

Simplifying,

$$\lim_{z_0 \to z_2} I = 2\pi \mathrm{i}\, \hat{\omega}_2 - (\bar{\omega}_3 - \bar{\omega}_1) + \bar{\omega}_2 \ln \left|\frac{z_1 - z_2}{z_3 - z_2}\right| - \mathrm{i}(2\pi - \theta)\, \bar{\omega}_2. \tag{21}$$

Hence, the nodal equation for arbitrary node z_2 in Γ is

$$2\pi \mathrm{i}\, \hat{\omega}_2 = \bar{\omega}_3 - \bar{\omega}_1 + \bar{\omega}_2 \left[\ln \left|\frac{z_3 - z_2}{z_1 - z_2}\right| + \mathrm{i}(2\pi - \theta)\right] + \sum_{z_i \neq z_1, z_2} \int_{z_i}^{z_{i+1}} \frac{G_1(\zeta)\, d\zeta}{\zeta - z_0}. \tag{22}$$

7.3 The Analytical Function Defined by the Approximator $\hat{\omega}(z)$

The approximation function $\hat{\omega}(z)$ is defined by the contour integral

$$\hat{\omega}(z) = \frac{1}{2\pi \mathrm{i}} \int_{\Gamma} \frac{G_1(\zeta)\, d\zeta}{\zeta - z}, \quad z \notin \Gamma$$

where Γ is a simple closed contour with simply connected interior Ω, and $z \in \Omega$. Specifically, the usual case is, Γ is a simple closed contour composed of straight line segments and its interior, Ω, has no holes in it (i.e. multiply connected). This type of domain $\Omega \cup \Gamma$ is said to be a member of the set of all such domains, P.

Since $G_1(\zeta)$ is a continuous function for $\zeta \in \Gamma$, then the $\hat{\omega}(z)$ function is analytic on Ω. This important result is proved in the following.

Theorem. Let Γ be a simple closed contour with finite length L. Let $h(\zeta)$ be a continuous function on Γ. Then $\hat{\omega}(z)$ is analytic in the interior of Γ where $\hat{\omega}(z)$ is defined by the contour integral

$$\hat{\omega}(z) = \frac{1}{2\pi \mathrm{i}} \int_\Gamma \frac{h(\zeta)}{\zeta - z}\, d\zeta$$

and $\hat{\omega}'(z)$ is given by the integral

$$\hat{\omega}'(z) = \frac{1}{2\pi \mathrm{i}} \int_\Gamma \frac{h(\zeta)}{(\zeta - z)^2}\, d\zeta\,.$$

Proof. Since $|h(\zeta)|$ is continuous on the compact set Γ there exists a real number M such that

$$|h(\zeta)| \leqq M \quad \text{for } \zeta \in \Gamma\,.$$

Let $z_0 \notin \Gamma$ be arbitrary and set $R = d(z_0, \Gamma)$ the distance from z_0 to the contour Γ. To show that $\hat{\omega}(z)$ is analytic at z_0, it will be shown that the derivative exists at z_0.

Let z be any complex number with $|z - z_0| < R/2$. Then

$$\begin{aligned}
\frac{\hat{\omega}(z) - \hat{\omega}(z_0)}{z - z_0} &= \frac{1}{z - z_0}\frac{1}{2\pi \mathrm{i}} \int_\Gamma \left[\frac{h(\zeta)}{\zeta - z} - \frac{h(\zeta)}{\zeta - z_0}\right] d\zeta \\
&= \frac{1}{z - z_0}\frac{1}{2\pi \mathrm{i}} \int_\Gamma \frac{z - z_0}{(\zeta - z)(\zeta - z_0)}\, h(\zeta)\, d\zeta \\
&= \frac{1}{2\pi \mathrm{i}} \int_\Gamma \frac{h(\zeta)}{(\zeta - z)(\zeta - z_0)}\, d\zeta\,.
\end{aligned}$$

Expanding these equations gives

$$\begin{aligned}
\left|\frac{\hat{\omega}(z) - \hat{\omega}(z_0)}{z - z_0} - \frac{1}{2\pi \mathrm{i}} \int_\Gamma \frac{h(\zeta)}{(\zeta - z_0)^2}\, d\zeta\right| &= \frac{1}{2\pi}\left|\int_\Gamma \left[\frac{h(\zeta)}{(\zeta - z)(\zeta - z_0)} - \frac{h(\zeta)}{(\zeta - z_0)^2}\right] d\zeta\right| \\
&= \frac{1}{2\pi}\left|\int_\Gamma \frac{z - z_0}{(\zeta - z)(\zeta - z_0)^2}\, h(\zeta)\, d\zeta\right| \\
&\leqq \frac{1}{2\pi} \int_\Gamma \frac{|z - z_0|}{|\zeta - z|\,|\zeta - z_0|^2}\, |h(\zeta)|\, |d\zeta| \\
&\leqq \frac{1}{2\pi} \int_\Gamma \frac{|z - z_0|}{R(R/2)^2}\, M\, |d\zeta| \\
&\leqq \frac{2M|z - z_0|}{\pi R^3} \int_\Gamma |d\zeta| \\
&\leqq \frac{2ML}{\pi R^3}\, |z - z_0|\,.
\end{aligned}$$

Since the right side goes to 0 as $z \to z_0$,

$$\lim_{z \to z_0} \frac{\hat{\omega}(z) - \hat{\omega}(z_0)}{z - z_0} = \frac{1}{2\pi i} \int_{\Gamma} \frac{h(\zeta)}{(\zeta - z_0)^2} d\zeta$$

and $\hat{\omega}(z_0)$ is differentiable at any point z_0 not on the contour Γ. Thus, $\hat{\omega}(z)$ is analytic for z_0 in the interior of Γ.

From the above important theorem, it is concluded that the $\hat{\omega}(z)$ approximation function contains real and imaginary functions $\hat{\omega}(z) = \bar{\phi}(z) + i\,\hat{\psi}(z)$ such that $\bar{\phi}(z)$ and $\hat{\psi}(z)$ are both harmonic in Ω. If $\hat{\omega}(z)$ satisfies the boundary conditions which are specified from the function $\omega(z)$, where $\omega(z)$ is analytic on $\Omega \cup \Gamma \in P$, then $\hat{\omega}(z) = \omega(z)$ for $z \in \Omega \cup \Gamma$. However, generally $\hat{\omega}(z) \neq \omega(z)$ for $z \in \Gamma$ and an error function $e(z)$ exists on $\Omega \cup \Gamma$ where

$$e(z) = \omega(z) - \hat{\omega}(z), \quad z \in \Gamma \tag{23}$$

and $e(z)$ is not identically zero on $\Omega \cup \Gamma$. Thus, the numerical modeling objective is to reduce $e(z)$ on $\Omega \cup \Gamma$.

7.4 A Constant Boundary Element Method

A simpler modeling approach than determining a continuous global interpolation function on Γ is to assume that the integration contribution from each element Γ_j is simply

$$\int_{\Gamma_j} \frac{\omega(\zeta)\, d\zeta}{\zeta - z_0} = \bar{\omega}_j \int_{\Gamma_j} \frac{d\zeta}{\zeta - z} \tag{24}$$

where $\bar{\omega}_j$ is a specified value for element Γ_j. In this modeling approach, the nodal points are assumed located at midelement (see Fig. 7).

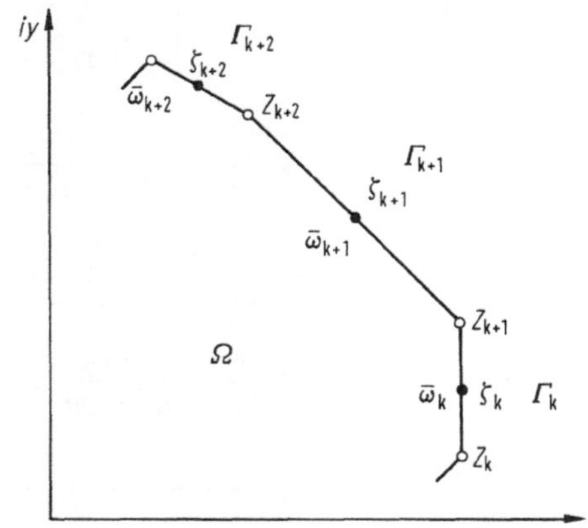

Fig. 7. Constant boundary element geometry. ○, element points; ● nodal point; ζ_k, midelement node on Γ_k

The nodal equation contributions are simply

$$\bar{\omega}_j \int_{\Gamma_j} \frac{d\zeta}{\zeta - \zeta_k} = \bar{\omega}_j \ln\left(\frac{z_{j+1} - \zeta_k}{z_j - \zeta_k}\right) \tag{25}$$

where ζ_k is the coordinate of node k. Letting $d(z_j, \zeta_k)$ be the usual distance function gives

$$\bar{\omega}_j \int \frac{d\zeta}{\zeta - \zeta_k} = \bar{\omega}_j \left[\ln\left(\frac{d(z_{j+1}, \zeta_k)}{d(z_j, \zeta_k)}\right) + \mathrm{i}\,\theta(j+1, j)\right] \tag{26}$$

where $\theta(j+1, j)$ is the central angle between coordinates z_{j+1}, z_j and ζ_k. Thus

$$\bar{\omega}_j \int_{\Gamma_j} \frac{d\zeta}{\zeta - \zeta_k} = \bar{\omega}_j\, h_j, \quad j \neq k. \tag{27}$$

The limiting value for $j = k$ is given by considering

$$\lim_{z \to \zeta_k} \bar{\omega}_k \int_{\Gamma_k} \frac{d\zeta}{\zeta - z} \tag{28}$$

where z is in the interior of Ω. But

$$\bar{\omega}_k \int_{\Gamma_k} \frac{d\zeta}{\zeta - z} = \bar{\omega}_k \ln\left(\frac{z_{k+1} - z}{z_k - z}\right) \tag{29}$$

where $\zeta_k = (z_{k+1} + z_k)/2$. Let $z = \zeta_k + \varepsilon_x + \mathrm{i}\,\varepsilon_y$, then

$$\bar{\omega}_k \int_{\Gamma_k} \frac{d\zeta}{\zeta - z} = \bar{\omega}_k \ln\left(\frac{(z_{k+1} - z_k)/2 - \varepsilon_x - \mathrm{i}\,\varepsilon_y}{(z_k - z_{k+1})/2 - \varepsilon_x - \mathrm{i}\,\varepsilon_y}\right)$$

and in the limit

$$\lim_{\substack{\varepsilon_x \to 0 \\ \varepsilon_y \to 0}} \int_{\Gamma_k} \frac{d\zeta}{\zeta - z} = \bar{\omega}_k \ln(-1) = \bar{\omega}_k\, \pi\, \mathrm{i}\,.$$

The constant element nodal equation is then

$$2\pi\,\mathrm{i}\,\hat{\omega}(\zeta_k) = \bar{\omega}_k\, \pi\, \mathrm{i} + \sum_{j \neq k} \bar{\omega}_j\, h_j; \quad k = 1, 2, \ldots, m \tag{30}$$

where it is assumed that m constant boundary elements are used in the model. The development of case I and case II matrix systems follows in an analogous fashion from Sect. 1.

7.5 The Complex Variable Boundary Element Method

The preceding discussions can be generalized to formulate a complex variable boundary element method, or simply the CVBEM.

Let $\Omega \cup \Gamma \in P$. Discretize Γ into m boundary elements Γ_j such that

$$\Gamma = \sum_{j=1}^{m} \Gamma_j$$

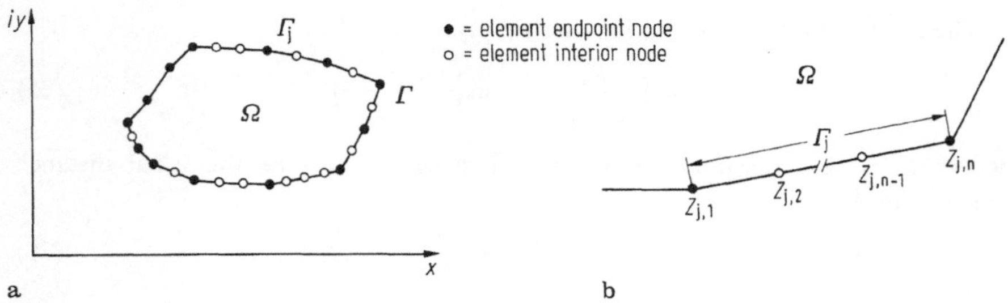

Fig. 8. **a** Boundary discretized into several n-node elements. **b** Detail of element Γ_j

and the intersection of $\Gamma_j \cap \Gamma_{j-1}$ is endpoint of Γ_j and Γ_{j-1}. On each element Γ_j define two or more evenly spaced nodal points. Number each boundary element sequentially from 1 to m along Γ in the counterclockwise direction. Similarly, number the nodal points sequentially from 1 as Γ is transversed in the counterclockwise direction.

A second nodal numbering scheme is as shown in Fig. 8 where the endpoints of an n-node element Γ_j are defined as $z_{j,1}$ and $z_{j,n}$.

Define on each element Γ_j a system of continuous basis functions $N_{j,k}(z)$ such that

$$N_{j,i}(z) = \begin{cases} 1, & z = z_{j,i} \\ 0, & z = z_{j,k}, \quad k \neq i \end{cases} \tag{31}$$

and $N_{j,k}(z) \equiv 0$ for $z \notin \Gamma_j$. The real valued interpolation functions $\alpha_{\phi_j}(z)$ and $\alpha_{\psi_j}(z)$ are defined on Γ_j by

$$\begin{aligned} \alpha_{\phi_j}(z) &= \sum_{k=1}^{n} N_{j,k}(z)\, \bar{\phi}_{j,k}, \quad z \in \Gamma_j, \\ \alpha_{\psi_j}(z) &= \sum_{k=1}^{n} N_{j,k}(z)\, \bar{\psi}_{j,k}, \quad z \in \Gamma_j \end{aligned} \tag{32}$$

where $\bar{\phi}_{j,k}$ and $\bar{\psi}_{j,k}$ are the specified nodal values at node $z_{j,k}$. For simplicity, the above function definitions can be written as

$$\alpha_{\phi_j}(z) = N_{j,k}\, \bar{\phi}_{j,k}$$

and

$$\alpha_{\psi_j}(z) = N_{j,k}\, \bar{\psi}_{j,k} \tag{33}$$

Although it is assumed that the same basis functions be used for both $\alpha_{\phi_j}(z)$ and $\alpha_{\psi_j}(z)$, this is not required. However, the continuity definition of the basis functions is required.

Define a continuous global function $G(z)$ on Γ by

$$G(z) = \sum_{j=1}^{m} (N_{j,k}\, \bar{\phi}_{j,k} + \mathrm{i}\, N_{j,k}\, \bar{\psi}_{j,k}), \quad z \in \Gamma_j \tag{34}$$

where

$$G(z_{j,k}) = \bar{\phi}_{j,k} + \mathrm{i}\, \bar{\psi}_{j,k}\,.$$

An approximation function $\hat{\omega}(z)$ can be defined on $\Omega \cup \Gamma$ by

$$\hat{\omega}(z) = \frac{1}{2\pi i} \int_\Gamma \frac{G(\zeta)\, d\zeta}{\zeta - z}, \quad z \in \Omega, \quad z \notin \Gamma. \tag{35}$$

Because $G(z)$ is a continuous function on Γ, then $\hat{\omega}(z)$ is an analytic function in Ω. Therefore, the function $\hat{\omega}(z)$ can be split into the real and imaginary components

$$\hat{\omega}(z) = \hat{\phi}(z) + i\,\hat{\psi}(z)$$

where $\hat{\phi}(z)$ and $\hat{\psi}(z)$ are real valued functions which satisfy the Laplace equation on Ω.

If $\phi(x, y)$ and $\psi(x, y)$ are known continuously on Γ and the boundary basis functions are chosen on each Γ_j such that $G(z) = \phi(z) + i\,\psi(z)$ on Γ, then $\hat{\omega}(z) = \omega(z)$ on $\Omega \cup \Gamma$. Generally, however, $\phi(z)$ and $\psi(z)$ are known only on portions of Γ such that $\phi(z)$ is given on Γ_a and $\psi(z)$ is given on Γ_b where $\Gamma_a \cup \Gamma_b = \Gamma$. Then half of the specified nodal values are unknown on Γ. Two methods of estimating values to be used for the unknown specific nodal values are described in the following:

Case I. Suppose there are N nodal points on Γ, and N_a values of $\phi(z)$ and N_b values of $\psi(z)$ are known on Γ where $N_a + N_b = N$ and $N_a, N_b > 0$. A set of values for the unknown specified nodal values can be developed by defining the matrix system

$$\begin{Bmatrix} \bar{\phi}_1 \\ \bar{\phi}_2 \\ \vdots \\ \bar{\phi}_{N_a} \\ \bar{\psi}_{N_a+1} \\ \vdots \\ \bar{\psi}_N \end{Bmatrix} = \begin{Bmatrix} \hat{\phi}_1 \\ \hat{\phi}_2 \\ \vdots \\ \hat{\phi}_{N_a} \\ \hat{\psi}_{N_a+1} \\ \vdots \\ \hat{\psi}_N \end{Bmatrix} \tag{36}$$

where $\bar{\phi}_1$ and $\hat{\phi}_1$ indicate specified and approximation functional nodal values respectively. Solving the above matrix system will determine values for the unknown specified nodal values $\{\bar{\psi}_1, \ldots, \bar{\psi}_{N_a}, \bar{\phi}_{N_a+1}, \ldots, \bar{\phi}_N\}$.

Case II. Analogous to case I, the unknown specified nodal values can be estimated by defining

$$\begin{Bmatrix} \bar{\psi}_1 \\ \bar{\psi}_2 \\ \vdots \\ \bar{\psi}_{N_a} \\ \bar{\phi}_{N_a+1} \\ \vdots \\ \bar{\phi}_N \end{Bmatrix} = \begin{Bmatrix} \hat{\psi}_1 \\ \hat{\psi}_2 \\ \vdots \\ \hat{\psi}_{N_a} \\ \hat{\phi}_{N_a+1} \\ \vdots \\ \hat{\phi}_N \end{Bmatrix} \tag{37}$$

This matrix system will also determine a set of values for $\{\bar{\psi}_1, \ldots, \bar{\psi}_{N_a}, \bar{\phi}_{N_a+1}, \ldots, \bar{\phi}_N\}$.

Using either set of values for the unknown specified nodal values on Γ will completely define an approximation function $\hat{\omega}(z)$ which is analytic on Ω. A property of $\hat{\omega}(z)$ is as follows:

Case I. The $\hat{\omega}(z)$ function values agree with the known specified nodal values.

Case II. The $\hat{\omega}(z)$ function values equal the estimated values for the unknown specified nodal values.

It can be seen that the Case I and Case II approximation functions differ and that a better notation for these two approximators is $\hat{\omega}_{\mathrm{I}}(z)$ and $\hat{\omega}_{\mathrm{II}}(z)$, respectively. A weighting factor $0 \leqq \eta \leqq 1$ can be introduced such that a more complete approximation function $\hat{\omega}(\eta)$ is defined by

$$\hat{\omega}(\eta) = \hat{\omega}_{\mathrm{I}}(1-\eta) + \hat{\omega}_{\mathrm{II}}(\eta). \tag{38}$$

Obviously, $\hat{\omega}(\eta)$ is also analytic on Ω.

Generally, it is preferable to use $\hat{\omega}(1)$ where all basis functions are linear polynomials on each Γ_j. Such a model utilizes m nodes and m boundary elements on Γ, and is very suitable for modeling error analysis and subsequent model refinements.

7.6 Approximation Error from the CVBEM

Let $\hat{\omega}(z)$ be an approximation function defined on $\Omega \cup \Gamma \in P$ such that

$$\hat{\omega}(z) = \frac{1}{2\pi\mathrm{i}} \int_\Gamma \frac{G(\zeta)\,d\zeta}{\zeta - z}; \quad z \in \Omega, \quad z \notin \Gamma$$

where $G(\zeta)$ is a continuous global trial function defined on Γ. Then $\hat{\omega}(z)$ is analytic in Ω. Let Γ^- be a simply connected contour which is a constant distance δ^* from boundary Γ such that Γ^- lies in the interior of Ω. Let $\omega(z) = \phi(z) + \mathrm{i}\,\psi(z)$ be an analytic function defined on $\Omega \cup \Gamma$ such that $\phi(z)$ and $\psi(z)$ solve the subject boundary value problem. The numerical modeling objective is to determine a $\hat{\omega}(z)$ such that for some $\varepsilon > 0$

$$|\omega(z) - \hat{\omega}(z)| < \varepsilon, \quad z \in \Gamma^-.$$

The boundary conditions which are part of the problem are assumed to be known of

$$\phi(z), \quad \text{for } z \in \Gamma_a$$

and

$$\psi(z), \quad \text{for } z \in \Gamma_b$$

where $\Gamma_a \cup \Gamma_b = \Gamma$, and Γ_a and Γ_b both have finite length. Both Γ_a and Γ_b may be composed of a finite number of line segments each of nonzero length.

In order to indicate what harmonic function $\phi(z)$ or $\psi(z)$ is known at a certain point $z \in \Gamma$, the following notation is used for $\omega(z)$:

$$\omega(z) = \Delta \xi_k + \Delta \xi_u \tag{39}$$

where the symbol Δ indicates

$$\Delta = \begin{cases} 1; & \text{if} \quad \xi_k \text{ (or } \xi_u\text{) are } \phi(z) \text{ values} \\ \mathrm{i}; & \text{if} \quad \xi_k \text{ (or } \xi_u\text{) are } \psi(z) \text{ values.} \end{cases}$$

The subscripts k and u indicate whether the nodal value is known or unknown, respectively.

For any z interior of Γ, there exists a Γ^- such that z is interior of Γ^- and

$$\omega(z) = \frac{1}{2\pi \mathrm{i}} \int_{\Gamma^-} \frac{\omega(\zeta)\, d\zeta}{\zeta - z}. \tag{40}$$

The approximation function $\hat{\omega}(z)$ is analytic on Ω and is therefore analytic on Γ^- and its interior, Ω^-. Then

$$\hat{\omega}(z) = \frac{1}{2\pi \mathrm{i}} \int_{\Gamma^-} \frac{\hat{\omega}(\zeta)\, d\zeta}{\zeta - z} + \frac{1}{2\pi \mathrm{i}} \int_{\Gamma} \frac{G(\zeta)\, d\zeta}{\zeta - z}. \tag{41}$$

An error function $e(z)$ is

$$e(z) = \omega(z) - \hat{\omega}(z), \quad z \in \Omega^- \cup \Gamma^-. \tag{42}$$

Then $e(z)$ is analytic on $\Omega^- \cup \Gamma^-$ and

$$\int_{\Gamma^-} e(\zeta)\, d\zeta = 0, \quad \zeta \in \Omega^- \cup \Gamma^-. \tag{43}$$

Let z_j^- be the closest point on Γ^- to $z_j \in \Gamma$. Using (39),

$$\hat{\omega}(z_j^-) = \bar{\phi}(z_j^-) + \mathrm{i}\, \hat{\psi}(z_j^-) \tag{44}$$

or

$$\hat{\omega}(z_j^-) = \Delta \hat{\xi}_k(z_j^-) + \Delta \hat{\xi}_u(z_j^-). \tag{45}$$

Then

$$e(z_j^-) = \Delta \xi_k(z_j^-) - \Delta \hat{\xi}_k(z_j^-) + \Delta \xi_u(z_j^-) - \Delta \hat{\xi}_u(z_j^-). \tag{46}$$

Equation (46) can be rewritten as

$$e(z_j^-) = \Delta\, (\xi_k(z_j^-) - \hat{\xi}_k(z_j^-)) + \Delta\, (\xi_u(z_j^-) - \hat{\xi}_u(z_j^-))$$

where the component $\Delta\, (\xi_k(z_j^-) - \hat{\xi}_k(z_j^-))$ is known from the given specified nodal values on Γ and from the $\hat{\omega}(z)$ approximation function nodal values. Thus for $\Delta\, (\xi_k(z_j^-) - \hat{\xi}_k(z_j^-))$ known continuously on Γ^-,

$$\int_{\Gamma^-} \Delta\, (\xi_k(\zeta) - \hat{\varepsilon}_k(\zeta))\, d\zeta = \int_{\Gamma^-} \Delta\, (\hat{\xi}_u(\zeta) - \xi_u(\zeta))\, d\zeta.$$

Let $\delta > 0$ be a constant distance between Γ and Γ^- such that $\min|z - z^-| = \delta$ for $z \in \Gamma$ and $z^- \in \Gamma^-$, then

$$\lim_{\delta \to 0} \int_{\Gamma^-} \Delta\,(\xi_k(\zeta) - \hat{\xi}_k(\zeta))\, d\zeta = \int_{\Gamma} \Delta\,(\xi_k(\zeta) - \hat{\xi}_k(\zeta))\, d\zeta \tag{47}$$

and

$$\lim_{\delta \to 0} \int_{\Gamma^-} \Delta\,(\hat{\xi}_u(\zeta) - \xi_u(\zeta))\, d\zeta = \int_{\Gamma} \Delta\,(\hat{\xi}_u(\zeta) - \xi_u(\zeta))\, d\zeta\,. \tag{48}$$

Thus an integrated error measure is determined on Γ which relates the integrated error of the unknown nodal values to the integrated error of the known nodal values,

$$\int_{\Gamma} \Delta\,(\xi_k(\zeta) - \hat{\xi}_k(\zeta))\, d\zeta = \int_{\Gamma} \Delta\,(\hat{\xi}_u(\zeta) - \xi_u(\zeta))\, d\zeta\,. \tag{49}$$

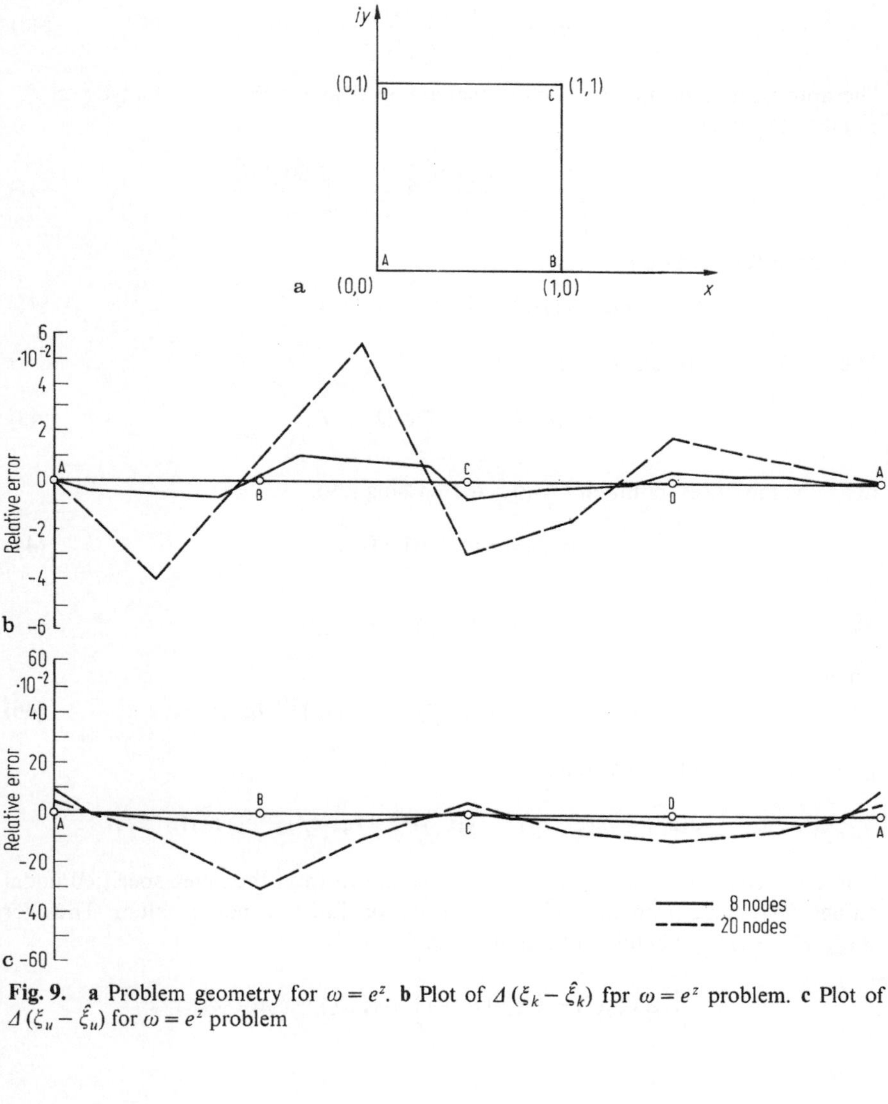

Fig. 9. **a** Problem geometry for $\omega = e^z$. **b** Plot of $\Delta\,(\xi_k - \hat{\xi}_k)$ fpr $\omega = e^z$ problem. **c** Plot of $\Delta\,(\xi_u - \hat{\xi}_u)$ for $\omega = e^z$ problem

7.7 A CVBEM Modeling Strategy to Reduce Approximation Error

In Sect. 5 the error $e(z)$ function was examined and it was concluded that an integrated error of the unknown nodal values can be calculated on Γ.

A strategy to reduce modeling error is as follows:

Step (1): Use a case II matrix system to estimate the unknown nodal variables. This system results in an approximator $\hat{\omega}(z)$ such that generally $\Delta(\xi_k(z_j) - \hat{\xi}_k(z_j)) \neq 0$,

Step (2): Using the defined approximator $\hat{\omega}(z)$, determine the $\Delta\hat{\xi}(z_j)$ values, $j = 1, 2, \ldots, m$,

Step (3): Determine $\Delta(\xi_k(z_j) - \hat{\xi}_k(z_j))$, $j = 1, 2, \ldots, m$,

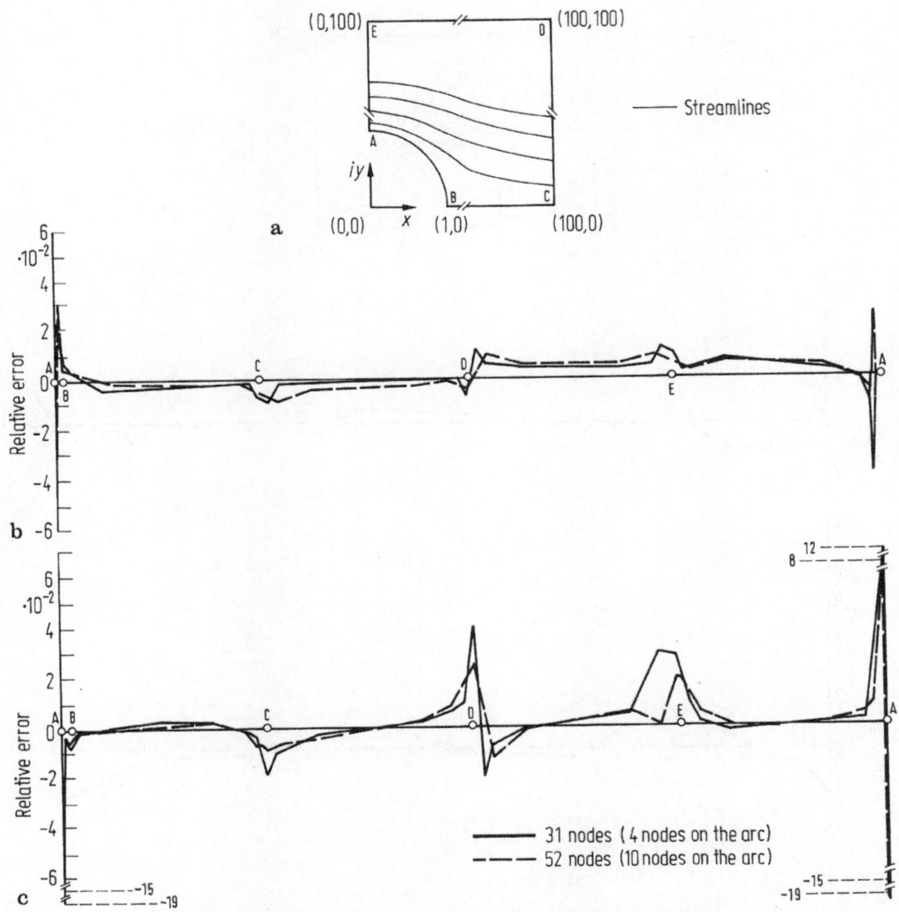

Fig. 10. **a** Problem geometry for $\omega = Z + Z^{-1}$ (ideal fluid over a cylinder). **b** Plot of $\Delta(\xi_k - \hat{\xi}_k)$ for $\omega = Z + Z^{-1}$ problem. **c** Plot of $\Delta(\xi_u - \hat{\xi}_u)$ for $\omega = Z + Z^{-1}$ problem

Step (4): Locate regions $\mathscr{D}_e$ on Γ where $\Delta\,(\xi_k(z_j) - \hat{\xi}\,(z_j))$ is large,
Step (5): Add nodal points to each $\mathscr{D}_e$,
Step (6): Return to step (1).

This modeling strategy is essentially an adaptive integration scheme which attempts to minimize $\Delta\,(\xi_k(z) - \hat{\xi}_k(z))$ on Γ by reducing the integration error on each $\mathscr{D}_e$ due to poor match of $G(z)$ to $\omega(z)$. To illustrate this procedure, several sample problems will be studied where the solution to the boundary value problem, $\omega(z)$, is known.

Example 2. Consider the simply connected domain

$$\Omega\colon\ \{z\,|\,0 \leqq x \leqq 1, 0 \leqq y \leqq 1\}$$

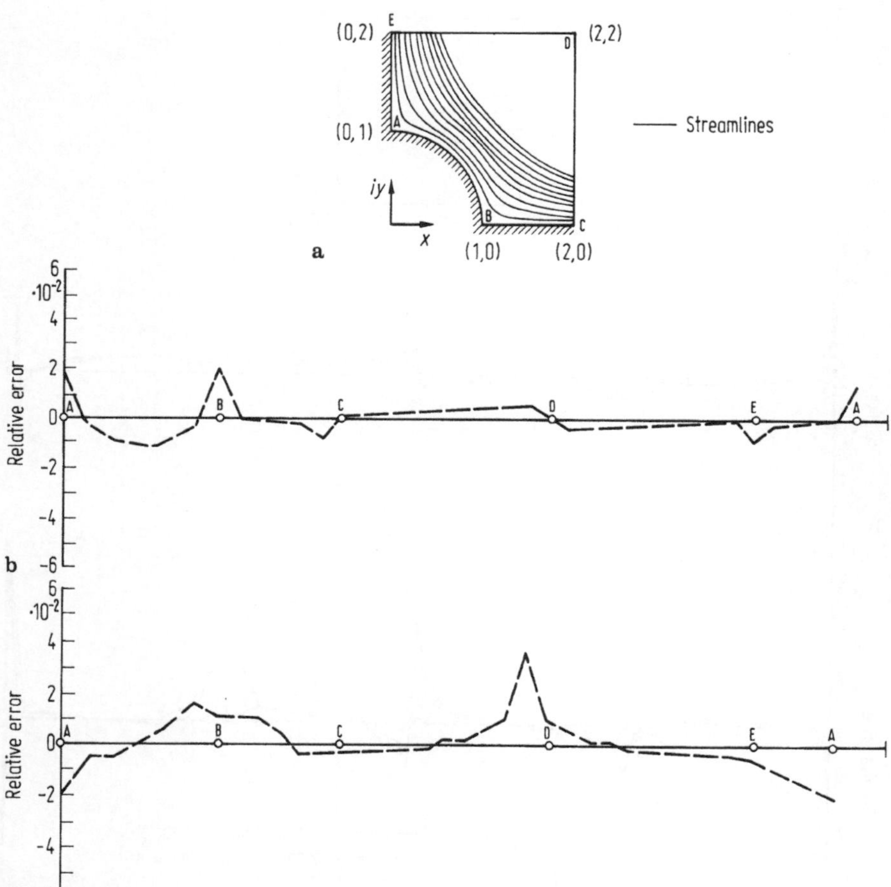

Fig. 11. **a** Problem geometry for $\omega = Z^2 + Z^{-2}$ (ideal fluid around a cylindrical corner). **b** Plot of $\Delta\,(\xi_k - \hat{\xi}_k)$ for $\omega = Z^2 + Z^{-2}$ problem. **c** Plot of $\Delta\,(\xi_u - \hat{\xi}_u)$ for $\omega = Z^2 + Z^{-2}$ problem

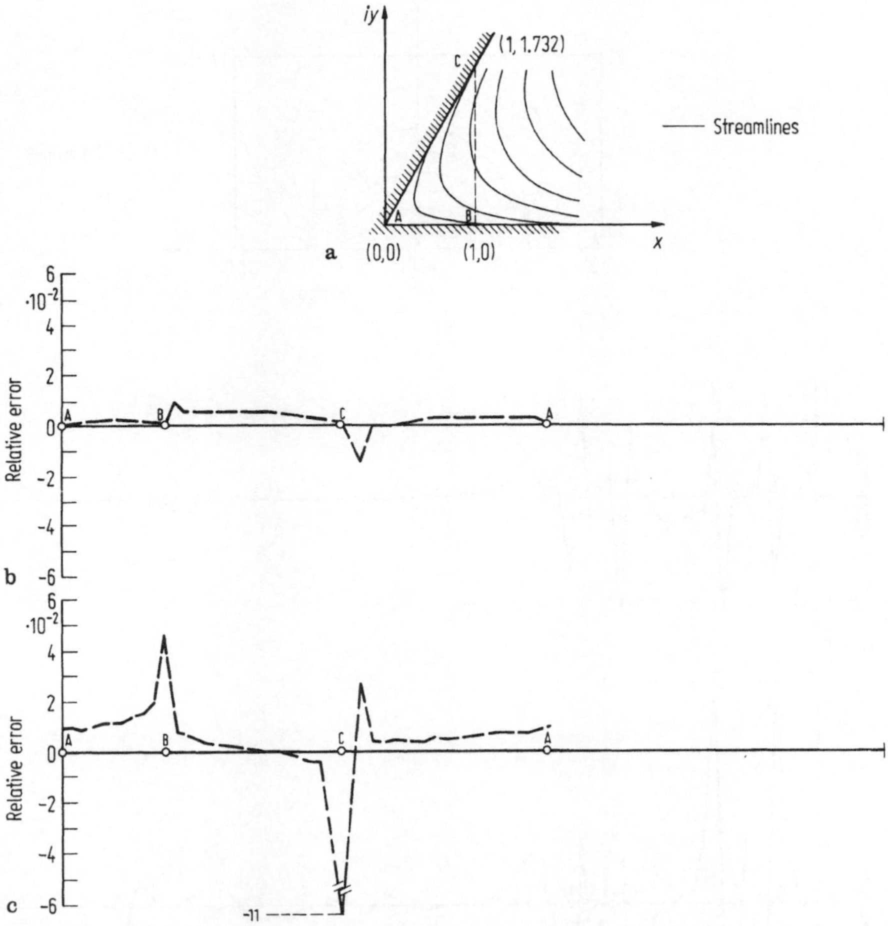

Fig. 12. **a** Problem geometry for $\omega = Z^3$ (ideal fluid around an angular region). **b** Plot of $\Delta(\xi_k - \hat{\xi}_k)$ for $\omega = Z^3$ problem. **c** Plot of $\Delta(\xi_u - \hat{\xi}_u)$ for $\omega = Z^3$ problem

and the analytic function $\omega = e^z$. Figure 9a shows the problem geometry (uniform nodal point placement). Several trials of approximation were made by adding nodal points according to Sect. 5. Figure 9b shows the relative error plots of $\Delta(\xi_k - \hat{\xi}_k)$ along Γ; and Fig. 9c shows the corresponding plot for $\Delta(\xi_u - \hat{\xi}_u)$ along Γ. From the figures, the approximation function $\hat{\omega}(z)$ better approximates $\omega(z) = e^z$ along Γ as the number of nodes are increased.

Example 3. Ideal fluid flow around a cylinder has the analytic model of $\omega(z) = A(z + z^{-1})$. Figure 10 shows the CVBEM results in modeling this problem.

Examples 4, 5, and 6. Ideal fluid flow around a cylindrical corner, around an angular region, and between a source and sink, are shown in the Figs. 11 through 13. Similar to the previous applications, plots of the known and unknown boundary condition CVBEM error distributions are shown.

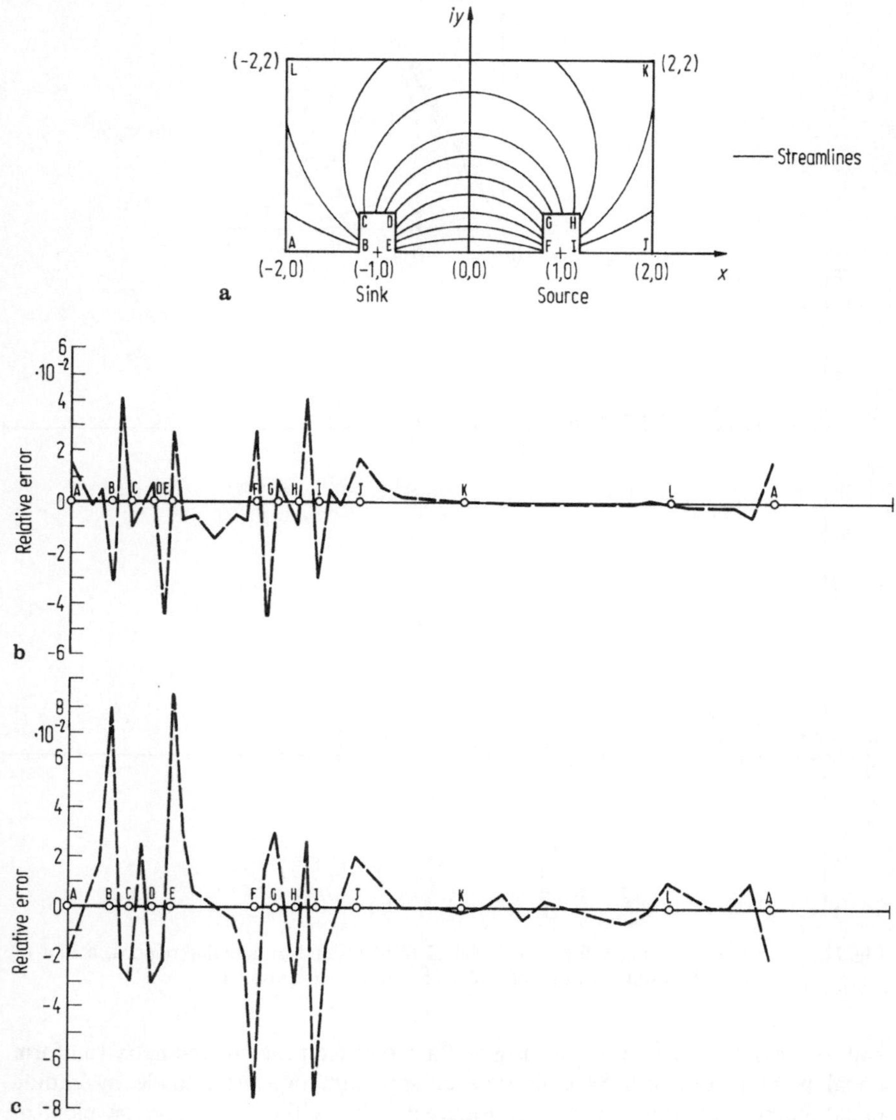

Fig. 13. **a** Problem geometry for $\omega = \log \dfrac{Z-1}{Z+1}$ (source and sink of equal strength). **b** Plot of $\Delta(\xi_k - \hat{\xi}_k)$ for $\omega = \log \dfrac{Z-1}{Z+1}$ problem. **c** Plot of $\Delta(\xi_u - \hat{\xi}_u)$ for $\omega = \log \dfrac{Z-1}{Z+1}$ problem

7.8 Expansion of the H_k Approximation Function

In this section, the CVBEM H_k approximation function $\hat{\omega}_k(z)$ will be expanded into the form

$$\hat{\omega}_k(z) = \sum_j P_j^k(z) \ln(z_j - z) + R_k(z) \tag{50}$$

where $P_j^k(z)$ is an order k complex polynomial on element Γ_j, and $R_k(z)$ is an order k reference complex polynomial. Should the solution to the boundary value problem $\omega(z)$, be an order k (or less) polynomial, then necessarily

$$\hat{\omega}_k(z) = \omega(z) = P_j^k(z). \tag{51}$$

The expansion of the H_1 and H_0 approximation functions will be developed first, with the results then generalized to the arbitrary H_k approximation function which is based on order k polynomial basis functions on each boundary element.

Let $\Omega \cup \Gamma \in P$ and $G_1(\zeta)$ be a linear global trial function. The CVBEM develops an H_1 approximation function $\hat{\omega}_1(z)$ by

$$2\pi \mathrm{i}\, \hat{\omega}_1(z) = \sum_{j=1}^{m} \int_{\Gamma_j} \frac{G_1(\zeta)\, d\zeta}{\zeta - z} \tag{52}$$

where

$$\Gamma = \bigcup_{j=1}^{m} \Gamma_j .$$

Solving each of the boundary element integrals gives

$$2\pi \mathrm{i}\, \hat{\omega}_1(z) = \sum_{j=1}^{m} (\bar{\omega}_{j+1} - \bar{\omega}_j) + \sum_{j=1}^{m} \left[\bar{\omega}_{j+1}\left(\frac{z - z_j}{z_{j+1} - z_j}\right) - \bar{\omega}_j\left(\frac{z - z_{j+1}}{z_{j+1} - z_j}\right)\right] h_j \tag{53}$$

where $h_j = \ln(z_{j+1} - z) - \ln(z_j - z)$, and $z \in \Omega$ but $z \notin \Gamma$.

Rewriting (53), the first summation term is zero and

$$2\pi \mathrm{i}\, \hat{\omega}_1(z) = \sum_{j=1}^{m} \left[\bar{\omega}_{j+1}\left(\frac{z - z_j}{z_{j+1} - z_j}\right) - \bar{\omega}_j\left(\frac{z - z_{j+1}}{z_{j+1} - z_j}\right)\right] [\ln(z_{j+1} - z) - \ln(z_j - z)] . \tag{54}$$

The transcendental $\ln(z_j - z)$ function is multivalued with the principal value assumed given by

$$\ln(z_j - z) = \ln|z_j - z| + \mathrm{i} \arg(z_j - z) \tag{55}$$

where $0 \leqq \arg(z_j - z) < 2\pi$, and $z_j \neq z$. Thus point $z \in \Omega$ is a branch point of the $\ln(z_j - z)$ function. For convenience, let the branch cut on $\ln(z_j - z)$ pass from point z through nodal point $z_1 \in \Gamma$. Then as the boundary integral on Γ is solved in the positive sense, it is noted that in the evaluation of the h_j terms,

$$\ln(z_1 - z) = \begin{cases} \ln|z_1 - z|, & \text{when evaluating the integral on element } \Gamma_1 \\ \ln|z_1 - z| + 2\pi \mathrm{i}, & \text{when evaluating the integral on element } \Gamma_m \end{cases} \tag{56}$$

Using the results of (56), Eq. (54) can be written as

$$2\pi\,\mathrm{i}\,\hat{\omega}_1(z) = -\left[\bar{\omega}_2\left(\frac{z-z_1}{z_2-z_1}\right) - \bar{\omega}_1\left(\frac{z-z_2}{z_2-z_1}\right)\right]\ln(z_1-z)$$

$$+\sum_{j=2}^{m}\left[-\bar{\omega}_{j+1}\left(\frac{z-z_j}{z_{j+1}-z_j}\right) + \bar{\omega}_j\left(\frac{z-z_{j+1}}{z_{j+1}-z_j}\right)\right.$$

$$\left.+\bar{\omega}_j\left(\frac{z-z_{j-1}}{z_j-z_{j-1}}\right) - \bar{\omega}_{j-1}\left(\frac{z-z_j}{z_j-z_{j-1}}\right)\right]\ln(z_j-z)$$

$$+\left[\bar{\omega}_1\left(\frac{z-z_m}{z_1-z_m}\right) - \bar{\omega}_m\left(\frac{z-z_1}{z_1-z_m}\right)\right][\ln(z_1-z)+2\pi\,\mathrm{i}]\,. \tag{57}$$

In (57), the $2\pi\,\mathrm{i}$ is added to the $\ln(z_1-z)$ term due to the complete circuit on Γ around branch point z. The above expansion is simplified by noting

$$\bar{\omega}_j\left(\frac{z-z_{j-1}}{z_j-z_{j-1}}\right) = \bar{\omega}_j\left(\frac{z-z_j}{z_j-z_{j-1}}\right) + \bar{\omega}_j \tag{58}$$

and

$$\bar{\omega}_j\left(\frac{z-z_{j+1}}{z_{j+1}-z_j}\right) = \bar{\omega}_j\left(\frac{z-z_j}{z_{j+1}-z_j}\right) - \bar{\omega}_j\,. \tag{59}$$

Substituting (58) and (59) into (57) and combining terms,

$$2\pi\,\mathrm{i}\,\hat{\omega}_1(z) = \left[\sum_{j=2}^{m}\left[\left(\frac{\bar{\omega}_j-\bar{\omega}_{j-1}}{z_j-z_{j-1}}\right) - \left(\frac{\bar{\omega}_{j+1}-\bar{\omega}_j}{z_{j+1}-z_j}\right)\right](z-z_j)\ln(z_j-z)\right]$$

$$-\left[\bar{\omega}_2\left(\frac{z-z_1}{z_2-z_1}\right) - \bar{\omega}_1\left(\frac{z-z_1}{z_2-z_1}\right) + \bar{\omega}_1\right]\ln(z_1-z)$$

$$+\left[\bar{\omega}_1\left(\frac{z-z_1}{z-z_m}\right) + \bar{\omega}_1 - \bar{\omega}_m\left(\frac{z-z_1}{z_1-z_m}\right)\right][\ln(z_1-z)+2\pi\,\mathrm{i}]\,. \tag{60}$$

Rearranging (60) gives the final form of the expansion

$$2\pi\,\mathrm{i}\,\hat{\omega}_1(z) = \left[\sum_{j=1}^{m}\left[\left(\frac{\bar{\omega}_j-\bar{\omega}_{j-1}}{z_j-z_{j-1}}\right) - \left(\frac{\bar{\omega}_{j+1}-\bar{\omega}_j}{z_{j+1}-z_j}\right)\right](z-z_j)\ln(z_j-z)\right]$$

$$+2\pi\,\mathrm{i}\left[\bar{\omega}_1 + \left(\frac{\bar{\omega}_1-\bar{\omega}_m}{z_1-z_m}\right)(z-z_1)\right]. \tag{61}$$

Thus, the H_1 approximation function can be written as the sum

$$\hat{\omega}_1(z) = \sum_{j=1}^{m}\gamma_j(z_j-z)\ln(z_j-z) + R_1(z) \tag{62}$$

where the γ_j are complex constants defined by

$$\gamma_j = \frac{1}{2\pi \mathrm{i}} \left[\left(\frac{\bar{\omega}_{j+1} - \bar{\omega}_j}{z_{j+1} - z_j} \right) - \left(\frac{\bar{\omega}_j - \bar{\omega}_{j-1}}{z_j - z_{j-1}} \right) \right] \tag{63}$$

and $R_1(z)$ is a reference polynomial (of order 1) given by

$$R_1(z) = \bar{\omega}_1 + \left(\frac{\bar{\omega}_1 - \bar{\omega}_m}{z_1 - z_m} \right) (z - z_1). \tag{64}$$

From the expansion of (62), it is easily shown that the limiting value of $\hat{\omega}_1(z)$ exists as z approaches node z_i

$$\lim_{z \to z_i} \hat{\omega}_1(z) = \left[\sum_{\substack{j=1 \\ j \neq i}}^{m} \gamma_j (z_j - z_i) \ln (z_j - z_i) \right] + R_1(z_i) \tag{65}$$

where for $j = i$ in (65)

$$\lim_{z \to z_i} \gamma_i (z_i - z) \ln (z_i - z) = 0. \tag{66}$$

Additionally, should $\omega(z)$ be a linear (or constant) polynomial on $\Omega \cup \Gamma$, then $\hat{\omega}(z)$ is exactly equal to $\omega(z)$.

In a similar fashion, the H_0 approximation function can be expanded as (for appropriate branches of the logarithm)

$$2\pi \mathrm{i}\, \hat{\omega}_0(z) = \sum_{j=1}^{m} \bar{\omega}_j \ln \left(\frac{z_{j+1} - z}{z_j - z} \right) \tag{67}$$

where again $\bar{\omega}_{m+1} \equiv \bar{\omega}_1$, $z_{m+1} \equiv z_1$. Rewriting (67),

$$\begin{aligned} 2\pi \mathrm{i}\, \hat{\omega}_0(z) = {} & \bar{\omega}_1 \ln (z_2 - z) - \bar{\omega}_1 \ln (z_1 - z) + \bar{\omega}_2 \ln (z_3 - z) \\ & - \bar{\omega}_2 \ln (z_2 - z) + \ldots + \bar{\omega}_m [\ln (z_1 - z) + 2\pi \mathrm{i}] - \bar{\omega}_m \ln (z_m - z). \end{aligned} \tag{68}$$

Thus using the notation that $\bar{\omega}_0 \equiv \bar{\omega}_m$,

$$2\pi \mathrm{i}\, \hat{\omega}_0(z) = \left[\sum_{j=1}^{m} (\bar{\omega}_{j-1} - \bar{\omega}_j) \ln (z_j - z) \right] + 2\pi \mathrm{i}\, \bar{\omega}_m \tag{69}$$

or in the form of (62),

$$\hat{\omega}_0(z) = \left[\sum_{j=1}^{m} \gamma_j \ln (z_j - z) \right] + R_0(z) \tag{70}$$

where in this case

$$\gamma_j = \frac{-1}{2\pi \mathrm{i}} (\bar{\omega}_j - \bar{\omega}_{j-1}) \tag{71}$$

and

$$R_0(z) = \bar{\omega}_m. \tag{72}$$

The above procedure can be extended to the H_k approximation function $\hat{\omega}_k(z)$. Let Γ be discretized into m $(k+1)$-node boundary elements, and assume that order k polynomial basis functions are used on each element Γ_j. Thus on Γ_j, the order k global trial function is

$$G_k(\zeta) = \sum_{j=1}^{k+1} N_{j,i}\,\bar{\omega}_{j,i}, \quad \zeta \in \Gamma_j. \tag{73}$$

The $\hat{\omega}_k(z)$ approximation function is

$$2\pi\,\mathrm{i}\,\hat{\omega}_k(z) = \sum_{j=1}^{m} \int_{\Gamma_j} \frac{G_k(\zeta)\,d\zeta}{\zeta - z}, \quad z \in \Omega,\ z \notin \Gamma. \tag{74}$$

The contribution to (74) from Γ_j is determined by

$$\int_{\Gamma_j} \frac{G_k(\zeta)\,d\zeta}{\zeta - z} = \sum_{j=1}^{k+1} \bar{\omega}_{j,i} \int_{\Gamma_j} \frac{N_{j,i}(\zeta)\,d\zeta}{\zeta - z}. \tag{75}$$

But each polynomial basis function is of the form

$$N_{j,i}(\zeta) = \begin{cases} a_{i,0} + a_{i,1}\zeta + \ldots + a_{i,k}\zeta^k, & \zeta \in \Gamma_j \\ 0, & \zeta \notin \Gamma_j \end{cases} \tag{76}$$

and for $z \in \Gamma_j$

$$N_{j,i}(z) = \begin{cases} 1, & z = z_{j,i} \\ 0, & z = z_{j,n}, \quad n \neq i. \end{cases} \tag{77}$$

Thus

$$\int_{\Gamma_j} \frac{N_{j,i}(\zeta)\,d\zeta}{\zeta - z} = \int_{\Gamma_j} \frac{(a_{i,0} + a_{i,1}\zeta + \ldots + a_{i,k}\zeta^k)\,d\zeta}{\zeta - z} \tag{78}$$

$$= \int_{\Gamma_j} [a_{i,k}\zeta^{k-1} + (z\,a_{i,k} + a_{i,k-1})\,\zeta^{k-2} + \ldots + (z^{k-1}a_{i,k} + \ldots + a_{i,1})]\,d\zeta$$
$$+ (z^k a_{i,k} + z^{k-1} a_{i,k-1} + \ldots + z\,a_{i,1} + a_{i,0}) \int_{\Gamma_j} \frac{d\zeta}{\zeta - z}. \tag{79}$$

The complex integral is evaluated by

$$(a_{i,k}z^k + \ldots + a_{i,1}z + a_{i,0}) \int_{\Gamma_j} \frac{d\zeta}{\zeta - z} = (a_{i,k}z^k + \ldots + a_{i,0}) \ln\left(\frac{z_{j,k+1} - z}{z_{j,1} - z}\right). \tag{80}$$

Comparing (80) to the results of the H_1 and H_0 approximation, function $\hat{\omega}_k(z)$ can be expanded into the form

$$\hat{\omega}_k(z) = \sum_{j=1}^{m} P_j^k(z) \ln(z_j - z) + R_k(z) \tag{81}$$

where $P_j^k(z)$ is an order k complex polynomial defined on element Γ_j, and $R_k(z)$ is a reference polynomial of order k which occurs due to the circuit on Γ about branch point z of the function $\ln(z_j - z)$.

7.9 Upper Half Plane Boundary Value Problems

Further insight into the CVBEM is gained by examining the approximation accuracy in modeling Dirichlet boundary value problems in the upper half plane. In this section, the Dirichlet problem is studied where $\psi(z)$ is known continuously on boundary Γ and a single reference value of $\phi(z)$ is known. The Dirichlet problem of $\phi(z)$ known on Γ is analogous to the above case due to $f(z) = \mathrm{i}\,\omega(z)$ being an analytic function (in which case $f(z) = -\psi(z) + \mathrm{i}\,\phi(z)$, and $\phi(z)$ is therefore the stream function of $f(z) = \mathrm{i}\,\omega(z)$).

By assumption Γ is a simple closed polygon. The Schwarz-Christoffel transformation $T(z)$ maps Γ onto the real axis $(-\infty < x < \infty)$, and domain Ω onto the upper half-plane $(y > 0)$. The transformation is

$$T(z) = A \int (z - x_1)^{\alpha_1/\pi - 1} \ldots (z - x_n)^{\alpha_n/\pi - 1}\, dz + B \tag{82}$$

where the α_i are angles (Fig. 14); A and B are constants.

Let Γ be discretized into m 2-node boundary elements Γ_j, $j = 1, 2, \ldots, m$. Then the real axis is also discretized into m boundary elements with one element, say Γ_t, being mapped by $T(z)$ into an infinite length boundary element on the real axis, Γ_t' (see Fig. 14). Assume the transformed boundary conditions on the real axis are linear distributions of $\psi'(z)$ on each Γ_j' except on Γ_t', where $\psi'(z)$ is a constant value. (All complex variables are denoted by prime to represent the transformed result.) Figure 15 shows the assumed boundary condition definition on Γ'.

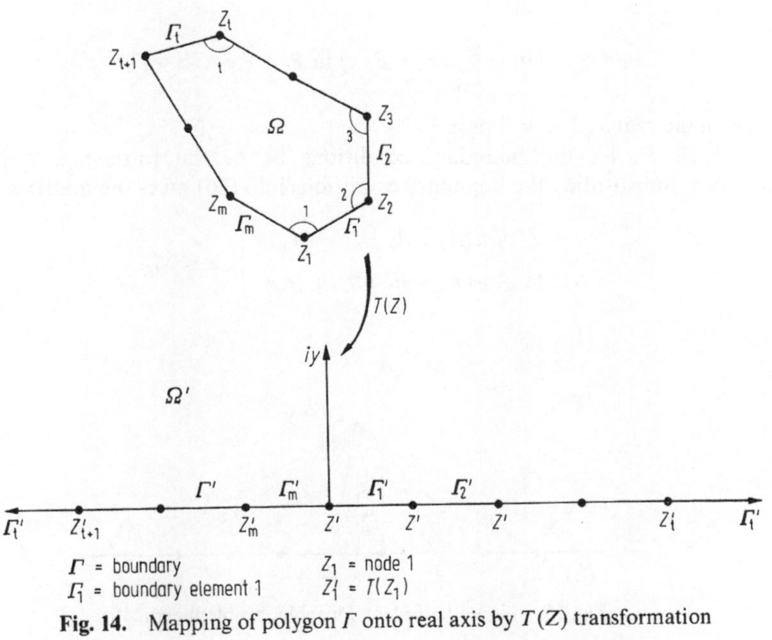

Fig. 14. Mapping of polygon Γ onto real axis by $T(Z)$ transformation

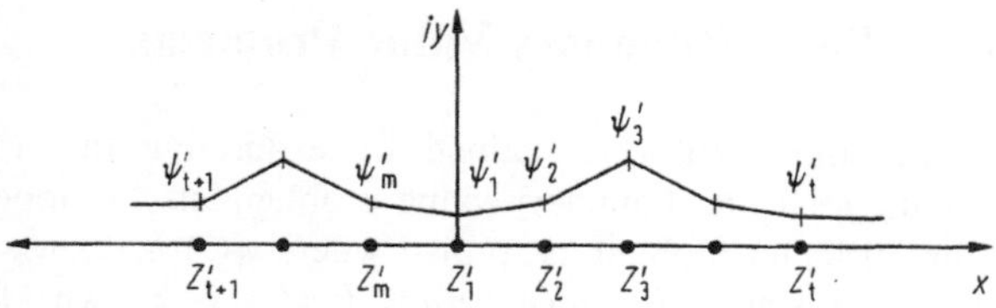

Fig. 15. Transformed boundary conditions on Γ'

For the case of $\psi'_t = \psi'_{t+1} = 0$, the solution of the boundary value problem is

$$\omega(z) = \sum_{j=1}^{m} (A_j + B_j z) \ln(z - z_j) + Cz + D, \quad z \in \Omega \cup \Gamma' \tag{83}$$

where the A_j, B_j and C, D are complex constants which have the form of the expanded H_1 approximation function $\hat{\omega}_1(z)$.

To use (83), let

$$\ln(z - z_j) = \ln R_j + \mathrm{i}\,\theta_j, \quad 0 \leqq \theta \leqq \pi, \quad R_j > 0 \tag{84}$$

where R_j and θ_j are defined in Fig. 16.

Then for every node z_j

$$(A_j + B_j z)(\ln R_j + \mathrm{i}\,\theta_j) = [(A_j + B_j x) \ln R_j - B_j y\, \theta_j] + \mathrm{i}\,[B_j y \ln R_j + (A_j + B_j x)\, \theta_j]. \tag{85}$$

On the real axis, y is identically zero and therefore for $x \neq x_j$, $j = 1, 2, \ldots, m$

$$\psi'(x, y = 0) = \sum_{j=1}^{m} (A_j + B_j x)\, \theta_j + \mathrm{Im}\,(Cz + D)$$

$$\phi'(x, y = 0) = \sum_{j=1}^{m} (A_j + B_j x) \ln R_j + \mathrm{Re}\,(Cz + D) \tag{86}$$

where on the real axis, $\theta_j = 0$ or π.

On each Γ_j, let the boundary conditions be of the form $\psi'(z \in \Gamma_j, y = 0) = a_j + b_j x$. Substituting the boundary conditions into (86) gives the matrix system

$$\left.\begin{aligned} 0 = a_1 &= \pi(A_1 + A_2 + \ldots + A_m) \\ 0 = b_1 &= \pi(B_1 + B_2 + \ldots + B_m) \end{aligned}\right\} \quad x < x'_{t+1}$$

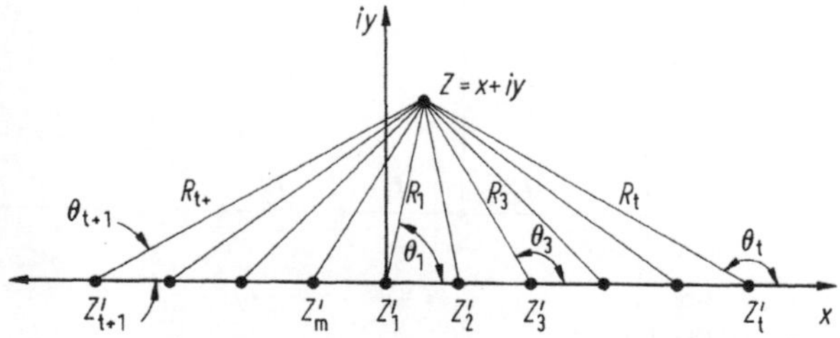

Fig. 16. Upper half plane Dirichlet problem geometry

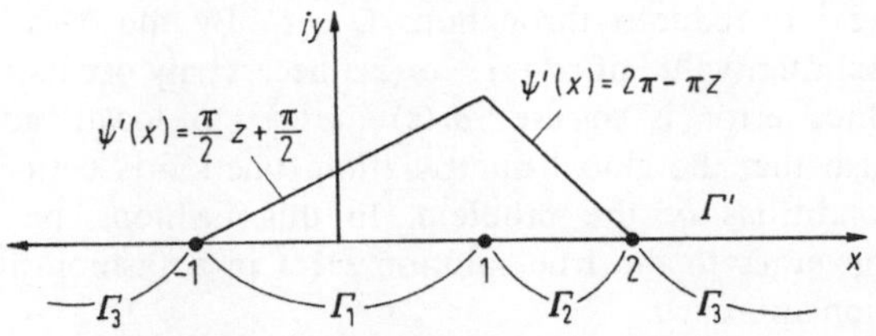

Fig. 17. Example 7 boundary conditions

$$\left.\begin{aligned} a_2 &= \pi(A_2 + \ldots + A_m) \\ &\vdots \\ b_2 &= \pi(B_2 + \ldots + B_m) \end{aligned}\right\} x'_{i+1} < x < x'_{i+2}$$

$$\left.\begin{aligned} a_m &= \pi A_m \\ b_m &= \pi B_m \end{aligned}\right\} x'_{i-1} < x < x'_i$$

Example 7. Let $\psi'(x)$ be given on a 3-element discretization of Γ' by (Fig. 17)

$$\psi'(x) = \begin{cases} 0, & x < -1 \\ \frac{\pi}{2}x + \frac{\pi}{2}, & -1 \leqq x \leqq 1 \\ 2\pi - \pi x, & 1 \leqq x \leqq 2 \\ 0, & x \geqq 2 \end{cases}$$

Then $\alpha_2 = \frac{\pi}{2}$, $b_2 = \frac{\pi}{2}$; $a_3 = 2\pi$, $b_3 = -\pi$. Using the relations given in (86), $m = 3$, and

$$A_2 = a_3/\pi = 2\pi/\pi = 2$$

$$B_3 = -\pi/\pi = -1\,.$$

Likewise,

$$A_2 = -1.5, \quad B_2 = 1.5; \quad A_1 = -0.5, \quad B_1 = -0.5$$

and the solution to this boundary value problem is

$$\omega(z) = (-0.5 - 0.5z)\ln(z+1) + (-1.5 + 1.5z)\ln(z-1) + (2-z)\ln(z-2),$$

$$\text{for} \quad x \neq -1, 2 \quad \text{and} \quad y \geqq 0.$$

It can be noted that the solution $\omega(z)$ is of a form analogous to the H_1 approximation function expansion of Eq. (62).

7.10 The Approximate Boundary for Error Analysis

The CVBEM is used to develop an analytic approximation function $\hat{\omega}(z)$ which exactly satisfies the Laplace equation throughout the interior of the problem domain, Ω. As the values of $\hat{\omega}(z)$ approach the values of the exact solution of the boundary value problem $\omega(z)$ for all points z on the problem boundary Γ, then the error $|\hat{\omega}(z)-\omega(z)|$ is reduced throughout $\Omega \cup \Gamma$. By the Maximum Modulus Theorem, the maximum value of $|\hat{\omega}(z)-\omega(z)|$ necessarily occurs for z on Γ. One approach to reduce error is to use $|\hat{\omega}(z)-\omega(z)|$ to locate additional nodal points z_j on Γ such that the global interpolation function is better adapted to fit the boundary conditions of the problem. In this fashion, the approximation function $\hat{\omega}(z)$ converges to the true solution $\omega(z)$ in a fashion analogous to an adaptive integration approach.

Rather than examining the above error values on the boundary, Γ, it is useful to determine an approximate boundary $\hat{\Gamma}$ upon which $\hat{\omega}(z)$ satisfies the given boundary conditions for $\omega(z)$ on $\hat{\Gamma}$. That is, given an approximator $\hat{\omega}(z)$, level curves of constant ϕ or ψ on Γ (where $\omega(z)=\phi+\mathrm{i}\,\psi$ and $\hat{\omega}(z)=\hat{\phi}+\mathrm{i}\,\hat{\psi}$) are compared to level curves of constant $\hat{\phi}$ or $\hat{\psi}$ on $\hat{\Gamma}$ where $\hat{\Gamma}$ is determined by setting the known $\phi=\hat{\phi}$ and $\psi=\hat{\psi}$. The resulting boundary $\hat{\Gamma}$ has the property that $\hat{\omega}(z)$ satisfies the specified boundary conditions on $\hat{\Gamma}$, and $\hat{\omega}(z)$ satisfies the governing Laplace equation in the interior, $\hat{\Omega}$. Consequently, $\hat{\omega}(z)$ is the exact solution to the boundary value problem with the true boundary Γ transformed into the approximate boundary $\hat{\Gamma}$.

The utilization of the approximate boundary provides the following features:

1. An exact solution of the subjection boundary value problem is provided for a transformation of the problem boundary (the complex variable transformation is undetermined).
2. The approximate boundary can be visually compared to the true boundary as to closeness of geometric fit.
3. Nodal points can be added to Γ to determine a more refined approximation $\hat{\omega}(z)$ so that $\hat{\Gamma}$ is geometrically closer to Γ in regions of high discrepancy.
4. The engineer works with a displacement of the problem boundary rather than examining a more abstract relative error propagation along the boundary.
5. The approximate boundary provides a direct visual representation of the sensitivity of the approximation $\hat{\omega}(z)$ in accommodation boundary conditions, variations in the Γ geometry, and the addition of nodal points.

In the following, several mixed boundary value problems of the Laplace equation are approximated by the CVBEM. The problems utilize boundaries Γ which geometrically coincide with lines of constant ϕ or ψ of $\omega(z)$. After developing a $\hat{\omega}(z)$, the approximate boundary $\hat{\Gamma}$ is determined by plotting the corresponding lines of constant $\hat{\phi}=\phi$ and $\hat{\psi}=\psi$ from $\hat{\omega}(z)$. In the accompanying figures, both Γ and the associated $\hat{\Gamma}$ are plotted so that a direct comparison is seen. Intuitively, as $\|\hat{\Gamma}-\Gamma\|$ becomes small then necessarily $|\hat{\omega}(z)-\omega(z)|$ is reduced and $\|\hat{\Gamma}-\Gamma\|=0$ implies $\hat{\omega}(z)=\omega(z)$.

A difficulty in using the $\hat{\omega}(z)$ function of (52) is that $\hat{\omega}(z)=0$ for z exterior of $\Omega \cup \Gamma$. Thus an analytic continuation of $\hat{\omega}(z)$ to the exterior of $\Omega \cup \Gamma$ is needed in order to determine $\hat{\Gamma}$. One procedure to develop this analytic continuation is to use the finite series expansion of (62) modified to have the angle of the form $\arg(z_j-z)$ to be measured with respect to branch cuts originating from each nodal point and extending outwards away from $\Omega \cup \Gamma$. Figure 18 illustrates the branch cut definition needed. In this fashion, the analytical continuation of $\hat{\omega}(z)$ is available everywhere except across each branch cut.

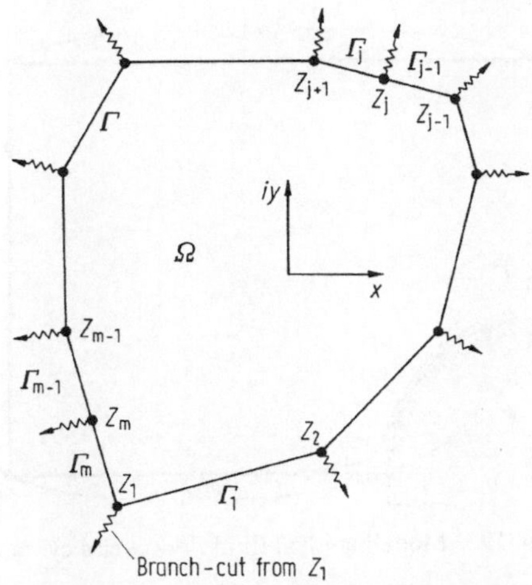

Fig. 18. The analytic continuation of $\hat{\omega}(z)$ to the exterior of $\Omega \cup \Gamma$. Note branch cuts along Γ at nodes z_j

The approximate boundary is developed by the following steps:

1. Use $\hat{\omega}(z)$ to estimate nodal values for the unknown nodal point state variable or stream function values, $\Delta\xi_U(z_j)$.
2. Using the analytic continuation of $\hat{\omega}(z)$ given by (62), determine constants γ_j by forcing $\hat{\omega}(z_j) = \Delta\xi_k(z_j) + \Delta\xi_U(z_j)$. In calculating the γ_j values, use branch cuts which lie exterior of $\Omega \cup \Gamma$ such as shown in Fig. 18.
3. Determine the approximate boundary $\hat{\Gamma}$ by locating (x, y) coordinates where level curves of $\hat{\omega}(z)$ match the boundary condition values. Because $\hat{\omega}(z_j)$ is now determined using $\Delta\xi_k(z_j)$ values, $\hat{\Gamma}$ necessarily intersects Γ at each nodal point location on Γ.

Example 7. This problem approximates the classic problem of ideal fluid flow over a cylinder. The exact solution is known to be $\omega(z) = z + 1/z$. The problem boundary Γ is specified to be the upper right quadrant as shown in the figure. Using a 47-node discretization, a $\hat{\omega}(z)$ approximator is developed. The corresponding approximate boundary $\hat{\Gamma}$ is plotted along with the true boundary Γ in Fig. 19.

Example 7 demonstrates the utility of determining an approximate boundary corresponding to CVBEM approximation functions. The approximate boundary $\hat{\Gamma}$ is developed by plotting the level curves of constant potential (or stream function) which match the boundary condition values on the problem boundary Γ. Consequently, this technique is directly applicable only to boundary value problems which have level curves for boundary conditions.

The error of approximation is manifested by the departure of the approximate boundary from the problem true boundary. Where large spatial discrepancies are observed, additional nodal points are needed to increase the approximation accuracy. The approximate boundary can often be argued to better represent the

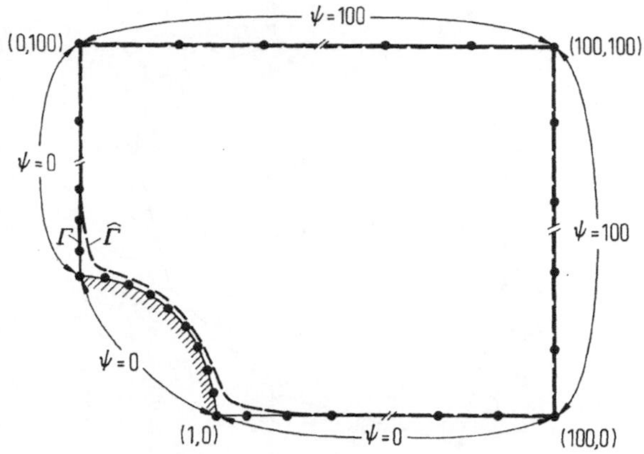

Fig. 19. Modeling ideal fluid flow over a cylinder

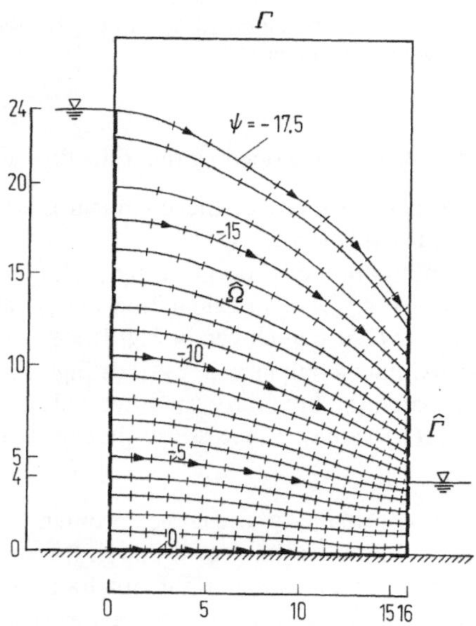

Fig. 20. Plot of streamlines and potentials for soil-water flow through a homogeneous soil

"as-built" or a more realistic problem boundary than the defined problem boundary. This latter idea is especially valid in large scale civil engineering studies where angle points are generally constructed as grossly rounded edges.

To illustrate the CVBEM approximation results within the interior of the problem domain, Figs. 20 and 21 show groundwater seepage problems with the approximate boundary, and several streamlines and lines of the constant potential plotted. Because the maximum error magnitude ε must occur on the boundary, interior values of $\hat{\omega}(z)$ necessarily differ from $\omega(z)$ in magnitude by less than ε.

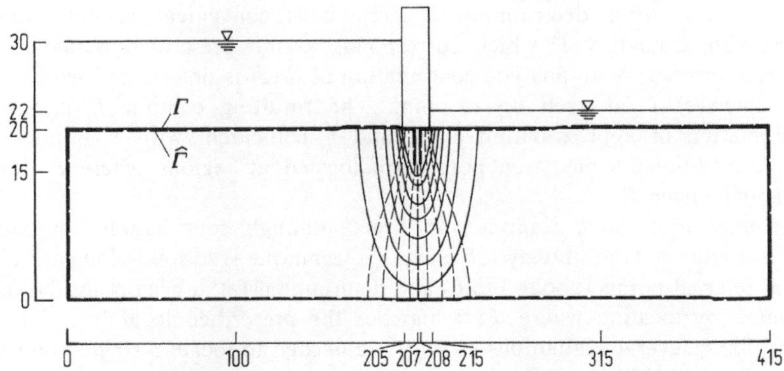

Fig. 21. Plot of streamlines and potentials for soil-water flow beneath a dam. (Note that the vertical and horizontal scales differ)

7.11 Locating Additional Nodal Points on Γ

The success of the CVBEM in developing solutions to the boundary value problem depends upon the accuracy of the trial function assumptions. Accuracy is improved by the addition of nodal points on Γ due to the subsequent reduction in the error of the trial function approximation.

The CVBEM offers a highly useful error analysis capability by simply examining the success of the CVBEM approximator in matching the boundary conditions. The usual approach to reduing error is to add nodal points on Γ where the CVBEM approximator performs poorly in meeting the specified boundary condition values.

In this section, four methods of examining CVBEM approximation error are compared as to their effectiveness in developing better $\hat{\omega}(z)$ functions by indicating where additional nodes should be added to Γ.

Method 1. A plot of relative error in matching boundary conditions continuously on Γ is obtained by subtracting the approximator function values (along Γ) from the known boundary condition values. Since only one of the conjugate functions (ϕ or ψ) is known as a boundary condition at a point, this error plot is a representation of the mixed boundary condition fit. From Sect. 4 it is noted that the unknown nodal point values can be estimated using a class I or class II CVBEM matrix system. If the class I system is used, further computation effort is needed due to this type of relative error being zero at nodal points. Thus, interior values of $\hat{\omega}(z)$ are computed on each Γ_j. If the class II system is used, this error is usually nonzero at nodal points, and is readily evaluated. After the determination of the error plot, additional nodal points are located where there is large error. Should the error be zero on each Γ_j, then $\hat{\omega}(z)$ satisfies the Laplace equation and also the prescribed boundary conditions, and $\hat{\omega}(z)$ is the exact solution.

Method 2. Generally, the prescribed boundary conditions are values of constant ϕ or ψ on each Γ_j. These values corresponds to level curves of the analytic function

$\omega(z) = \phi + i\,\psi$. After determining a $\hat{\omega}(z)$, it is convenient to determine an approximate boundary $\hat{\Gamma}$ which corresponds to the prescribed boundary conditions. From Sect. 9 an analytic continuation of $\hat{\omega}(z)$ is determined which forces $\hat{\Gamma}$ to intersect Γ at each nodal point. The resulting contour $\hat{\Gamma}$ is a visual representation of approximation error, and $\hat{\Gamma}$ coincident with Γ implies $\hat{\omega}(z) = \omega(z)$. Additional collocation points are located at regions where $\hat{\Gamma}$ deviates substantially from Γ.

Implementation on a computer is direct although considerable computation effort is required. One strategy for using this technique is to subdivide each Γ_j with several internal points (about 4 to 6) and determine $\hat{\omega}(z)$ at each point. Next, $\hat{\Gamma}$ is estimated by locating where $\hat{\omega}(z)$ matches the prescribed local boundary condition. Thus, several evaluations of $\hat{\omega}(z)$ are needed to locate a single point on $\hat{\Gamma}$. The end product, however, is very useful since it can be argued that $\hat{\omega}(z)$ is the exact solution to the boundary value problem with Γ transformed to $\hat{\Gamma}$. Thus $\hat{\Gamma}$ is a visual indication of approximation error.

Method 3. This technique includes features from both Methods 1 and 2, and yet involves a significant reduction in computer effort over Method 2 alone. First, the error distribution of Method 1 is determined along Γ between the known function (ϕ or ψ) of $\omega(z)$ and the corresponding approximation of $\hat{\omega}(z)$. The next step is to weight the error determined above (designated as $e(z)$ for $z \in \Gamma$) by the tangential gradient of the function conjugate to the local boundary condition variable. For example, if ϕ is known on Γ_n, then for $z \in \Gamma_n$ we have $e(z) = \phi - \hat{\phi}$. This relative error is weighted by $\partial\hat{\psi}/\partial s$ which is determined directly by finite differences of $\hat{\omega}(z)$. Using the Cauchy-Riemann relations, an estimated distance of departure $d(\hat{\Gamma}, \Gamma, z)$ between the approximate boundary $\hat{\Gamma}$ and the problem boundary Γ at point $z \in \Gamma$ is given by

$$d(\hat{\Gamma}, \Gamma, z) \cong \left| e_\phi(z) \Big/ \frac{\partial\hat{\psi}}{\partial s} \right| . \tag{87}$$

In (87) the error $e_\phi(z)$ has a subscript notation for the ϕ function. A similar relationship holds for the $e_\psi(z)$ error

$$d(\hat{\Gamma}, \Gamma, z) \cong \left| e_\psi(z) \Big/ \frac{\partial\hat{\phi}}{\partial s} \right| .$$

The final form of error used, $V(z)$, is the ratio

$$V(z) = \begin{cases} e_\phi^2(z) \Big/ \left| \dfrac{\partial\hat{\psi}}{\partial s} \right| , & \text{if } \phi \text{ is known at } z \\[2ex] e_\psi^2(z) \Big/ \left| \dfrac{\partial\hat{\phi}}{\partial s} \right| , & \text{if } \psi \text{ is known at } z . \end{cases} \tag{88}$$

Additional nodal points are defined at locations on Γ where $V(z)$ is large.

An advantage of Method 3 over Method 1 is that more weight is given to the error which also has a large distance of departure between Γ and $\hat{\Gamma}$. Similarly, Method 3 provides an improved definition of the error associated with the approximate boundary of Method 2 by including the description of whether $e(z)$ is large or small and $\hat{\Gamma}$ has a large departure from Γ simply due to a small normal gradient of the specified boundary condition variable. Figure 22 illustrates a geometric interpretation of $V(z)$ as a "point area of error" in the CVBEM approximation. From the figure, the positive area at point z_0 equals one-half of the quantity defined in the relations of (88). Also shown in the figure is the actual approximation value $\hat{\phi}(n)$ as a function of normal distance (n) from point $z_0 \in \Gamma$.

Method 4. Because $G(\zeta)$ is continuous on Γ, $\hat{\omega}(z)$ is analytic in Ω. Thus for $z \in \Omega$ and $z \notin \Gamma$

$$\hat{\omega}(z) = \frac{1}{2\pi i} \int_\Gamma \frac{G(\zeta)\, d\zeta}{\zeta - z} = \frac{1}{2\pi i} \int_\Gamma \frac{\hat{\omega}(\zeta)\, d\zeta}{\zeta - z}. \tag{89}$$

But for $z_0 \in \Gamma$, the limit as $z \to \Gamma$ (where $z \in \Omega$) can be determined and an error $E(z_0)$ is defined by

$$E(z_0) = \lim_{z \to z_0} \frac{1}{2\pi i} \int_\Gamma \frac{G(z_0)\, d\zeta}{\zeta - z_0} - \frac{1}{2\pi i} \int_\Gamma \frac{\hat{\omega}(\zeta)\, d\zeta}{\zeta - z_0} \tag{90}$$

or simply

$$E(z_0) = \lim_{z \to z_0} \frac{1}{2\pi i} \int_\Gamma \frac{[G(z_0) - \hat{\omega}(\zeta)]\, d\zeta}{\zeta - z_0}. \tag{91}$$

Setting $E(z_j) = 0$ for $j = 1, 2, \ldots, m$ determines a class I or II system of equations which are used to estimate values for the unknown nodal variable function. The objective in this method is to obtain a global trial function such that in the limit $G(z_0) = \hat{\omega}(z_0)$ for all $z_0 \in \Gamma$. Thus additional nodal points are located on Γ where $|G(z_0) - \hat{\omega}(z_0)|$ is a maximum.

A comparison of Method 4 to Method 1 indicates that Method 4 involves approximately the same computational effort as Method 1, yet includes an error contribution for both the potential and stream functions. Thus a total error magnitude is provided by this technique which is not immediately available by the other three approaches.

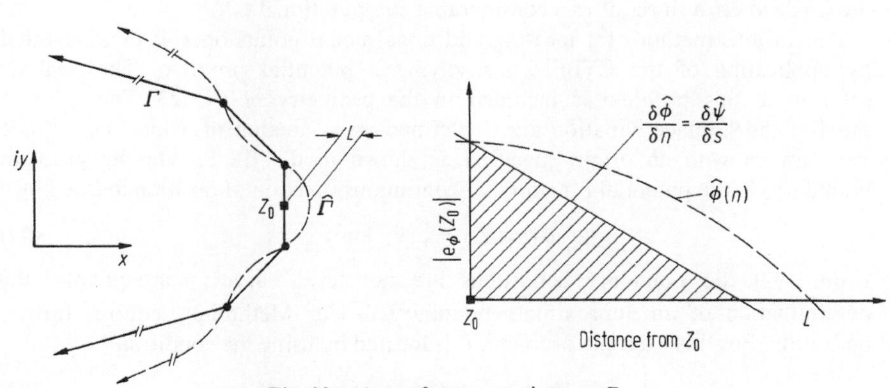

Fig. 22. Area of error at point $z_0 \in \Gamma$

Table 1. Comparison of CVBEM error evaluation methods

Method number	1	2	3	4
Error analysis approach	Relative error of known boundary condition	Approximate boundary	Error area	Total relative error
Collocation point locating criteria	Maximum value of error	Maximum departure between $\hat{\Gamma}$ and Γ	Maximum point area of error	Maximum value of error
Computational effort	Single point evaluation of $\hat{\omega}(z)$	Iteration of $\hat{\omega}(z)$ for each point of $\hat{\Gamma}$	4 to 6 evaluations of $\hat{\omega}(z)$ for Γ_j	Single point evaluation of $\hat{\omega}(z)$
Representation of error	Relative error plot of boundary condition match	Plot of $\hat{\Gamma}$ for comparison with Γ	Plot of relative error area along Γ	Plot of total relative error $\lvert G(z)-\hat{\omega}(z)\rvert$
CVBEM class type used of estimate of unknown nodal values	II	I	II	II
Evaluation of $\hat{\omega}(z)$ at nodes	Yes	No	Yes	Yes
Evaluation of $\hat{\omega}(z)$ at points within Γ_j	No	About 4 to 6	No	No
Includes contributions of both harmonic functions	No	No	Yes	Yes
Approximate ratio of computational effort with Method 1	100%	1700%	120%	110%

Table 1 summarizes the main features of the four methods presented. Included in the table are estimates of the computational effort (in CPU time) expressed as a ratio of the considered technique versus Method 1. It is noted that although Method 2 (approximate boundary) generates an easy-to-interpret representation of the CVBEM error, it requires a considerable computational effort.

The various methods for locating additional nodal points on Γ is demonstrated by application of the CVBEM for solving a potential problem. The analytic solution to this problem is included in the geometry of Fig. 23. The solution satisfies the Laplace equation and is defined as a function of a local coordinate $x-y$ system with an origin specified as shown in the figure. On the problem boundary, Γ, the potential function is a continuous function of position defined by

$$\phi(z \in \Gamma) = \tfrac{1}{2}(x^2+y^2). \tag{92}$$

From (92) the boundary conditions are not level curves; consequently, the determination of an approximate boundary $\hat{\Gamma}$ (for Method 2) requires further definition. For this example problem, $\hat{\Gamma}$ is located by using the condition

$$\hat{\Gamma} \equiv \{z : \hat{\phi}(z) = \tfrac{1}{2}(x^2+y^2) = \tfrac{1}{2}|z|^2\}. \tag{93}$$

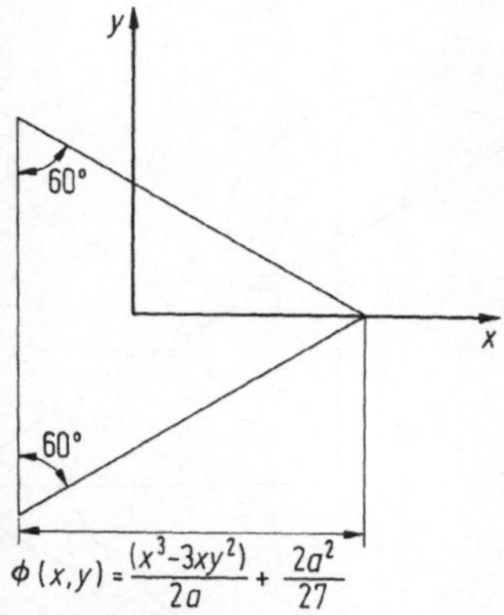

Fig. 23. Application problem geometrics and exact solutions for temperature, $\phi(x, y)$

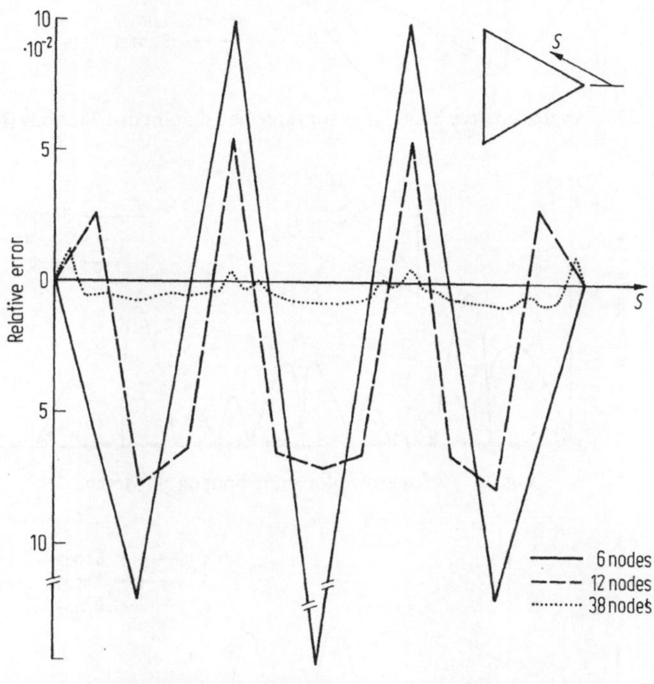

Fig. 24. Boundary relative error plot (Method 1)

Figures 24 through 27 illustrate the several error evaluation methods for 3 nodal placements (evenly spaced).

From the figures, Methods 1, 3, and 4 provide similar abstract representations of the CVBEM modeling error. However, Method 2 results in a visual representation of approximation error which is easily interpretable. Often it can be argued that the precise mathematical description of the problem boundary is not achieved due to

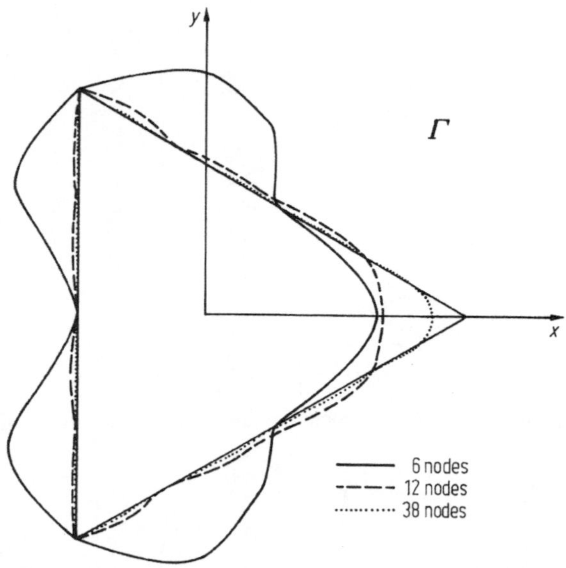

Fig. 25. Approximative boundaries for three nodal point distributions (Method 2)

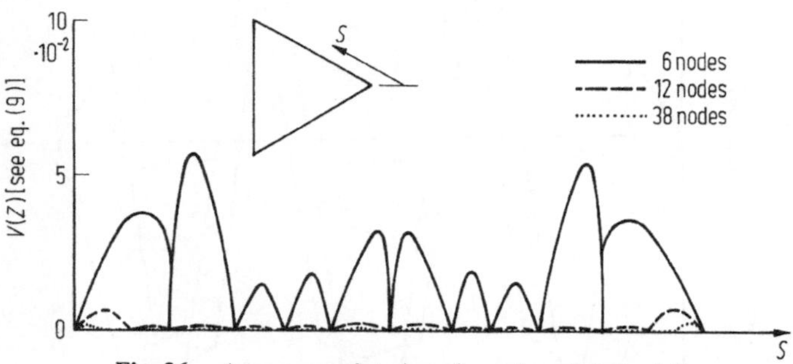

Fig. 26. Area error plot along boundary (Method 3)

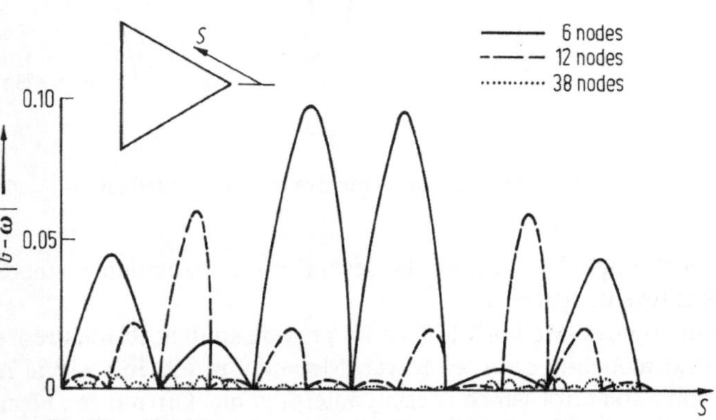

Fig. 27. Trial function error plot along boundary (Method 4)

the construction of the prototype, and that the approximate boundary $\hat{\Gamma}$ may actually represent a more probable end product. Because $\hat{\omega}(z)$ is the exact solution of the boundary value problem with Γ transformed into $\hat{\Gamma}$, then the selection of a $\hat{\Gamma}$ has the advantage of also being associated with the generating $\hat{\omega}(z)$ solution.

7.12 Sources and Sinks

Let $\Omega \cup \Gamma \in P$. Then the CVBEM develops a function $\hat{\omega}_k(z)$ analytic on Ω and continuous on Γ (for $k \geqq 0$). Let $f_i(z)$ be analytic on Ω. Then

$$F(z) = \hat{\omega}_k(z) = \sum_{i=1}^{k} f_i(z) \tag{94}$$

is analytic on Ω for k, a finite integer.

Let $\omega(z)$ be an analytic function on $\Omega \cup \Gamma$ except at points $z_i \in \Omega$ where there exist sources or sinks. A CVBEM approximation of $\omega(z)$ is determined by

$$F(z) = \hat{\omega}_k^*(z) + \sum_{i=1}^{k} (-S) \ln(z - z_i) \tag{95}$$

where $z_i \in \Omega$, S is the strength of the source, and $\hat{\omega}_k^*(z)$ is a CVBEM approximation function determined by approximating the modified boundary condition values of $\omega(z) + \sum S \ln(z - z_i)$ on Γ.

To illustrate the source (sink) function, let point $z_i = x_i + \mathrm{i}\, y_i \in \Omega$. Let z be a point in Ω and define the radial coordinates

$$z - z_i = R\mathrm{e}^{\mathrm{i}\theta}; \quad R > 0, \quad 0 \leqq \theta < 2\pi . \tag{96}$$

At a radial distance R_0 from z_i, the circumference is $2\pi R_0$. The unit flux (unit flow per unit cross-section length) in the R-direction is given by

$$q = -K \frac{\partial \phi}{\partial R} \tag{97}$$

where K is a transport coefficient. Then the total flow away from point z_i at a distance of R_0 is

$$Q(R_0) = \left(-K \frac{\partial \phi}{\partial R} \bigg|_{R_0} \right) (2\pi R_0) . \tag{98}$$

For $Q(R_0)$ assume a constant value $Q(R_0) = Q$,

$$-\frac{Q}{2\pi K} = \left(R \frac{\partial \phi}{\partial R} \right) \bigg|_{R_0}, \quad R_0 > 0 . \tag{99}$$

Let $S = Q/(2\pi K)$. Then

$$-S \frac{dR}{R} = d\phi \tag{100}$$

where it is noted that $\phi(\theta, R) = \phi(R)$ due to symmetry of flow from point z_i. Then

$$-S \ln R = \phi(R). \tag{101}$$

Similarly, a sink is defined as a negative source by

$$S \ln R = \phi(R). \tag{102}$$

Thus a flow field containing k sources and sinks described by the analytic function

$$F^*(z) = -\sum_{i=1}^{k} S \ln(z - z_i). \tag{103}$$

The function $F^*(z)$ of (103) imposes complex values on boundary Γ. The objective is to approximate $\omega(z)$ on $\Omega \cup \Gamma$ where $\omega(z)$ contains $F^*(z)$. For instance, the domain $\Omega \cup \Gamma$ may also be subjected to other effects such as linear flow, corners, and other possibilities. However we do know values of $\omega(z)$ along the boundary Γ which include the effects of $F^*(z)$. Thus, to approximate $\omega(z)$ on $\Omega \cup \Gamma$, the CVBEM is used to determine a $\hat{\omega}_k^*(z)$ which approximates $[\omega(z) - F^*(z)]$ on Γ.

7.13 Regional Inhomogeneity

Figure 28 illustrates the case of two dissimilar materials with conductivities K_1 and K_2. for steady state conditions, two conditions are satisfied along the interface; namely 1) the potential ϕ is a boundary condition for both domains Ω_1 and Ω_2, and 2) the normal flux $|\partial\phi/\partial n|$ values are equal for Ω_1 and Ω_2.

The CVBEM is used for the problem of Fig. 28 by developing two approximations, one for each of Ω_1 and Ω_2, such that the specified and interface boundary conditions are both satisfied. The error of the approximations can be evaluated using the approximate boundary approach where special attention is paid towards the two approximate boundaries (from Ω_1 and Ω_2) developed along the interface.

Because the CVBEM results in square, fully-occupied matrix systems, the analysis of multiple regions can result in very large matrix systems. Consequently, the CVBEM may become computationally inefficient when dealing with domains composed of several dissimilar materials.

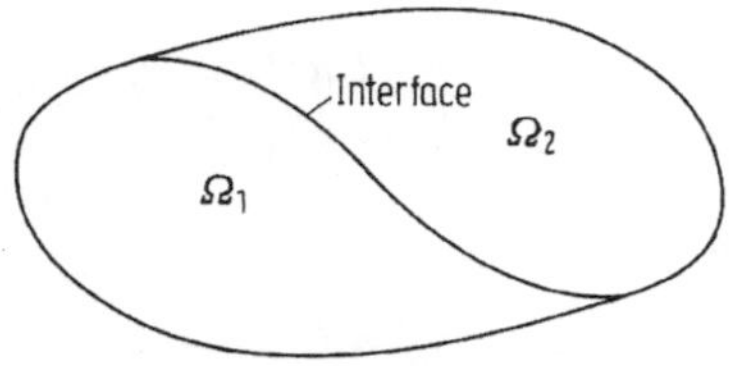

Fig. 28. Dissimilar materials problem

7.14 The Poisson Equation

The two-dimensional Poisson equation on domain Ω is given by

$$\nabla^2 \phi = f(x, y), \quad (x, y) \in \Omega \tag{104}$$

where $\phi(x, y)$ is a potential function and $f(x, y)$ is a prescribed function of (x, y) coordinates. Given boundary conditions on the simple closed boundary Γ (enclosing the simply connected domain Ω), the CVBEM can be extended to approximate the boundary value problem of (104).

Let $\phi_p(x, y)$ be a particular solution of (104). Let $\hat{\omega}^*(z)$ be a CVBEM approximation of the Laplace equation $\nabla^2 \phi = 0$ where boundary conditions on Γ are specified by subtracting the value of $\phi_p(x, y)$ for $(x, y) \in \Gamma$. That is, determine $\hat{\omega}^*(z)$ on $\Omega \cup \Gamma$ such that $\Delta \xi_k^*$ are the boundary conditions given by

$$\Delta \xi_k^* = \Delta \xi_k - \phi_p|_\Gamma . \tag{105}$$

Then necessarily $\hat{\omega}^*(z) = \hat{\phi}^*(z) + \mathrm{i}\, \hat{\psi}^*(z)$, and the CVBEM solution to the boundary value problem of (104) with boundary conditions $\Delta \xi_k$ on Γ is

$$\phi(z) = \phi_p(z) + \hat{\phi}^*(z), \quad z \in \Omega . \tag{106}$$

The above modeling approach is outlined by the following steps:

1. Find $\phi_p(z) = \phi_p(x, y)$ such that $\nabla^2 \phi_p = f(x, y)$.
2. Evaluate $\Delta \xi_k^* = \Delta \xi_k - \phi_p(x, y)$ for $(x, y) \in \Gamma$.
3. Develop $\hat{\omega}^*(z)$ based on $\Delta \xi_k^*$ boundary conditions.
4. Develop error analysis based on the solution of step 3.
5. Construct the CVBEM solution $\omega(z)$ by adding, $\omega(z) = \omega^*(z) + \phi_p(z)$.
6. CVBEM solution to (104) is $\phi(z) = \hat{\phi}^*(z) + \phi_p(z)$.

It is seen from the above methodology that the approximation of the Poisson equation is simply the application of the CVBEM to a Laplace problem with modified boundary conditions. Consequently, an important step to this solution technique is the development of a particular solution, $\phi_p(x, y)$. The following Table 2 provides a few basic particular solutions. It shold be noted that an infinity of particular solutions are possible for each $f(x, y)$.

Table 2. Particular solutions of the Poisson equation

$f(x, y)$	$\phi_p(x, y)$
k	$k\, x^2/2$
k	$k\, y^2/2$
k	$k\,(x^2 + y^2)/4$
$k\, x$	$k\, x^3/6$
$k\, y$	$k\, y^3/6$
$a\, p\,(p-1)\, x^{p-2} + b\, q\,(q-1)\, y^{q-2}$	$a\, x^p + b\, y^q$

7.15 Computer-Aided-Analysis and the CVBEM

The CVBEM can be used in a computer-aided-design environment where the engineer or scientist idenifies additional boundary element nodal point locations based on computer errors in satisfying the known boundary conditions. In this fashion, the analyst develops a problem geometry which is acceptable for prototype construction, and the CVBEM approximation determines the exact solution for the potential problem defined over this prototype geometry. Because the computer interactive technique uses graphical displays, the approach is efficient and easy to use.

In the previous chapter, the CVBEM has been shown to be a powerful tool for the numerical analysis of Laplace or Poisson equation boundary value problems. The numerical approach is to discretize the boundary Γ by nodal points into boundary elements (Fig. 29), and then specify a continuous global trial function $G(\zeta)$ on Γ as a function of the nodal values. Using the Cauchy integral, the resulting integral equation is

$$\hat{\omega}(z_0) = \frac{1}{2\pi i} \int_{\Gamma} \frac{G(\zeta)\, dz}{\zeta - z_0} \tag{107}$$

where $\hat{\omega}(z_0)$ is the CVBEM approximation for $z_0 \in \Omega$; and Ω is a two-dimensional simply connected domain enclosed by the simple closed contour Γ.

Because $G(\zeta)$ is continuous on Γ, then $\hat{\omega}(z)$ is analytic over Ω and can be written as the sum of two harmonic functions

$$\hat{\omega}(z) = \hat{\phi}(z) + i\,\hat{\psi}(z)\,. \tag{108}$$

Thus both $\hat{\phi}(z)$ and $\hat{\psi}(z)$ exactly satisfy the Laplace equation over Ω.

Approximation error occurs due to $\hat{\omega}(z)$ not satisfying the boundary conditions on Γ exactly. However, an approximate boundary $\hat{\Gamma}$ can be developed which represents the location where $\hat{\omega}(z)$ equals the specified boundary conditions such as level curves. Consequently, the CVBEM approximation error can be interpreted as a transformation of $\Gamma \to \hat{\Gamma}$ where the ultimate objective is to have $\hat{\Gamma}$ coincide

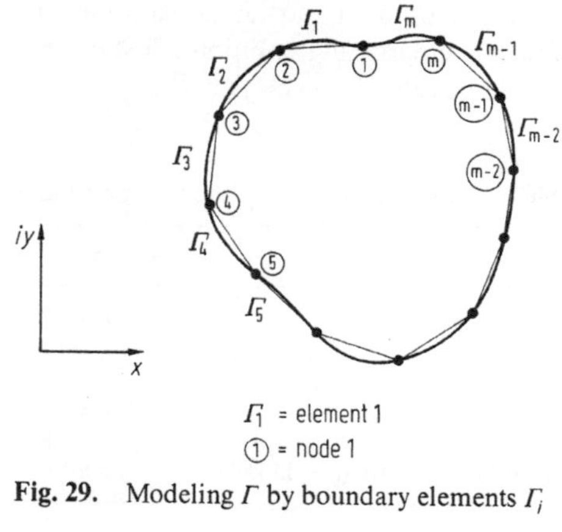

Fig. 29. Modeling Γ by boundary elements Γ_i

with Γ. Because all the error of approximation is due to incorrect boundary element trial functions, accuracy is increased by the addition of boundary nodal points where approximation error is identified to be large (i.e., adaptive integration).

As $\hat{\Gamma}$ approaches Γ geometrically, the analyst is assured by the Maximum Modulus Theorem that the maximum approximation error occurs on Γ and that the governing partial differential equation (Laplace) is solved exactly. Consequently, the final product is the exact solution for a problem geometry which is the construction tolerance for the prototype construction.

Generally, the types of numerical approximation errors in solving potential problems is of two forms: (i) errors due to not satisfying the governing equation over Ω, and (ii) errors due to not satisfying the boundary conditions continuously on Γ. For the CVBEM (and for other boundary integral equation methods), the first type of approximation error is eliminated due to both $\hat{\phi}$ and $\hat{\psi}$ being potential functions. But $\hat{\omega}(z)$ does not usually satisfy the boundary conditions continuously on Γ (if it did, then $\hat{\omega}(z) = \omega(z)$). The next step in the CVBEM analysis is to work with $\hat{\omega}(z)$ in order that $\hat{\omega}(z) \rightarrow \omega(z)$.

This step in the analysis of approximation error provides a significant advantage over domain numerical methods such as finite elements or finite differences. In the domain methods, the analyst examines error with a form of vector space Cauchy convergence criteria by arbitrarily increasing the domain nodal densities and comparing the resulting change in estimated nodal values. Whereas with the CVBEM, the analysis has several forms of the approximation error to work with.

The easiest form of error to study is the development of an approximate boundary $\hat{\Gamma}$ which represents the location where $\hat{\omega}(z)$ achieves the problem boundary values of $\omega(z)$. Generally, the boundary conditions are constant values of ϕ or ψ along boundary elements, i.e. $\phi = \phi_j$ for $z \in \Gamma_j$ or $\psi = \psi_k$, for $z \in \Gamma_k$. This set of m nodal values $\{\phi_j, \psi_k\}$ are level curves of $\omega(z)$. The approximate boundary $\hat{\Gamma}$ is determined by locating those points where $\hat{\phi} = \phi_j$ and $\hat{\psi} = \psi_k$. Due to the collocation process, $\hat{\Gamma}$ intersects Γ at least at each nodal location, z_j, $j = 1, 2, \ldots, m$.

To determine $\hat{\Gamma}$, each element Γ_j is further subdivided by interior points where $\hat{\omega}(z)$ is to be evaluated. At each element interior point, $\hat{\omega}(z)$ is calculated from the Cauchy line integral and the values of $\hat{\phi}$ and $\hat{\psi}$ are determined. If the appropriate $\hat{\phi}$ (or $\hat{\psi}$) matches the boundary condition on Γ_j, then $\hat{\Gamma}$ intersects Γ at that point. Otherwise, subsequent points are evaluated by marching pointwise along a line perpendicular to Γ_j until the boundary condition value is reached. For point locations interior of $\tilde{\Omega} \cup \Gamma$, an analytic continuation of $\hat{\omega}(z)$ is used.

In this fashion, a set of points is determined where $\hat{\omega}(z)$ equals the desired ϕ_j or ψ_k values. The contour $\hat{\Gamma}$ is estimated by connecting these points with straight lines. Because $\hat{\Gamma}$ and Γ intersect at least at nodal point locations, $\hat{\Gamma}$ appears as a plot which oscillates about the Γ contour.

It is convenient to use a graphical display of both Γ and $\hat{\Gamma}$ superimposed on the CRT. By magnification of the departure between Γ and $\hat{\Gamma}$, the analyst can easily inspect the performance of the CVBEM approximation. Because the approximation error is due to the assumed basis function assumptions, the integration error is reduced by the addition of nodal points on Γ, similar to an adaptive integration technique.

The addition of nodal points can be made directly via the CRT screen and a "locating the closest boundary coordinate" computergraphics subroutine. After the nodal additions are completed, a new $\hat{\omega}(z)$ is determined and the revised $\hat{\Gamma}$ plotted on Γ. By the addition (and deletion) of nodal points from Γ, the analyst is able to quickly evaluate the quality of the CVBEM model. Because the addition of a nodal point can be interpreted as the addition of an approximation error sink term, the geometric representation of error by means of $\hat{\Gamma}$ provides a mathematically sophisticated yet easy-to-use modeling tool.

References

1 Hromadka II, T.V. and Guymon, G.L., Application of a Boundary Integral Equation to Prediction of Freezing Fronts in Soils. In: Cold Regions Science and Technology **6,** 115–121 (1982).

2 Hromadka II, T.V. and Guymon, G.L., Complex Polynomial Approximation of the Laplace Equation. In: ASCE Hydraulics Division, Vol. 110, No. 3, March 1984

3 Hromadka II, V.T. and Guymon, G.L., A Complex Variable Boundary Element Method: Development. International Journal of Numerical Methods in Engineering, April 1983

4 Hromadka II, T.V., The Complex Variable Boundary Element Method. Springer-Verlag, Berlin, Heidelberg, New York 1984

Chapter 8

Potential Problems*

by L.C. Wrobel

8.1 Introduction

A large number of engineering problems can be mathematically described by Laplace's equation. Since functions satisfying Laplace's equation are generally called potential functions, these engineering problems are referred to as potential problems. Under this classification, we can include heat conduction problems, where the potential function is the temperature; flow of ideal fluids, formulated either with a velocity potential or a stream function; groundwater flow, where the function is the piezometric head; torsion of prismatic shafts, with a warping function; and many others.

Since the pioneering works of Hess and Smith [1], Jaswon [2] and Symm [3] the boundary element method, both in its direct and indirect forms, has found wide-spread application for the solution of potential problems (for a more detailed history and further references see [4]).

Clearly, the efficiency of any numerical method is directly related to the way the method is implemented in the computer. Thus, a careful programming is essential. Furthermore, a potential user with a particular problem would be faced with enormous difficulties if forced to program from scratch for each new class of problems. It is therefore important that the programming effort of previous works should be utilized.

This chapter describes an efficient and reliable boundary element code for the solution of potential problems, employing linear boundary elements. Its starting point was the simple program described by Brebbia [5] but it was almost completely reformulated in order to increase its efficiency without losing its easy-to-follow character. In addition, several other features were included to make the program more general.

The chapter starts with a brief review of the boundary element theory. Then, some numerical and computational aspects are discussed. A detailed description of the code follows, together with its application to a number of simple, practical problems.

8.2 Review of the Theory

The problem we are dealing with is governed by the following equation and boundary conditions:

$$\nabla^2 u = \frac{\partial^2 u}{\partial x_1^2} + \frac{\partial^2 u}{\partial x_2^2} = 0 \quad \text{in } \Omega \tag{1}$$

$$u = \bar{u} \quad \text{on } \Gamma_1 \tag{2}$$

$$q = \frac{\partial u}{\partial n} = \bar{q} \quad \text{on} \quad \Gamma_2 \tag{3}$$

where u is the potential function, $\bar{u}$ and $\bar{q}$ are prescribed values, $\Gamma = \Gamma_1 + \Gamma_2$ is the contour of domain Ω and n is the unit outward normal to surface Γ (see Fig. 1).

An integral equation, equivalent to (1)–(3) above, can be derived by using Green's third identity or a weighted residual formulation [4], and reads

$$u(\xi) = \int_\Gamma q(x)\, u^*(\xi, x)\, d\Gamma(x) - \int_\Gamma u(x)\, q^*(\xi, x)\, d\Gamma(x)\,. \tag{4}$$

The function u^* in Eq. (4) is the free space Green's function, or fundamental solution, to Laplace's equation, and represents the solution to Eq. (1) with a discrete singularity, in the form

$$\nabla^2 u^*(\xi, x) = -\Delta(\xi, x) \tag{5}$$

where Δ is the Dirac delta function, ξ and x are the source and field points, respectively. The function q^* in Eq. (4) is the outward normal derivative of u^*, along the contour Γ, with respect to the field point, i.e. $q^*(\xi, x) = \partial u^*(\xi, x)/\partial n(x)$.

For two-dimensional problems, u^* and q^* are of the form

$$u^* = \frac{1}{2\pi} \ln \frac{1}{r}, \quad q^* = -\frac{1}{2\pi r} r_n\,,$$

$$r = (X_1^2 + X_2^2)^{1/2}, \quad r_n = \frac{\partial r(\xi, x)}{\partial n(x)} = -\frac{1}{r}(X_1 x_{1,n} + X_2 x_{2,n})\,,$$

$$X_i = x_i(\xi) - x_i(x); \quad i = 1, 2$$

and $x_{i,n}$ are the direction cosines of the outward normal at x.

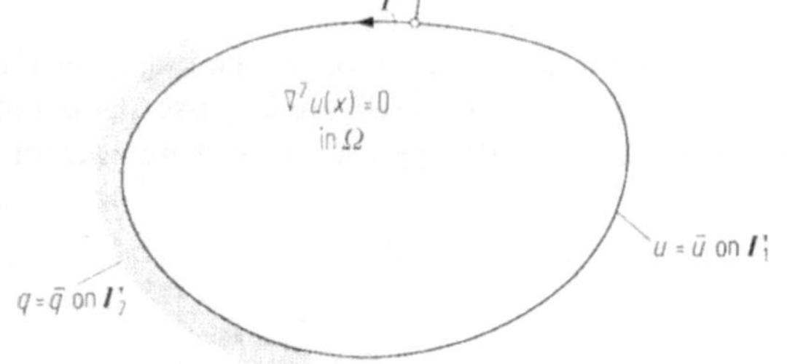

Fig. 1. Definition of the problem

Equation (4) permits computing the potential value at internal points once all values of u and q are known along the boundary. Fluxes at internal points can be computed by taking the derivative of Eq. (4) with respect to the source point,

$$\frac{\partial u(\xi)}{\partial x_i(\xi)} = \int_\Gamma q(x) \frac{\partial u^*(\xi, x)}{\partial x_i(\xi)} d\Gamma(x) - \int_\Gamma u(x) \frac{\partial q^*(\xi, x)}{\partial x_i(\xi)} d\Gamma(x) \tag{6}$$

in which

$$\frac{\partial u^*(\xi, x)}{\partial x_i(\xi)} = -\frac{X_i}{2\pi r^2},$$

$$\frac{\partial q^*(\xi, x)}{\partial x_i(\xi)} = \frac{1}{\pi r^2}\left[\frac{x_{i,n}}{2} - \frac{X_i}{r^2}(X_1 x_{1,n} + X_2 x_{2,n})\right].$$

Since in a well-posed problem only the value of u or q is prescribed at each boundary point, we can use Eq. (4) to compute the remaining unknowns. For this, we make the source point ξ tend to Γ and take into account the discontinuity of the second integral, obtaining

$$c(\xi)\, u(\xi) = \int_\Gamma q(x)\, u^*(\xi, x)\, d\Gamma(x) - \int_\Gamma u(x)\, q^*(\xi, x)\, d\Gamma(x) \tag{7}$$

in which the second integral is computed in the Cauchy principal value sense. The coefficient c, which is a function of the internal angle the boundary Γ makes at point ξ, can be calculated analytically [4] or, alternatively, by prescribing a constant potential, an idea analogous to the rigid-body translations of the theory of elasticity. This procedure will be discussed in detail in the next section.

We now discretize Eq. (7) by dividing the boundary Γ into a series of boundary elements. Furthermore, we assume that the value of u and q at a point within each element is related to its values at the element nodal points through some interpolation functions. In the present work, discretization is carried out by using M straight segments with linear interpolation functions. Thus, Eq. (7) becomes

$$c_i u_i = \sum_{j=1}^{M} \int_{\Gamma_j} (\psi_1 q_1 + \psi_2 q_2)\, u^*\, d\Gamma - \sum_{j=1}^{M} \int_{\Gamma_j} (\psi_1 u_1 + \psi_2 u_2)\, q^*\, d\Gamma \tag{8}$$

where

$$\psi_1 = \tfrac{1}{2}(1-\eta); \quad \psi_2 = \tfrac{1}{2}(1+\eta)$$

and η is a dimensionless coordinate equal to $2y/l$, l being the element length and y a local coordinate (Fig. 2).

Calling

$$g_{ij}^k = \int_{\Gamma_j} \psi_k u^*\, d\Gamma; \quad h_{ij}^k = \int_{\Gamma_j} \psi_k q^*\, d\Gamma$$

and adding the contribution from two adjoining elements, $(j-1)$ and (j), into one term, we can rewrite Eq. (8) in the form

$$c_i u_i = \sum^{N} G_{ij} q_j - \sum^{N} \hat{H}_{ij} u_j \tag{9}$$

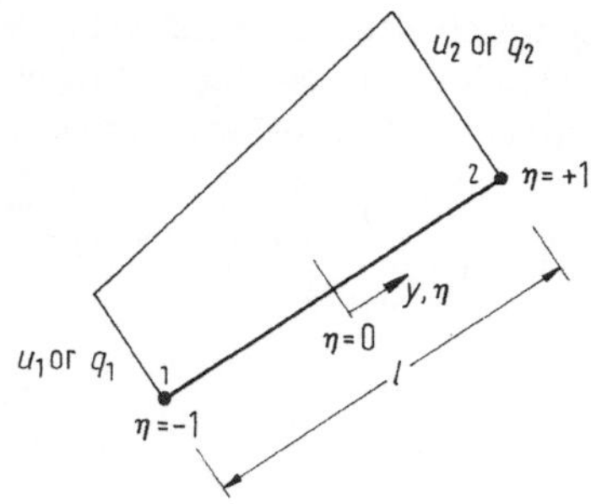

Fig. 2. Linear element

where N is the total number of boundary nodes and each nodal coefficient G_{ij} is equal to the term g_{ij}^2 of element $(j-1)$ plus the term g_{ij}^1 of element (j), for an anticlockwise numbering system; the same applies to $\hat{H}_{ij}$.

Equation (9) can be further rewritten as

$$\sum_{j=1}^{N} H_{ij} u_j = \sum_{j=1}^{N} G_{ij} q_j \tag{10}$$

with $H_{ij} = \hat{H}_{ij} + \delta_{ij} c_i$, where δ_{ij} is the Kronecker delta.

Equation (10) can now be applied, using a collocation technique, to all nodal points, producing a system of equations of order $N \times N$ which can be represented in matrix form as

$$\boldsymbol{HU} = \boldsymbol{GQ}. \tag{11}$$

By applying the prescribed boundary conditions of the problem, Eq. (11) can be reordered in such a way that a final system of equations is obtained

$$\boldsymbol{KY} = \boldsymbol{F} \tag{12}$$

where $\boldsymbol{K}$ is a fully populated square matrix of order N and $\boldsymbol{Y}$ is a vector containing all boundary unknowns. From the computational viewpoint, the coefficients H_{ij} $(i \neq j)$ and G_{ij} can be assembled directly into matrix $\boldsymbol{K}$ so that Eq. (11) need not be formed.

8.3 Numerical and Computational Aspects

8.3.1 Integration

Calculation of the g_{ij}^k and h_{ij}^k terms $(i \neq j)$ is carried out by evaluating the integrals numerically, using Gauss quadrature. The number of integration points employed depends on the distance from the source point to the element under consideration, such that the farther the element lies from the source point the fewer the number of integration points used in the calculation. The following criterion was adopted [4]:

$s \leqq 1.5$: 6 integration points
$1.5 < s < 5.5$: 4 integration points
$5.5 \leqq s$: 2 integration points

$$s = \frac{1}{2l} \{[2x_1(\xi) - x_1(1) - x_1(2)]^2 + [2x_2(\xi) - x_2(1) - x_2(2)]^2\}^{1/2}$$

where $x_i(\xi)$ are the coordinates of the source point, $x_i(1)$ and $x_i(2)$ the coordinates of the extreme points of the element to be integrated, and l its length. This criterion was found, by trial and error, to be both economical and accurate.

The terms g^1_{ii} and g^2_{ii} are calculated analytically by the following expressions:

$$g^1_{ii} = \frac{l_i}{8\pi} \int_{-1}^{1} \ln\left[\frac{2}{l_i(1+\eta)}\right](1-\eta)\, d\eta = \frac{l_i}{4\pi}\left[\tfrac{3}{2} - \ln(l_i)\right],$$
$$g^2_{ii} = \frac{l_i}{8\pi} \int_{-1}^{1} \ln\left[\frac{2}{l_i(1+\eta)}\right](1+\eta)\, d\eta = \frac{l_i}{4\pi}\left[\tfrac{1}{2} - \ln(l_i)\right], \tag{13}$$

in which the source point coincides with node 1, and the integral for g^1_{ii} has a logarithmic singularity at $\eta = -1$.

Computation of the terms H_{ii}, which include the coefficients c_i, is carried out by noting the fact that when a uniform potential is applied over the whole body the values of the derivatives must be zero. Under this condition Eq. (11) for closed domains produces

$$\boldsymbol{H}\boldsymbol{I} = \boldsymbol{0}. \tag{14}$$

where $\boldsymbol{I}$ is a vector of unit values. The above equation indicates that the sum of all the elements of $\boldsymbol{H}$ in a row should be zero, hence the values of the coefficients in the diagonal can be easily calculated once the off-diagonal coefficients are all known, i.e.

$$H_{ii} = -\sum_{\substack{j=1\\(j \neq i)}}^{N} H_{ij}; \quad i = 1, 2, \ldots, N. \tag{15}$$

8.3.2 Infinite Regions

The boundary integral equation (7) is also valid for infinite regular regions in the sense defined by Kellogg [6], provided certain regularity conditions concerning the behaviour of the functions in (7) on an infinitely distant surface are fulfilled.

Let $\bar{\Gamma}$ be the contour of a circle of radius R surrounding the boundary Γ and centered at point ξ (Fig. 3). A boundary integral equation similar to (7) for the finite domain Ω enclosed by the actual contour Γ and the fictitious contour $\bar{\Gamma}$ can be written as

$$c(\xi)\, u(\xi) + \int_{\Gamma} u(x)\, q^*(\xi, x)\, d\Gamma(x) + \int_{\bar{\Gamma}} u(x)\, q^*(\xi, x)\, d\bar{\Gamma}(x)$$
$$= \int_{\Gamma} q(x)\, u^*(\xi, x)\, d\Gamma(x) + \int_{\bar{\Gamma}} q(x)\, u^*(\xi, x)\, d\bar{\Gamma}(x) \tag{16}$$

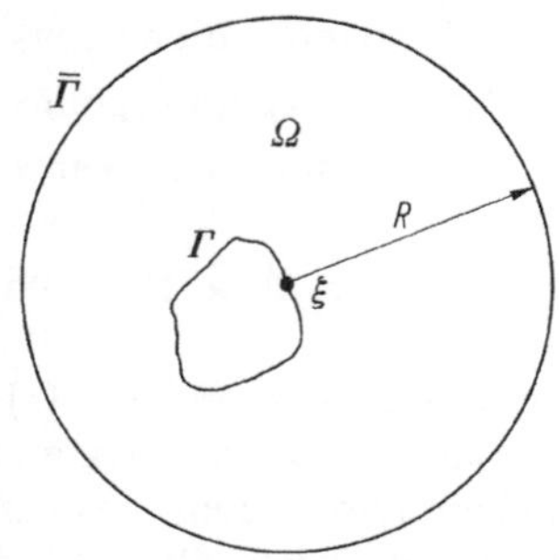

Fig. 3. Infinite region with cavity

If we let the radius $R \to \infty$; (16) will only be expressed in terms of boundary integrals over Γ alone if

$$\lim_{R \to \infty} \int_{\bar{\Gamma}} [q(x)\, u^*(\xi, x) - u(x)\, q^*(\xi, x)]\, d\bar{\Gamma}(x) = 0\,. \tag{17}$$

Since the function u^* behaves as the logarithm of R and its derivative q^* behaves as $O(R^{-1})$ as $R \to \infty$, the regularity conditions at infinity for this case imply that u behaves at most as $\ln R$, such that $q = O(R^{-1})$. As $d\Gamma = |J|\, d\theta$ and $|J| = O(R)$, the terms in the integral in Eq. (17) do not approach zero separately as $R \to \infty$ but they cancel each other, thus fulfilling condition (17).

Therefore, applying condition (17) into Eq. (16) yields Eq. (7), what means that the boundary integral equation obtained for points on the internal contour Γ of an infinite regular region is of the same form as the one for finite regions. The same is also valid for the integral equation for points inside the infinite region.

Special care has to be taken on the evaluation of the diagonal coefficients of matrix $\boldsymbol{H}$ through condition (14). In this case, the regularity conditions at infinity are violated as the function u is now assumed to be constant everywhere in Ω. Since it can easily be shown that

$$\lim_{R \to \infty} \int_{\bar{\Gamma}} q^*\, d\bar{\Gamma} = 1$$

we obtain instead of (15) the following relation

$$H_{ii} = -\sum_{\substack{j=1 \\ (j \neq i)}}^{N} H_{ij} + 1; \quad i = 1, 2, \ldots, N\,. \tag{18}$$

8.3.3 Robin Boundary Conditions

Other types of boundary conditions that frequently appear on practical problems can easily be incorporated into the formulation.

The Robin condition prescribes a linear relation between the potential and its normal derivative at points along the boundary as follows:

$$q = a\,u + b \tag{19}$$

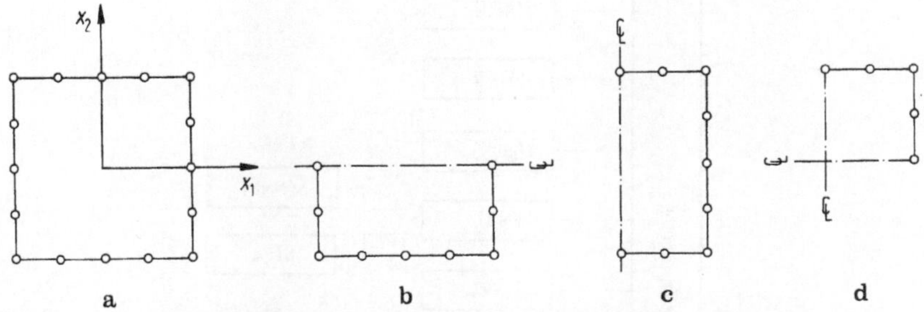

Fig. 4a–d. Types of symmetry. **a** No symmetry. **b** Symmetry with respect to x_1. **c** Symmetry with respect to x_2. **d** Symmetry with respect to x_1 and x_2

where a and b are known functions of the position. This is the case, for instance, of the convection boundary condition of heat conduction problems, the impedance condition of electromagnetic problems, etc.

If condition (19) is applied at all boundary nodes, we can write in matrix form

$$\boldsymbol{Q} = \boldsymbol{A}\boldsymbol{U} + \boldsymbol{B} \qquad (20)$$

where $\boldsymbol{A}$ is a diagonal matrix and $\boldsymbol{B}$ a vector containing the values of a and b, respectively, at each boundary node.

Substituting (20) into Eq. (11) yields the system of equations

$$(\boldsymbol{H} - \boldsymbol{G}\boldsymbol{A})\,\boldsymbol{U} = \boldsymbol{G}\boldsymbol{B}\,. \qquad (21)$$

After its solution, the normal derivatives of potentials along the boundary can be evaluated pointwise by using condition (19).

8.3.4 Symmetry

In the computer program to be described in the next section, symmetry is taken into account by a direct condensation process with integration over reflected elements such that no discretization of the symmetry axes is necessary. The cases that can be considered, together with the corresponding discretization, are presented in Fig. 4.

8.4 Description of the Computer Program

8.4.1 Main Program and Data Structure

The macro flow chart for the program can be seen in Fig. 5. The main program defines the maximum dimension of the system of equations, which is 100 in the present case. It also allocates channel 5 for the input (LEC) and channel 6 for the output (IMP).

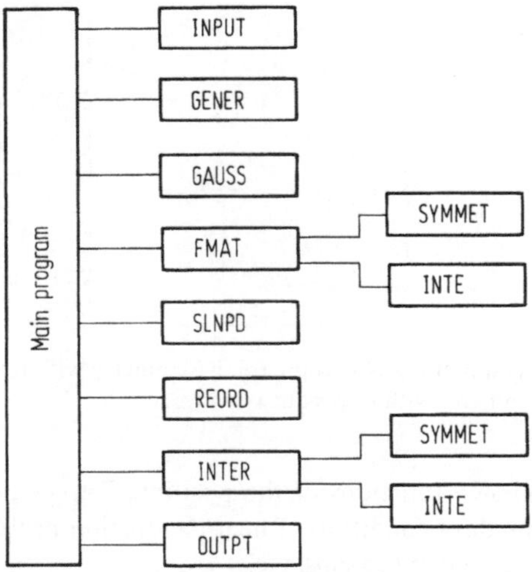

Fig. 5. Programm flow chart

The variables used in the program, together with their meaning, are listed below:

N Number of nodes
NE Number of elements
L Number of internal points
N1 Number of nodal coordinates cards
N2 Number of boundary conditions cards
NS Symmetry parameter (0 for no-symmetry; 1 for symmetry with respect to x_1; 2 with respect to x_2; 3 with respect to x_1 and x_2)
ND Parameter to indicate the type of analysis (0 for running the job; 1 for data check only)
M Number of different surfaces
MC Parameter to indicate the type of domain (0 bounded; 1 unbounded).

The following arrays are used in the program, the problem-dependent ones being indicated with the required dimensions for possible modifications (note that herein the listings are presented with N = 100, NE = 100, L = 50, M = 6):

X(N), Y(N) Coordinates x_1 and x_2 of boundary nodes
CX(L), CY(L) Coordinates x_1 and x_2 of internal points
FI(N) Vector to store prescribed boundary conditions. After subroutine REORD is called it contains the nodal potentials
DFI(N) Vector of independent coefficients in the system of equations. After subroutine SLNPD is called it contains the boundary solution. After subroutine REORD is called it contains the nodal fluxes
TO(N) Vector to store the free parameter if a linear relation between potential and flux is prescribed
SOL(L) Vector of internal potentials
QX(L), QY(L) Vectors of internal fluxes

SR(NE)	Vector of elements length
A(N, N)	System matrix
IC1(NE), IC2(NE)	Boundary elements connectivity
KODE(N)	Boundary conditions code vector (KODE = 0 if potential is prescribed; = 1 if flux is prescribed; = 2 if a linear relation is prescribed)
NL(M)	Vector containing the number of last node in each surface
IPR(M)	Vector containing parameters to indicate the type of each surface (IPR = 0 if surface is closed; = 1 if it is open)
TITLE(18)	Title of the problem
XI(6, 3)	Abscissas of Gaussian integration points
W(6, 3)	Weights of Gaussian integration points
G(2), H(2), GX(2), HX(2), GY(2), HY(2)	Auxiliary vectors.

```
      PROGRAM WROBEL1
C
C  PROGRAM FOR SOLUTION OF TWO-DIMENSIONAL POTENTIAL PROBLEMS
C  BY THE BOUNDARY ELEMENT METHOD USING LINEAR ELEMENTS
C
      COMMON /CONO/X(100),Y(100)
      COMMON /BOCO/DFI(100),FI(100),TO(100)/SMAT/A(100,100)
      COMMON /COIN/CX(50),CY(50),SOL(50),QX(50),QY(50)/ALENG/SR(100)
      COMMON /NCONE/IC1(100),IC2(100)/NST/KODE(100)
      COMMON /IPAR/N,NE,L,LEC,IMP,MC,M,NS,NL(6),IPR(6)
C
C  ASSIGN DATA SET NUMBERS FOR INPUT,LEC, AND OUTPUT,IMP
C
      LEC=5
      IMP=6
C
C  INPUT
C
      CALL INPUT(ND)
C
C  GENERATE ELEMENTS DATA
C
      CALL GENER
      IF(ND.EQ.1) STOP
C
C  SET UP THE GAUSS INTEGRATION POINTS
C
      CALL GAUSS
C
C  FORM SYSTEM OF EQUATIONS
C
      CALL FMAT
C
C  SOLUTION OF THE SYSTEM OF EQUATIONS
C
      CALL SLNPD
C
C  REORDER FI AND DFI ARRAYS
C
      CALL REORD
C
C  COMPUTE VALUES AT INTERNAL POINTS
C
      IF(L.GT.0) CALL INTER
C
C  OUTPUT
C
      CALL OUTPT
C
C  END OF PROGRAM
C
      STOP
      END
```

8.4.2 Subroutine INPUT

The input required by the program is read in this subroutine. The input data consists of the following groups of cards:

a) *Title Card:* One card containing the title of the problem
   FORMAT (18A4)
b) *Basic Parameters:* One card containing the following parameters:
   N, NE, L, M, N1, N2, NS, ND, MC
   FORMAT (9I5)
c) *Surfaces Parameters:* M cards indicating for each surface K:
   IPR(K), NL(K)
   FORMAT (2I5)
d) *Nodal Coordinates:* N1 cards indicating for each node K:
   K, X(K), Y(K)
   FORMAT (I5, 2F10.0)

Note: (i) In the case of equally spaced nodes only the extreme ones need be read, the others being generated by the program using linear interpolation.

(ii) In the case of double nodes [4], the coordinates of both should be given with coincident values.

(iii) The numbering of the nodes should be in the clockwise sense for unbounded domains and in the anticlockwise sense for bounded ones (see Fig. 6).

e) *Boundary Conditions:* N2 cards indicating for each node K:
   K, KODE(K), FI(K), TO(K)
   FORMAT (2I5, 2F10.0)

Note: In the case when the boundary conditions are constant or vary linearly over part of the boundary, only the values at the extreme nodes of this part need be read, the others being generated by the program using linear interpolation.

f) *Internal Points:* L cards indicating for each point K:
   CX(K), CY(K)
   FORMAT (2F10.0).

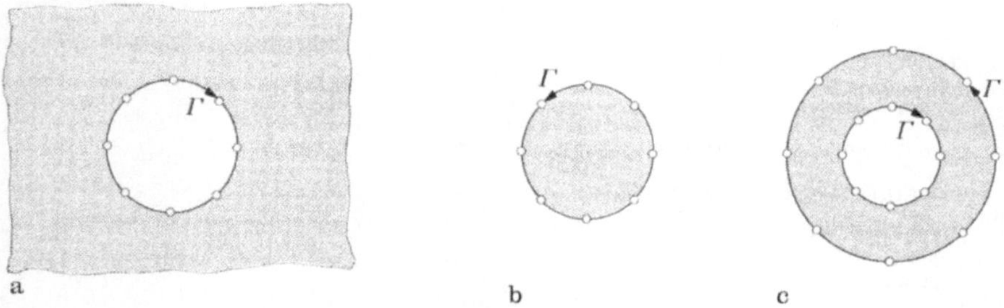

Fig. 6a–c. Element numbering sense. **a** Unbounded case. **b** Bounded case. **c** Multi-surface case

```
      SUBROUTINE INPUT(ND)
C
C  THIS SUBROUTINE READS ALL THE INPUT DATA
C
      COMMON /COND/X(100),Y(100)
      COMMON /BOCO/DFI(100),FI(100),TO(100)
      COMMON /COIN/CX(50),CY(50),SOL(50),QX(50),QY(50)
      COMMON /NST/KODE(100)
      COMMON /IPAR/N,NE,L,LEC,IMP,MC,M,NS,NL(6),IPR(6)
      DIMENSION TITLE(18)
      WRITE(IMP,100)
  100 FORMAT(' ',120('*'))
C
C  READ NAME OF THE JOB
C
      READ(LEC,150)TITLE
  150 FORMAT(18A4)
      WRITE(IMP,200)TITLE
  200 FORMAT(25X,18A4)
C
C  READ BASIC PARAMETERS
C
      READ(LEC,250)N,NE,L,M,N1,N2,NS,ND,MC
  250 FORMAT(9I5)
      WRITE(IMP,300)N,NE,L,M,NS
  300 FORMAT(//2X,'DATA',//2X,'NUMBER OF BOUNDARY NODES=',I3/2X,
     *'NUMBER OF BOUNDARY ELEMENTS=',I3/2X,
     *'NUMBER OF INTERNAL POINTS WHERE VALUES ARE REQUIRED=',I3/2X,
     *'NUMBER OF SURFACES=',I3/2X,'SYMMETRY PARAMETER=',I3)
C
C  READ CHARACTERISTICS OF EACH SURFACE
C
      WRITE(IMP,350)
  350 FORMAT(/2X,'SURFACE',10X,'TYPE',10X,'LAST NODE')
      DO 10 I=1,M
      READ(LEC,400)IPR(I),NL(I)
  400 FORMAT(2I5)
   10 WRITE(IMP,450)I,IPR(I),NL(I)
  450 FORMAT(3X,I3,2(13X,I3))
C
C  TEST FOR UNBOUNDED DOMAIN
C
      IF(MC.EQ.0) GO TO 20
      WRITE(IMP,500)
  500 FORMAT(/2X,'UNBOUNDED DOMAIN: ADD 1. TO THE DIAGONAL COEFFICIENTS
     *OF MATRIX H')
C
C  READ AND GENERATE NODAL COORDINATES
C
   20 WRITE(IMP,550)
  550 FORMAT(//2X,'NODAL COORDINATES',//5X,'NODE',10X,'X',18X,'Y')
      J=1
      DO 40 I=1,N1
      I1=J
      READ(LEC,600) J,X(J),Y(J)
  600 FORMAT(I5,2F10.0)
      J1=J-I1
      IF(J1.LE.1) GO TO 40
      J2=J1-1
      AJ1=FLOAT(J1)
      AUX=(X(J)-X(I1))/AJ1
      AUY=(Y(J)-Y(I1))/AJ1
      DO 30 K=1,J2
      I2=I1+K
      I3=I2-1
      X(I2)=X(I3)+AUX
   30 Y(I2)=Y(I3)+AUY
   40 CONTINUE
      DO 50 I=1,N
   50 WRITE(IMP,650) I,X(I),Y(I)
  650 FORMAT(5X,I3,2(5X,E14.7))
C
C  READ AND GENERATE BOUNDARY CONDITIONS
C
C  FI(J)= VALUE OF THE POTENTIAL IN THE NODE IF KODE=0,
C         VALUE OF THE POTENTIAL DERIVATIVE IF KODE=1.
C
```

```
C   IF KODE(J)= 2, THE BOUNDARY CONDITION PRESCRIBES A LINEAR
C   RELATION BETWEEN POTENTIAL AND POTENTIAL DERIVATIVE IN THE FORM:
C
C             DER = A * POT + B
C
C   FI(J) = VALUE OF COEFFICIENT A
C   TO(J) = VALUE OF COEFFICIENT B
C
      WRITE(IMP,700)
  700 FORMAT(//2X,'BOUNDARY CONDITIONS'//5X,'NODE',6X,'CODE',6X,'PRESCRI
     *BED VALUE',6X,'FREE COEFFICIENT')
      J=1
      DO 70 I=1,N2
      I1=J
      READ(LEC,750) J,KODE(J),FI(J),TO(J)
  750 FORMAT(2I5,2F10.0)
      J1=J-I1
      IF(J1.LE.1) GO TO 70
      J2=J1-1
      AJ1=FLOAT(J1)
      AUX=(FI(J)-FI(I1))/AJ1
      AUY=(TO(J)-TO(I1))/AJ1
      DO 60 K=1,J2
      I2=I1+K
      I3=I2-1
      KODE(I2)=KODE(I3)
      FI(I2)=FI(I3)+AUX
   60 TO(I2)=TO(I3)+AUY
   70 CONTINUE
      DO 80 I=1,N
   80 WRITE(IMP,800) I,KODE(I),FI(I),TO(I)
  800 FORMAT(5X,I3,9X,I1,2(8X,E14.7))
C
C   READ INTERNAL POINTS COORDINATES
C
      IF(L.LE.0) RETURN
      DO 90 I=1,L
   90 READ(LEC,850)CX(I),CY(I)
  850 FORMAT(2F10.0)
      RETURN
      END
```

8.4.3 Subroutine GENER

This subroutine generates the elements connectivity, calculates their length and sets a parameter for the symmetry loop, according to the type of the symmetry of the problem. It has as input the variables N, NE, NS, M and the arrays X, Y, NL, IPR; the routine outputs the arrays SR, IC1, IC2 and parameter M, which is now used as a parameter for the symmetry loop.

```
      SUBROUTINE GENER
C
C   THIS SUBROUTINE GENERATES THE ELEMENTS CONNECTIVITY,
C   CALCULATES THEIR LENGTH AND SETS UP A PARAMETER FOR THE SYMMETRY LOOP
C
      COMMON /CONO/X(100),Y(100)
      COMMON /ALENG/SR(100)
      COMMON /NCONE/IC1(100),IC2(100)
      COMMON /IPAR/N,NE,L,LEC,IMP,MC,M,NS,NL(6),IPR(6)
C
C   GENERATE BOUNDARY ELEMENTS CONNECTIVITY
C
      WRITE(IMP,100)
  100 FORMAT(//2X,'ELEMENT',6X,'NODE 1',7X,'NODE 2',10X,'LENGTH')
      I=0
      DO 30 J=1,N
      J1=J+1
      DO 10 K=1,M
      IF(J.NE.NL(K)) GO TO 10
      IF(IPR(K).EQ.1) GO TO 30
```

```
      IF(K.EQ.1) J1=1
      IF(K.GT.1) J1=NL(K-1)+1
      GO TO 20
   10 CONTINUE
   20 TEMP1=ABS(X(J)-X(J1))
      TEMP2=ABS(Y(J)-Y(J1))
      IF(TEMP1.LE.1.E-10.AND.TEMP2.LE.1.E-10) GO TO 30
      I=I+1
      IC1(I)=J
      IC2(I)=J1
C
C  CALCULATE BOUNDARY ELEMENTS LENGTH
C
      SR(I)=SQRT(TEMP1*TEMP1+TEMP2*TEMP2)
      WRITE(IMP,200) I,IC1(I),IC2(I),SR(I)
  200 FORMAT(3X,I3,2(10X,I3),8X,E14.7)
   30 CONTINUE
C
C  SET UP PARAMETER FOR SYMMETRY LOOP
C
      M=1
      IF(NS.EQ.1.OR.NS.EQ.2) M=2
      IF(NS.EQ.3) M=4
      RETURN
      END
```

8.4.4 Subroutine SYMMET

This subroutine reflects the singular node according to the type of symmetry of the problem.

Input: parameters XC, YC, IS, NS. Output: parameters XP, YP.

```
      SUBROUTINE SYMMET(XC,YC,XP,YP,IS,NS)
C
C  THIS SUBROUTINE REFLECTS THE SINGULAR NODE
C  ACCORDING TO THE TYPE OF SYMMETRY
C
      GO TO (10,20,30,40),IS
   10 XP=XC
      YP=YC
      RETURN
   20 IF(NS.EQ.2) GO TO 30
      XP=XC
      YP=-YC
      RETURN
   30 XP=-XC
      YP=YC
      RETURN
   40 XP=-XC
      YP=-YC
      RETURN
      END
```

8.4.5 Subroutine GAUSS

This subroutine attributes the abscissas and weights of the Gaussian integration points.

Output: arrays XI, W.

```
      SUBROUTINE GAUSS
C
C  THIS SUBROUTINE SETS UP THE GAUSSIAN INTEGRATION POINTS
C
      COMMON /PINT/XI(6,3),W(6,3)
C
C  ABCISSAS OF GAUSSIAN POINTS
C
```

```
C  TWO-POINTS SCHEME
C
      XI(1,1)=0.577350269189626
      XI(2,1)=-XI(1,1)
C
C  FOUR-POINTS SCHEME
C
      XI(1,2)=0.339981043584856
      XI(2,2)=-XI(1,2)
      XI(3,2)=0.861136311594053
      XI(4,2)=-XI(3,2)
C
C  SIX-POINTS SCHEME
C
      XI(1,3)=0.238619186083197
      XI(2,3)=-XI(1,3)
      XI(3,3)=0.661209386466265
      XI(4,3)=-XI(3,3)
      XI(5,3)=0.932469514203152
      XI(6,3)=-XI(5,3)
C
C  WEIGHTS OF GAUSSIAN POINTS
C
C  TWO-POINTS SCHEME
C
      W(1,1)=1.
      W(2,1)=W(1,1)
C
C  FOUR-POINTS SCHEME
C
      W(1,2)=0.652145154862546
      W(2,2)=W(1,2)
      W(3,2)=0.347854845137454
      W(4,2)=W(3,2)
C
C  SIX-POINTS SCHEME
C
      W(1,3)=0.467913934572691
      W(2,3)=W(1,3)
      W(3,3)=0.360761573048139
      W(4,3)=W(3,3)
      W(5,3)=0.171324492379170
      W(6,3)=W(5,3)
      RETURN
      END
```

8.4.6 Subroutine FMAT

This subroutine computes matrix $\boldsymbol{K}$ and vector $\boldsymbol{F}$ of Eq. (12) without previously forming matrices $\boldsymbol{H}$ and $\boldsymbol{G}$ of Eq. (11). This is done by taking into account the prescribed boundary conditions for the node under consideration at element level.

The off-diagonal coefficients of $\boldsymbol{G}$ and $\boldsymbol{H}$ are computed in an auxiliary routine, INTE. The G_{ii} and H_{ii} coefficients are computed directly in this routine using Eqs. (13) and (15) or (18), respectively.

Input: Parameters N, NE, NS, MC, M and arrays X, Y, FI, TO, SR, IC1, IC2, KODE. Output: arrays DFI, A.

```
      SUBROUTINE FMAT
C
C  THIS SUBROUTINE COMPUTES THE SYSTEM MATRIX AND VECTOR
C  OF INDEPENDENT COEFFICIENTS
C
      COMMON /CONO/X(100),Y(100)
      COMMON /BOCO/DFI(100),FI(100),TO(100)/SMAT/A(100,100)
      COMMON /ALENG/SR(100)
```

```
      COMMON /NCONE/IC1(100),IC2(100)/NST/KODE(100)
      COMMON /IPAR/N,NE,L,LEC,IMP,MC,M,NS,NL(6),IPR(6)
      DIMENSION G(2),H(2),GX(2),HX(2),GY(2),HY(2)
C
C  CLEAR THE SYSTEM MATRIX AND VECTOR OF INDEPENDENT COEFFICIENTS
C
      DO 10 I=1,N
      DFI(I)=0.
      DO 10 J=1,N
   10 A(I,J)=0.
C
C  COMPUTE THE G(I,J) AND H(I,J) COEFFICIENTS AND ARRENGE
C  THE SYSTEM OF EQUATIONS READY TO BE SOLVED. DFI ORIGINALLY
C  CONTAINS THE INDEPENDENT COEFFICIENTS, AFTER SOLUTION IT
C  WILL CONTAIN THE VALUES OF THE SYSTEM UNKNOWNS (THE POTENTIAL
C  FOR POINTS WHERE A LINEAR RELATION IS PRESCRIBED)
C
      TPI=2.*3.1415926535 8979
      FPI=2.*TPI
C
C  LOOP OVER BOUNDARY NODES
C
      DO 80 I=1,N
C
C  SET DIAGONAL TERM OF MATRIX H
C
      DIAG=0.
C
C  LOOP OVER SYMMETRY AXES
C
      DO 70 IS=1,M
      CALL SYMMET(X(I),Y(I),XP,YP,IS,NS)
C
C  LOOP OVER BOUNDARY ELEMENTS
C
      DO 70 J=1,NE
      J1=IC1(J)
      J2=IC2(J)
      TEMP1=ABS(XP-X(J1))
      TEMP2=ABS(YP-Y(J1))
      TEMP3=ABS(XP-X(J2))
      TEMP4=ABS(YP-Y(J2))
C
C  TEST IF SOURCE POINT BELONGS TO PRESENT ELEMENT
C
      IF(TEMP1.LT.1.E-10.AND.TEMP2.LT.1.E-10) GO TO 20
      IF(TEMP3.LT.1.E-10.AND.TEMP4.LT.1.E-10) GO TO 20
C
C  CALCULATION OF OFF-DIAGONAL TERMS OF MATRICES G AND H
C
      CALL INTE(XP,YP,X(J1),Y(J1),X(J2),Y(J2),SR(J),H(1),H(2),G(1),
     *G(2),HX(1),HX(2),GX(1),GX(2),HY(1),HY(2),GY(1),GY(2),FPI,1)
      GO TO 30
C
C  CALCULATION OF DIAGONAL TERMS OF MATRICES G AND H
C
   20 H(1)=0.
      H(2)=0.
      CH=SR(J)*ALOG(SR(J))/2.
      G(1)=(0.75*SR(J)-CH)/TPI
      G(2)=(0.25*SR(J)-CH)/TPI
      IF(TEMP1.LT.1.E-10.AND.TEMP2.LT.1.E-10) GO TO 30
      CH=G(1)
      G(1)=G(2)
      G(2)=CH
   30 DIAG=DIAG-H(1)-H(2)
C
C  INTRODUCTION OF BOUNDARY CONDITIONS
C
      DO 70 K=1,2
      IF(K.EQ.1) K1=J1
      IF(K.EQ.2) K1=J2
      IF(KODE(K1)-1) 40,50,60
   40 CH=H(K)
      H(K)=-G(K)
      G(K)=-CH
```

```
   50 DFI(I)=DFI(I)+G(K)*FI(K1)
      A(I,K1)=A(I,K1)+H(K)
      GO TO 70
   60 DFI(I)=DFI(I)+G(K)*TO(K1)
      A(I,K1)=A(I,K1)+H(K)-G(K)*FI(K1)
   70 CONTINUE
C
C  CONSIDERATION OF DIAGONAL TERM OF MATRIX H
C
      IF(KODE(I).EQ.0) DFI(I)=DFI(I)-DIAG*FI(I)
      IF(KODE(I).GT.0) A(I,I)=A(I,I)+DIAG
   80 CONTINUE
      IF(MC.EQ.0) RETURN
C
C  COMPLEMENTARY TERM FOR UNBOUNDED DOMAINS
C
      DO 90 I=1,N
      IF(KODE(I).EQ.0) DFI(I)=DFI(I)-FI(1)
      IF(KODE(I).GT.0) A(I,I)=A(I,I)+1.
   90 CONTINUE
      RETURN
      END
```

8.4.7 Subroutine INTE

This subroutine computes the values of the off-diagonal coefficients of matrices ***G*** and ***H*** using numerical integration. It also computes the integrals necessary for the evaluation of potential and fluxes at internal points. The number of integration points adopted in the calculations is a function of the distance from the source point to the extreme points of the element under consideration, as discussed in Sect. 3.1.

Input: parameters XP, YP, X1, Y1, X2, Y2, TR, FPI, LF and arrays XI, W. Output: parameters A1, A2, B1, B2, C1, C2, D1, D2, C3, C4, D3, D4.

```
      SUBROUTINE INTE(XP,YP,X1,Y1,X2,Y2,TR,A1,A2,B1,B2,C1,C2,D1,D2,
     *C3,C4,D3,D4,FPI,LF)
C
C  THIS SUBROUTINE COMPUTES THE VALUES OF THE G AND H MATRICES
C  OFF-DIAGONAL COEFFICIENTS BY MEANS OF NUMERICAL INTEGRATION
C
      COMMON /PINT/XI(6,3),W(6,3)
C
C  CLEAR PARAMETERS
C
      A1=0.
      A2=0.
      B1=0.
      B2=0.
      C1=0.
      C2=0.
      D1=0.
      D2=0.
      C3=0.
      C4=0.
      D3=0.
      D4=0.
C
C  AUXILIARY PARAMETERS
C
      AX=(X1-X2)/2.
      BX=(X1+X2)/2.
      AY=(Y2-Y1)/2.
      BY=(Y2+Y1)/2.
      SR=TR/2.
C
C  DIRECTION COSINES
C
      CA=AY/SR
      SA=AX/SR
C
```

```
C   SELECT NUMBER OF INTEGRATION POINTS
C
      SEL=0.5*SQRT((2.*XP-X1-X2)**2+(2.*YP-Y1-Y2)**2)/TR
      NPI=4
      IF(SEL.LE.1.5) NPI=6
      IF(SEL.GT.3.5) NPI=2
      INP=NPI/2
C
C   LOOP OVER INTEGRATION POINTS
C
      DO 10 I=1,NPI
      GI=XI(I,INP)
      OME=W(I,INP)
C
C   INTERPOLATION FUNCTIONS (DIVIDED BY 2.*PI)
C
      GIP=(1.+GI)/FPI
      GIL=(1.-GI)/FPI
C
C   GLOBAL COORDINATES OF INTEGRATION POINTS
C
      XCO=BX-AX*GI
      YCO=BY+AY*GI
C
C   CALCULATION OF G(I,J) AND H(I,J)
C
      XC=XP-XCO
      YC=YP-YCO
      CT=XC*CA+YC*SA
      RA=XC*XC+YC*YC
      CT1=SR*OME/RA
      G=-0.5*ALOG(RA)*CT1*RA
      H=CT*CT1
      A1=A1+GIL*H
      A2=A2+GIP*H
      B1=B1+GIL*G
      B2=B2+GIP*G
      IF(LF.EQ.1) GO TO 10
C
C   CALCULATION OF INTEGRALS NECESSARY FOR INTERNAL FLUXES
C
      CT2=2.*CT/RA
      GX=-XC*CT1
      GY=-YC*CT1
      HX=(CA-CT2*XC)*CT1
      HY=(SA-CT2*YC)*CT1
      C1=C1+GIL*HX
      C2=C2+GIP*HX
      D1=D1+GIL*GX
      D2=D2+GIP*GX
      C3=C3+GIL*HY
      C4=C4+GIP*HY
      D3=D3+GIL*GY
      D4=D4+GIP*GY
   10 CONTINUE
      RETURN
      END
```

8.4.8 Subroutine SLNPD

This is a standard subroutine, taken from [7], for the solution of systems of equations with a nonpositive definite matrix using Gauss elimination. If the matrix has a zero in the diagonal it will interchange rows, deciding that the system matrix is singular only when no row interchange produces a nonzero diagonal coefficient for a fixed row position.

The final result is a vector of unknown boundary potentials and fluxes which is overwritten in B, which locally corresponds to DFI in the main program.

Input: parameter N and arrays A, B. Output: array B.

```
      SUBROUTINE SLNPD
C
C  SOLUTION OF THE LINEAR SYSTEM OF EQUATIONS
C  BY THE GAUSS ELIMINATION METHOD PROVIDING
C  FOR INTERCHANGING ROWS WHEN ENCOUNTERING A
C  ZERO DIAGONAL COEFFICIENT
C
C  A : SYSTEM MATRIX
C  B : ORIGINALLY IT CONTAINS THE INDEPENDENT
C      COEFFICIENTS, AFTER SOLUTION IT CONTAINS
C      THE VALUES OF THE SYSTEM UNKNOWNS
C
C  N : ACTUAL NUMBER OF EQUATIONS
C
      COMMON /BOCO/B(100),FI(100),TO(100)/SMAT/A(100,100)
      COMMON /IPAR/N,NE,L,LEC,IMP,MC,M,NS,NL(6),IPR(6)
      N1=N-1
      DO 100 K=1,N1
      K1=K+1
      C=A(K,K)
      IF(ABS(C).GT.0.) GO TO 3
      DO 7 J=K1,N
C
C  TRY TO INTERCHANGE ROWS TO GET NON-ZERO DIAGONAL COEFFICIENT
C
      IF(ABS(A(J,K)).LE.0.) GO TO 7
      DO 6 L1=K,N
      C=A(K,L1)
      A(K,L1)=A(J,L1)
    6 A(J,L1)=C
      C=B(K)
      B(K)=B(J)
      B(J)=C
      C=A(K,K)
      GO TO 3
    7 CONTINUE
    8 WRITE(IMP,2) K
    2 FORMAT(' **** SINGULARITY IN ROW',I5)
      STOP
C
C  DIVIDE ROW BY DIAGONAL COEFFICIENT
C
    3 DO 4 J=K1,N
    4 A(K,J)=A(K,J)/C
      B(K)=B(K)/C
C
C  ELIMINATE UNKNOWN X(K) FROM ROW I
C
      DO 10 I=K1,N
      C=A(I,K)
      DO 9 J=K1,N
    9 A(I,J)=A(I,J)-C*A(K,J)
   10 B(I)=B(I)-C*B(K)
  100 CONTINUE
C
C  COMPUTE LAST UNKNOWN
C
      IF(ABS(A(N,N)).LE.0.) GO TO 8
      B(N)=B(N)/A(N,N)
C
C  APPLY BACKSUBSTITUTION PROCESS TO COMPUTE REMAINING UNKNOWNS
C
      DO 200 L1=1,N1
      K=N-L1
      K1=K+1
      DO 200 J=K1,N
  200 B(K)=B(K)-A(K,J)*B(J)
      RETURN
      END
```

8.4.9 Subroutine REORD

This subroutine reorders the FI and DFI arrays in order to put all the potential values in FI and all the normal derivatives in DFI. It also calculates the derivatives at nodes where a Robin condition is prescribed.

Input: parameter N and arrays FI, DFI, TO, KODE. Output: arrays FI, DFI.

```
      SUBROUTINE REORD
C
C  THIS SUBROUTINE REORDERS FI AND DFI ARRAYS TO PUT ALL POTENTIAL
C  VALUES IN FI AND ALL DERIVATIVE VALUES IN DFI.
C  IT ALSO CALCULATES THE DERIVATIVES AT POINTS WHERE A
C  ROBIN CONDITION IS PRESCRIBED
C
      COMMON /BOCO/DFI(100),FI(100),TO(100)
      COMMON /NST/KODE(100)
      COMMON /IPAR/N,NE,L,LEC,IMP,MC,M,NS,NL(6),IPR(6)
      DO 10 I=1,N
C
C  COMPUTE DERIVATIVES USING ROBIN CONDITION
C
      IF(KODE(I).EQ.2) FI(I)=FI(I)*DFI(I)+TO(I)
      IF(KODE(I).LE.0) GO TO 10
C
C  REORDER FI AND DFI ARRAYS
C
      CH=FI(I)
      FI(I)=DFI(I)
      DFI(I)=CH
   10 CONTINUE
      RETURN
      END
```

8.4.10 Subroutine INTER

This subroutine computes the potential and its cartesian derivatives at selected internal points, using Eqs. (4) and (6). It uses the auxiliary routine INTE for the calculation of the integrals.

Input: parameters NE, L, NS, M and arrays X, Y, CX, CY, SR, IC1, IC2, FI, DFI. Output: arrays SOL, QX, QY.

```
      SUBROUTINE INTER
C
C  THIS SUBROUTINE COMPUTES THE POTENTIAL
C  AND ITS DERIVATIVES AT INTERNAL POINTS
C
      COMMON /COND/X(100),Y(100)
      COMMON /BOCO/DFI(100),FI(100),TO(100)
      COMMON /COIN/CX(50),CY(50),SOL(50),QX(50),QY(50)
      COMMON /ALENG/SR(100)
      COMMON /NCONE/IC1(100),IC2(100)
      COMMON /IPAR/N,NE,L,LEC,IMP,MC,M,NS,NL(6),IPR(6)
      FPI=4.*3.14159265358979
C
C  LOOP OVER INTERNAL POINTS
C
      DO 50 I=1,L
      SOL(I)=0.
      QX(I)=0.
      QY(I)=0.
C
C  LOOP OVER SYMMETRY AXES
C
      DO 50 IS=1,M
      CALL SYMMET(CX(I),CY(I),XP,YP,IS,NS)
C
```

```
C   LOOP OVER BOUNDARY ELEMENTS
C
      DO 50 J=1,NE
      J1=IC1(J)
      J2=IC2(J)
      CALL INTE(XP,YP,X(J1),Y(J1),X(J2),Y(J2),SR(J),A1,A2,B1,B2,C1,C2,
     *D1,D2,C3,C4,D3,D4,FPI,0)
      GO TO (40,10,20,20),IS
   10 IF(NS.NE.2) GO TO 30
   20 C1=-C1
      C2=-C2
      D1=-D1
      D2=-D2
      IF(IS.NE.4) GO TO 40
   30 C3=-C3
      C4=-C4
      D3=-D3
      D4=-D4
   40 SOL(I)=SOL(I)+DFI(J1)*B1+DFI(J2)*B2-FI(J1)*A1-FI(J2)*A2
      QX(I)=QX(I)+DFI(J1)*D1+DFI(J2)*D2-FI(J1)*C1-FI(J2)*C2
      QY(I)=QY(I)+DFI(J1)*D3+DFI(J2)*D4-FI(J1)*C3-FI(J2)*C4
   50 CONTINUE
      RETURN
      END
```

8.4.11 Subroutine OUTPT

This subroutine outputs the results for the potential and its derivatives at boundary nodes and selected internal points.

Input: parameters N, L and arrays X, Y, CX, CY, FI, DFI, SOL, QX, QY.

```
      SUBROUTINE OUTPT
C
C   THIS SUBROUTINE OUTPUTS THE RESULTS
C
      COMMON /CONO/X(100),Y(100)
      COMMON /BOCO/DFI(100),FI(100),TO(100)
      COMMON /COIN/CX(50),CY(50),SOL(50),QX(50),QY(50)
      COMMON /IPAR/N,NE,L,LEC,IMP,MC,M,NS,NL(6),IPR(6)
C
C   PRINT RESULTS FOR BOUNDARY NODES
C
      WRITE(IMP,100)
  100 FORMAT(' ',120('*')//1X,'RESULTS'//2X,'BOUNDARY NODES'//16X,'X',
     *23X,'Y',20X,'POTENTIAL',9X,'POTENTIAL DERIVATIVE'/)
      DO 10 I=1,N
   10 WRITE(IMP,200)X(I),Y(I),FI(I),DFI(I)
  200 FORMAT(4(10X,E14.7))
      IF(L.LE.0) GO TO 30
C
C   PRINT RESULTS AT INTERNAL POINTS
C
      WRITE(IMP,300)
  300 FORMAT(//,2X,'INTERNAL POINTS',//11X,'X',18X,'Y',15X,'POTENTIAL',
     *11X,'FLUX QX',12X,'FLUX QY'/)
      DO 20 K=1,L
   20 WRITE(IMP,400) CX(K),CY(K),SOL(K),QX(K),QY(K)
  400 FORMAT(5(5X,E14.7))
   30 WRITE(IMP,500)
  500 FORMAT(' ',120('*'))
      RETURN
      END
```

8.5 Applications

In what follows, three examples of application are discussed. In order to illustrate the input data required by the program, listings of the data files are presented together with complete printouts corresponding to the solution of the problems.

8.5.1 Pipe of Elliptical Cross-Section

The equation of motion of a uniform incompressible viscous fluid in steady unidirection flow (in the x_3 direction) is [8]

$$-\frac{\partial p}{\partial x_3}+\mu\left(\frac{\partial^2 u}{\partial x_1^2}+\frac{\partial^2 u}{\partial x_2^2}\right)=0$$

where μ is the viscosity of the fluid, $\partial p/\partial x_3=-G$ is a constant pressure gradient and u is the velocity component in the x_3 direction. This equation can be rewritten as

$$\nabla^2 u=-\frac{G}{\mu}.$$

For a pipe of elliptical cross-section the velocity distribution is of the form [8]

$$u=\frac{G}{2\mu(a^{-2}+b^{-2})}\left(1-\frac{x_1^2}{a^2}-\frac{x_2^2}{b^2}\right)$$

where a and b are the semiaxes of the ellipse. Taking the value of the constant $G/\mu=2$ and the semiaxes $a=2$ and $b=1$, the problem to be solved is

$$\nabla^2 u=-2$$

with boundary condition $u=0$ on Γ.

The solution of the above Poisson's equation can be divided into two parts, $u=u_1+u_2$, where $u_1=-(x_1^2+x_2^2)/2$ is a particular solution and u_2 a complementary one. Thus, the problem is reduced to the solution of Laplace's equation

$$\nabla^2 u_2=0$$

with boundary condition $u_2=-u_1$ on Γ.

Results for the velocity u and the derivatives $\partial u/\partial x_1$ and $\partial u/\partial x_2$ (necessary for the evaluation of the tangential stresses $\tau_{x_1x_3}$ and $\tau_{x_2x_3}$) are presented in Table 1, compared to the analytical solution. These results were obtained with a discretiza-

Table 1. Results for pipe of elliptical cross-section

Point	Variable	BEM	Exact
A	u	0.351	0.350
(1.50; 0)	$\partial u/\partial x_1$	− 0.603	− 0.600
	$\partial u/\partial x_2$	0.0	0.0
B	u	0.568	0.566
(0.60; 0.45)	$\partial u/\partial x_1$	− 0.240	− 0.240
	$\partial u/\partial x_2$	− 0.719	− 0.720
C	u	0.640	0.638
(0; 0.45)	$\partial u/\partial x_1$	0.0	0.0
	$\partial u/\partial x_2$	− 0.720	− 0.720

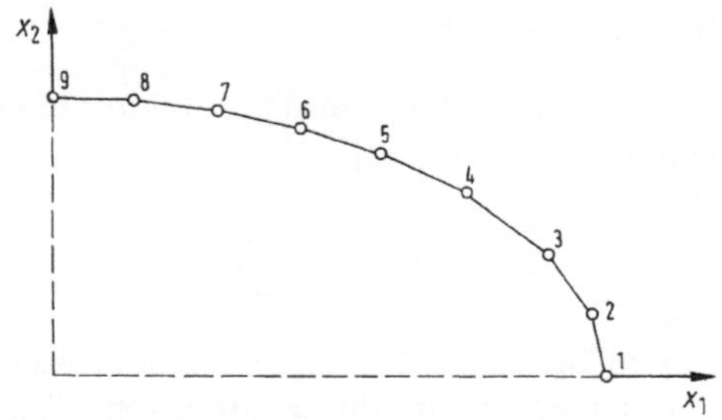

Fig. 7. Pipe of elliptical cross-section: discretization

Table 2. Data file for elliptical pipe

```
FLOW IN PIPE OF ELLIPTICAL CROSS-SECTION
    9     8     3     1     9     9     3     0     0
    1     9
    1  2.          0.
    2  1.95        0.222
    3  1.8         0.436
    4  1.5         0.661
    5  1.2         0.800
    6  0.9         0.893
    7  0.6         0.954
    8  0.3         0.989
    9  0.          1.
    1     0  2.
    2     0  1.926
    3     0  1.715
    4     0  1.343
    5     0  1.040
    6     0  0.804
    7     0  0.635
    8     0  0.534
    9     0  0.500
    1.5        0.
    0.6        0.45
    0.         0.45
```

tion of eight elements for one-quarter of the ellipse, taking advantage of the double symmetry of the problem (Figure 7). The data file and complete printout are presented in Tables 2 and 3, respectively.

8.5.2 Circular Cavity in Infinite Medium

Let us consider the problem of a circular cavity of unit radius in an infinite two-dimensional region with Neumann boundary conditions, i.e. a constant radial influx $\bar{q}$ specified along the cavity surface.

Although the potential is not specified at any point this problem has a unique solution, $u = -\bar{q} \ln r$, which shows a logarithmic behaviour at infinity [4].

Results for the potential and its radial derivative at points on the boundary and inside the infinite region are given in Tables 4 and 5, respectively, compared to the exact solution. Data file and printout of the problem for $N = 5$ (four boundary elements in a quarter of the region, as shown in Fig. 8) are presented in Tables 6 and 7.

Table 3. Printout for elliptical pipe

FLOW IN PIPE OF ELLIPTICAL CROSS-SECTION

DATA

NUMBER OF BOUNDARY NODES= 9
NUMBER OF BOUNDARY ELEMENTS= 8
NUMBER OF INTERNAL POINTS WHERE VALUES ARE REQUIRED= 3
NUMBER OF SURFACES= 1
SYMMETRY PARAMETER= 3

SURFACE	TYPE	LAST NODE
1	1	9

NODAL COORDINATES

NODE	X	Y
1	0.2000000E+01	0.0000000E+00
2	0.1950000E+01	0.2220000E+00
3	0.1800000E+01	0.4360000E+00
4	0.1500000E+01	0.6610000E+00
5	0.1200000E+01	0.8000000E+00
6	0.9000000E+00	0.8930000E+00
7	0.6000000E+00	0.9540000E+00
8	0.3000000E+00	0.9890000E+00
9	0.0000000E+00	0.1000000E+01

BOUNDARY CONDITIONS

NODE	CODE	PRESCRIBED VALUE	FREE COEFFICIENT
1	0	0.2000000E+01	0.0000000E+00
2	0	0.1926000E+01	0.0000000E+00
3	0	0.1715000E+01	0.0000000E+00
4	0	0.1343000E+01	0.0000000E+00
5	0	0.1040000E+01	0.0000000E+00
6	0	0.8040000E+00	0.0000000E+00
7	0	0.6350000E+00	0.0000000E+00
8	0	0.5340000E+00	0.0000000E+00
9	0	0.5000000E+00	0.0000000E+00

ELEMENT	NODE 1	NODE 2	LENGTH
1	1	2	0.2275610E+00
2	2	3	0.2613351E+00
3	3	4	0.3750000E+00
4	4	5	0.3306372E+00
5	5	6	0.3140844E+00
6	6	7	0.3061388E+00
7	7	8	0.3020348E+00
8	8	9	0.3002016E+00

RESULTS

BOUNDARY NODES

X	Y	POTENTIAL	POTENTIAL DERIVATIVE
0.2000000E+01	0.0000000E+00	0.2000000E+01	0.1210856E+01
0.1950000E+01	0.2220000E+00	0.1926000E+01	0.1020629E+01
0.1800000E+01	0.4360000E+00	0.1715000E+01	0.5749767E+00
0.1500000E+01	0.6610000E+00	0.1343000E+01	0.1070103E+00
0.1200000E+01	0.8000000E+00	0.1040000E+01	-0.1972276E+00
0.9000000E+00	0.8930000E+00	0.8040000E+00	-0.3880150E+00
0.6000000E+00	0.9540000E+00	0.6350000E+00	-0.5126085E+00
0.3000000E+00	0.9890000E+00	0.5340000E+00	-0.5771257E+00
0.0000000E+00	0.1000000E+01	0.5000000E+00	-0.6042791E+00

INTERNAL POINTS

X	Y	POTENTIAL	FLUX QX	FLUX QY
0.1500000E+01	0.0000000E+00	0.1475513E+01	0.8966051E+00	0.6065238E-07
0.6000000E+00	0.4500000E+00	0.8494186E+00	0.3595006E+00	-0.2694215E+00
0.0000000E+00	0.4500000E+00	0.7415615E+00	0.5064066E-08	-0.2696070E+00

Table 4. Potential at points on infinite region, for $\bar{q} = 10$

R	BEM ($N = 5$)	BEM ($N = 9$)	Exact
1.0	0.06	0.02	0
1.5	− 4.03	− 4.05	− 4.06
2.0	− 6.89	− 6.92	− 6.93
3.0	− 10.92	− 10.97	− 10.99
5.0	− 15.99	− 16.07	− 16.09
10.0	− 22.88	− 22.99	− 23.03
100.0	− 45.76	− 45.98	− 46.05
1000.0	− 68.63	− 68.97	− 69.08

Table 5. Radial flux for points on infinite region, for $\bar{q} = 10$

R	BEM ($N = 5$)	BEM ($N = 9$)	Exact
1.0	− 10.00	− 10.00	− 10.00
1.5	− 6.62	− 6.66	− 6.67
2.0	− 4.97	− 4.99	− 5.00
3.0	− 3.31	− 3.33	− 3.33
5.0	− 1.99	− 2.00	− 2.00
10.0	− 0.99	− 1.00	− 1.00
100.0	− 0.10	− 0.10	− 0.10
1000.0	− 0.01	− 0.01	− 0.01

Table 6. Data file for circular cavity

```
CIRCULAR CAVITY IN INFINITE MEDIUM
    5    4    7    1    5    2    3    0    1
    1    5
    1 0.        1.
    2 0.382683  0.923880
    3 0.707107  0.707107
    4 0.923880  0.382683
    5 1.        0.
    1    1   10.
    5    1   10.
   1.5
   2.
   3.
   5.
  10.
 100.
1000.
```

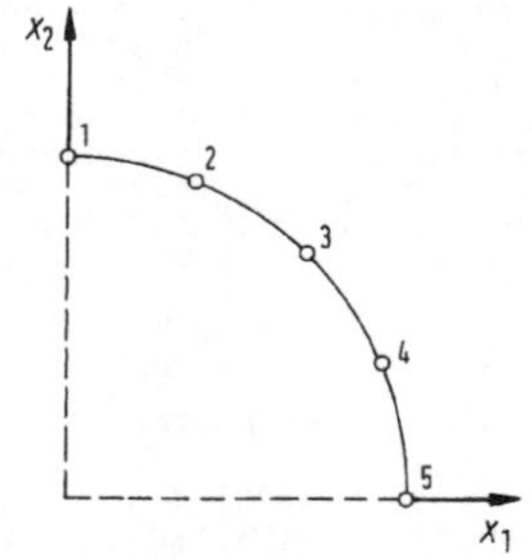

Fig. 8. Circular cavity in infinite region: discretization

Table 7. Printout for circular cavity

CIRCULAR CAVITY IN INFINITE MEDIUM

DATA

NUMBER OF BOUNDARY NODES= 5
NUMBER OF BOUNDARY ELEMENTS= 4
NUMBER OF INTERNAL POINTS WHERE VALUES ARE REQUIRED= 7
NUMBER OF SURFACES= 1
SYMMETRY PARAMETER= 3

SURFACE	TYPE	LAST NODE
1	1	5

UNBOUNDED DOMAIN: ADD 1. TO THE DIAGONAL COEFFICIENTSOF MATRIX H

NODAL COORDINATES

NODE	X	Y
1	0.0000000E+00	0.1000000E+01
2	0.3826830E+00	0.9238800E+00
3	0.7071070E+00	0.7071070E+00
4	0.9238800E+00	0.3826830E+00
5	0.1000000E+01	0.0000000E+00

BOUNDARY CONDITIONS

NODE	CODE	PRESCRIBED VALUE	FREE COEFFICIENT
1	1	0.1000000E+02	0.0000000E+00
2	1	0.1000000E+02	0.0000000E+00
3	1	0.1000000E+02	0.0000000E+00
4	1	0.1000000E+02	0.0000000E+00
5	1	0.1000000E+02	0.0000000E+00

ELEMENT	NODE 1	NODE 2	LENGTH
1	1	2	0.3901801E+00
2	2	3	0.3901813E+00
3	3	4	0.3901813E+00
4	4	5	0.3901801E+00

RESULTS

BOUNDARY NODES

X	Y	POTENTIAL	POTENTIAL DERIVATIVE
0.0000000E+00	0.1000000E+01	0.6108193E-01	0.1000000E+02
0.3826830E+00	0.9238800E+00	0.6108002E-01	0.1000000E+02
0.7071070E+00	0.7071070E+00	0.6107893E-01	0.1000000E+02
0.9238800E+00	0.3826830E+00	0.6107866E-01	0.1000000E+02
0.1000000E+01	0.0000000E+00	0.6108143E-01	0.1000000E+02

INTERNAL POINTS

X	Y	POTENTIAL	FLUX QX	FLUX QY
0.1500000E+01	0.0000000E+00	-0.4028593E+01	-0.6624506E+01	-0.1620047E-06
0.2000000E+01	0.0000000E+00	-0.6887018E+01	-0.4967941E+01	0.1395838E-06
0.3000000E+01	0.0000000E+00	-0.1091367E+02	-0.3311958E+01	0.1930130E-07
0.5000000E+01	0.0000000E+00	-0.1599117E+02	-0.1987174E+01	0.1980334E-07
0.1000000E+02	0.0000000E+00	-0.2287819E+02	-0.9935871E+00	0.1448143E-08
0.1000000E+03	0.0000000E+00	-0.4575637E+02	-0.9935872E-01	-0.2701839E-10
0.1000000E+04	0.0000000E+00	-0.6863454E+02	-0.9735870E-02	0.3539669E-12

8.5.3 Rectangular Concrete Column

This application deals with a concrete column of rectangular cross-section where part of the boundary surface is subjected to an interior ambient condition, another part is subjected to outside weather conditions, and the remainder is in contact

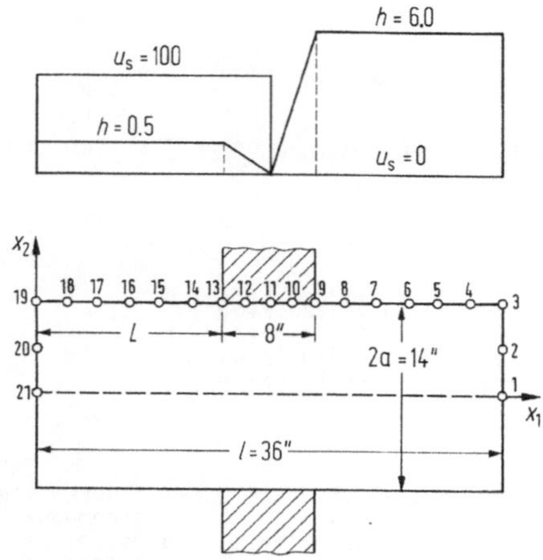

Fig. 9. Geometry, discretization and boundary conditions for concrete column

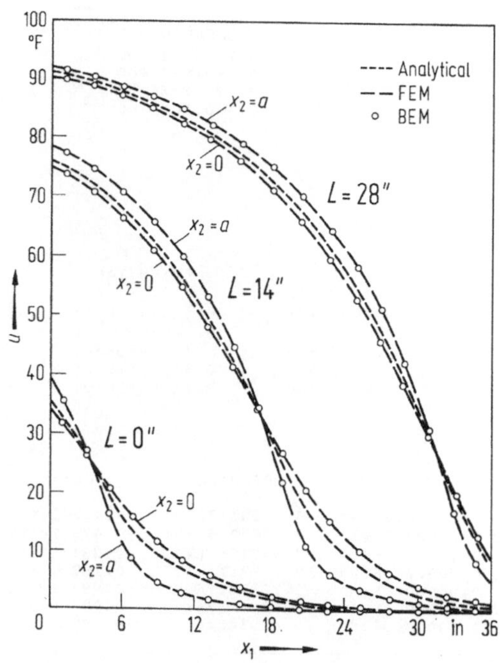

Fig. 10. Temperature distribution for three different positions of the abutting wall

Table 8. Data file for concrete column

```
CONCRETE COLUMN OF RECTANGULAR CROSS-SECTION
   21    20    15     1     6     3     1     0     0
    1    21
    1 3.          0.
    3 3.          0.583333
    9 1.833333    0.583333
   13 1.166667    0.583333
   19 0.          0.583333
   21 0.          0.
    1     2     -6.        0.
    9     2     -6.        0.
   11     2      0.        0.
   13     2     -0.5      50.
   21     2     -0.5      50.
  2.805556
  2.611111
  2.416667
  2.222222
  2.027778
  1.833333
  1.666667
  1.5
  1.333333
  1.166667
  0.972222
  0.777778
  0.583333
  0.388889
  0.194444
```

with an abutting wall which separates both. The boundary conditions of the problem are of the convective type, $q + hu = hu_s$, where h is the heat transfer coefficient and u_s the temperature of the surrounding medium.

The temperature and surface heat transfer coefficient on the interior face ($x_1 = 0$) are 100 °F and 0.5 Btu/(h ft^2 °F), respectively, and at the exterior face ($x_1 = l$), are 0 °F and 6.0 Btu/(h ft^2 °F). The variation of the temperature and heat transfer coefficient along the faces $x_2 = \pm a$ is indicated in Fig. 9. Note that the thermal conductivity was assumed to be 1.0 Btu/(h ft °F).

Results corresponding to three different positions for the abutting wall are presented in Fig. 10, compared with finite element results [9] and an analytical solution [10] (in terms of a mean temperature over the width of the cross-section). The boundary element analyses were performed by discretizing one-half of the column into 20 elements (see Fig. 9) taking into account symmetry with respect to the x_1 axis, while the finite element ones employed 252 quadrilateral elements.

The data file and printout for the case depicted in Fig. 9 are presented in Tables 8 and 9.

Table 9. Printout for concrete column

```
                    CONCRETE COLUMN OF RECTANGULAR CROSS-SECTION

DATA

NUMBER OF BOUNDARY NODES= 21
NUMBER OF BOUNDARY ELEMENTS= 20
NUMBER OF INTERNAL POINTS WHERE VALUES ARE REQUIRED= 15
NUMBER OF SURFACES=  1
SYMMETRY PARAMETER=  1


SURFACE          TYPE           LAST NODE
   1               1               21


NODAL COORDINATES

   NODE           X                   Y
     1     0.3000000E+01       0.0000000E+00
     2     0.3000000E+01       0.2916665E+00
     3     0.3000000E+01       0.5833330E+00
     4     0.2805556E+01       0.5833330E+00
     5     0.2611111E+01       0.5833330E+00
     6     0.2416667E+01       0.5833330E+00
     7     0.2222222E+01       0.5833330E+00
     8     0.2027778E+01       0.5833330E+00
     9     0.1833333E+01       0.5833330E+00
    10     0.1666667E+01       0.5833330E+00
    11     0.1500000E+01       0.5833330E+00
    12     0.1333333E+01       0.5833330E+00
    13     0.1166667E+01       0.5833330E+00
    14     0.9722225E+00       0.5833330E+00
    15     0.7777780E+00       0.5833330E+00
    16     0.5833336E+00       0.5833330E+00
    17     0.3888891E+00       0.5833330E+00
    18     0.1944446E+00       0.5833330E+00
    19     0.0000000E+00       0.5833330E+00
    20     0.0000000E+00       0.2916665E+00
    21     0.0000000E+00       0.0000000E+00


BOUNDARY CONDITIONS
   NODE      CODE      PRESCRIBED VALUE      FREE COEFFICIENT
     1         2        -0.6000000E+01         0.0000000E+00
     2         2        -0.6000000E+01         0.0000000E+00
     3         2        -0.6000000E+01         0.0000000E+00
     4         2        -0.6000000E+01         0.0000000E+00
     5         2        -0.6000000E+01         0.0000000E+00
     6         2        -0.6000000E+01         0.0000000E+00
     7         2        -0.6000000E+01         0.0000000E+00
     8         2        -0.6000000E+01         0.0000000E+00
     9         2        -0.6000000E+01         0.0000000E+00
    10         2        -0.3000000E+01         0.0000000E+00
    11         2         0.0000000E+00         0.0000000E+00
    12         2        -0.2500000E+00         0.2500000E+02
    13         2        -0.5000000E+00         0.5000000E+02
    14         2        -0.5000000E+00         0.5000000E+02
    15         2        -0.5000000E+00         0.5000000E+02
    16         2        -0.5000000E+00         0.5000000E+02
    17         2        -0.5000000E+00         0.5000000E+02
    18         2        -0.5000000E+00         0.5000000E+02
    19         2        -0.5000000E+00         0.5000000E+02
    20         2        -0.5000000E+00         0.5000000E+02
    21         2        -0.5000000E+00         0.5000000E+02
```

ELEMENT	NODE 1	NODE 2	LENGTH
1	1	2	0.2916665E+00
2	2	3	0.2916665E+00
3	3	4	0.1944444E+00
4	4	5	0.1944444E+00
5	5	6	0.1944444E+00
6	6	7	0.1944444E+00
7	7	8	0.1944444E+00
8	8	9	0.1944449E+00
9	9	10	0.1666665E+00
10	10	11	0.1666665E+00
11	11	12	0.1666665E+00
12	12	13	0.1666665E+00
13	13	14	0.1944445E+00
14	14	15	0.1944445E+00
15	15	16	0.1944445E+00
16	16	17	0.1944445E+00
17	17	18	0.1944445E+00
18	18	19	0.1944446E+00
19	19	20	0.2916665E+00
20	20	21	0.2916665E+00

RESULTS

BOUNDARY NODES

X	Y	POTENTIAL	POTENTIAL DERIVATIVE
0.3000000E+01	0.0000000E+00	0.8431733E+00	-0.5059040E+01
0.3000000E+01	0.2916665E+00	0.6868079E+00	-0.4120847E+01
0.3000000E+01	0.5833330E+00	0.2662871E+00	-0.1597723E+01
0.2805556E+01	0.5833330E+00	0.6177006E+00	-0.3706203E+01
0.2611111E+01	0.5833330E+00	0.1079539E+01	-0.6477236E+01
0.2416667E+01	0.5833330E+00	0.1737004E+01	-0.1042202E+02
0.2222222E+01	0.5833330E+00	0.2731186E+01	-0.1638712E+02
0.2027778E+01	0.5833330E+00	0.4345937E+01	-0.2607562E+02
0.1833333E+01	0.5833330E+00	0.7664944E+01	-0.4598967E+02
0.1666667E+01	0.5833330E+00	0.1524610E+02	-0.4573831E+02
0.1500000E+01	0.5833330E+00	0.2822847E+02	0.0000000E+00
0.1333333E+01	0.5833330E+00	0.3996932E+02	0.1500767E+02
0.1166667E+01	0.5833330E+00	0.4964356E+02	0.2517822E+02
0.9722225E+00	0.5833330E+00	0.5771670E+02	0.2114165E+02
0.7777780E+00	0.5833330E+00	0.6389421E+02	0.1805289E+02
0.5833336E+00	0.5833330E+00	0.6883279E+02	0.1558360E+02
0.3888891E+00	0.5833330E+00	0.7279564E+02	0.1360218E+02
0.1944446E+00	0.5833330E+00	0.7593585E+02	0.1203207E+02
0.0000000E+00	0.5833330E+00	0.7829480E+02	0.1085260E+02
0.0000000E+00	0.2916665E+00	0.7589204E+02	0.1205398E+02
0.0000000E+00	0.0000000E+00	0.7507751E+02	0.1246125E+02

INTERNAL POINTS

X	Y	POTENTIAL	FLUX QX	FLUX QY
0.2805556E+01	0.0000000E+00	0.1898019E+01	-0.6138177E+01	-0.7124618E-06
0.2611111E+01	0.0000000E+00	0.3286980E+01	-0.8340743E+01	-0.1681969E-05
0.2416667E+01	0.0000000E+00	0.5228067E+01	-0.1187889E+02	0.9983778E-06
0.2222222E+01	0.0000000E+00	0.8017490E+01	-0.1712302E+02	0.9499490E-07
0.2027778E+01	0.0000000E+00	0.1200878E+02	-0.2421731E+02	-0.1821667E-05
0.1833333E+01	0.0000000E+00	0.1751132E+02	-0.3243056E+02	-0.4060566E-06
0.1666667E+01	0.0000000E+00	0.2346731E+02	-0.3877129E+02	-0.1080334E-05
0.1500000E+01	0.0000000E+00	0.3029973E+02	-0.4270615E+02	-0.1862645E-06
0.1333333E+01	0.0000000E+00	0.3751920E+02	-0.4338378E+02	0.1594424E-05
0.1166667E+01	0.0000000E+00	0.4460075E+02	-0.4120081E+02	0.6154180E-05
0.9722220E+00	0.0000000E+00	0.5217900E+02	-0.3649817E+02	0.4351139E-05
0.7777780E+00	0.0000000E+00	0.5874506E+02	-0.3100594E+02	-0.9536743E-05
0.5833330E+00	0.0000000E+00	0.6424661E+02	-0.2564428E+02	0.6675720E-05
0.3888890E+00	0.0000000E+00	0.6874868E+02	-0.2074984E+02	0.1430511E-05
0.1944440E+00	0.0000000E+00	0.7234904E+02	-0.1635710E+02	0.1144409E-04

References

1 Hess, J.L. and Smith, A.M.O., Calculation of Potential Flow about Arbitrary Bodies. Progress in Aeronautical Sciences, Vol. 8 (D. Küchemann, ed.). Pergamon Press, London 1967

2 Jaswon, M.A., Integral Equation Methods in Potential Theory, I. Proc. Royal Soc. Ser. A **275,** 23–32 (1963)

3 Symm, G.T., Integral Equation Methods in Potential Theory, II. Proc. Royal Soc. Ser. A **275,** 33–46 (1963)

4 Brebbia, C.A., Telles, J.C.F., and Wrobel, L.C., Boundary Element Techniques: Theory and Applications in Engineering. Springer-Verlag, Berlin 1984

5 Brebbia C.A., The Boundary Element Method for Engineers. Pentech Press, London 1978

6 Kellogg, O.D., Foundations of Potential Theory. Springer-Verlag, Berlin 1929

7 Brebbia, C.A. and Ferrante, A. J., Computational Methods for the Solution of Engineering Problems. Pentech Press, London 1978

8 Batchelor, G.K., An Introduction to Fluid Dynamics. Cambridge University Press, Cambridge 1967

9 Wilson, E.L. and Nickell, R.E., Application of the Finite Element Method to Heat Conduction Analysis. Nuclear Engng. Design **4,** 276–286 (1966)

10 Peavy, B.A., Steady-State Heat Conduction in an Exposed Exterior Column of Rectangular Cross-Section. J. Res. Nat. Bur. Stand. **69 C,** 145–151 (1965)

Chapter 9

Elastostatic Problems*

by J.C.F. Telles

9.1 Introduction

In this chapter a simple, yet efficient, FORTRAN computer program for two-dimensional (plane Strain/Stress) problems is described. The code is implemented with linear boundary elements, i.e. linear interpolation functions for boundary tractions and displacements. In order to keep the chapter self-contained, a summary of the theory required is presented together with a complete description of each subroutine, so that the interested reader will find useful means of getting started with the technique and in the future will also be able to modify or adapt the code according to his/her own needs.

Despite its simplicity, the program presents some interesting features, such as selective numerical integration, stresses on the boundary and symmetry consideration without boundary discretization of symmetry axes; these are going to be explained in the text. In addition, any one contemplating using the program can try some of the examples found at the end of the chapter, they are here included to serve as bench-marks for debugging any newly typed version of the code.

The presentation starts with a review of the main theoretical aspects of the boundary element formulation for 2-D elasticity problems and before launching into the program itself, introduces some comments on the numerical characteristics of the method. A fully documented version of the code is then presented and this is followed by some simple examples solved by the program.

9.2 Outline of Theory

Throughout this chapter the so-called Cartesian tensor notation is used. Therefore, the repetition of any letter subscript in a term will imply summation. Herein, since we are only dealing with two-dimensional elasticity problems, unless locally stated, the subscripts are assumed to have a range of two. Furthermore, space derivatives will be indicated by a comma, i.e. $\partial f_i/\partial x_j = f_{i,j}$.

The infinitesimal strain tensor ε_{ij} can be expressed in terms of the derivatives of the displacement components (u_i) in the form

$$\varepsilon_{ij} = \tfrac{1}{2}(u_{i,j} + u_{j,i}) . \quad (1)$$

* *Important Notice:* The copyright of the computer program described in this chapter remains the property of the author (Dr. J.C.F. Telles) and the editor of this book (Dr. C.A. Brebbia).

If an infinitesimal rectangular parallelepiped is assumed to suround a given point within a stressed body, it readily follows that the static internal equilibrium inside the domain Ω of the region is given by

$$\sigma_{ij,i} + b_j = 0 \tag{2}$$

where the components of the stress tensor are represented by σ_{ij} and the body forces are represented by b_j.

Equation (2) is valid in the interior of the body. The equilibrium condition when applied to the boundary Γ of the region yields

$$p_i - \sigma_{ij} n_j = 0 \tag{3}$$

in which the surface tractions p_i are equilibrated by the stresses on the boundary and n_j represents the direction cosines of the outward normal to Γ.

For an isotropic elastic material in which there is no change in temperature, Hooke's law relating stresses and strains can be stated as follows

$$\sigma_{ij} = C_{ijkl} \varepsilon_{kl} \tag{4}$$

where C_{ijkl} can be written as

$$C_{ijkl} = \frac{2G\nu}{1-2\nu} \delta_{ij} \delta_{kl} + G(\delta_{ik} \delta_{jl} + \delta_{il} \delta_{jk}) \tag{5}$$

in which G is the shear modulus, ν is the Poisson ratio and δ_{ij} is the Kronecker delta symbol.

An important remark is now due; all the expressions presented here are assumed valid for plane strain problems, the plane stress case can be dealt with by the same equations, providing ν is replaced by $\bar{\nu} = \nu/(1+\nu)$.

The substitution of Eq. (1) into Eq. (4) leads to the stresses in terms of displacement derivatives. These can then be substituted into Eq. (2) to obtain two second-order partial differential equations for the displacements:

$$G u_{j,kk} + \frac{G}{1-2\nu} u_{k,kj} + b_j = 0. \tag{6}$$

These equations are the well-known Navier equilibrium equations and are subject to the following boundary conditions:

$$u_i = \bar{u}_i \quad \text{on } \Gamma_1 \qquad \text{and} \qquad p_i = \bar{p}_i \quad \text{on } \Gamma_2. \tag{7}$$

where $\bar{u}_i$ and $\bar{p}_i$ are prescribed values and the total boundary of the body is $\Gamma = \Gamma_1 + \Gamma_2$. Note that the subdivision of Γ into two parts is to be interpreted as a concept, since one can have at the same physical point of Γ the two types of boundary conditions in different directions.

An integral equation, equivalent to Eqs. (6) and (7), can now be obtained through a weighted residual formulation or a simple reciprocity principle based on the symmetry of C_{ijkl} [1]. This equation, which is also known as Somigliana identity for displacements [2], can be written as

$$u_i(\xi) = \int_\Gamma u^*_{ij}(\xi, x)\, p_j(x)\, d\Gamma(x) - \int_\Gamma p^*_{ij}(\xi, x)\, u_j(x)\, d\Gamma(x), \quad \xi \in \Omega \tag{8}$$

where $b_j = 0$ was assumed for simplicity and the starred tensors represent a closed form solution (fundamental solution) to Eq. (6) in which the body force components correspond to a unit point load applied at ξ in "i" direction. Notice that u^*_{ij} and p^*_{ij} represent the component in "j" direction of displacement and traction respectively at point x of the fundamental problem.

For plane strain applications where the fundamental solution corresponds to unit point loads applied within an infinite plane (Kelvin), the components of the fundamental displacements and tractions are given by

$$u^*_{ij}(\xi, x) = \frac{-1}{8\pi(1-\nu)G}\{(3-4\nu)\ln(r)\,\delta_{ij} - r_{,i}\,r_{,j}\} \tag{9}$$

and

$$p^*_{ij}(\xi, x) = \frac{-1}{4\pi(1-\nu)r}\left\{[(1-2\nu)\,\delta_{ij} + 2r_{,i}\,r_{,j}]\frac{\partial r}{\partial n} - (1-2\nu)(r_{,i}\,n_j - r_{,j}\,n_i)\right\} \tag{10}$$

where $r = r(\xi, x)$ represents the distance between the load point ξ and the field point x and its derivatives are taken with reference to the coordinates of the latter, i.e.

$$\begin{aligned} r &= (r_i\,r_i)^{1/2}, \\ r_i &= x_i(x) - x_i(\xi), \\ r_{,i} &= \frac{\partial r}{\partial x_i(x)} = \frac{r_i}{r}. \end{aligned} \tag{11}$$

Since Somigliana's identity provides a continuous representation of displacements at any point $\xi \in \Omega$ is can also be used to generate the internal stresses. To this end one can find the derivatives of Eq. (8) with respect to the coodinates of ξ and then substitute in Eqs. (1) and (4). The resulting equation is of the form

$$\sigma_{ij}(\xi) = \int_\Gamma u^*_{ijk}(\xi, x)\,p_k(x)\,d\Gamma(x) - \int_\Gamma p^*_{ijk}(\xi, x)\,u_k(x)\,d\Gamma(x) \tag{12}$$

in which the new fundamental tensor components read (plane strain)

$$u^*_{ijk}(\xi, x) = \frac{1}{4\pi(1-\nu)r}\{(1-2\nu)(r_{,j}\,\delta_{ik} + r_{,i}\,\delta_{jk} - r_{,k}\,\delta_{ij}) + 2r_{,i}\,r_{,j}\,r_{,k}\} \tag{13}$$

$$\begin{aligned} p^*_{ijk}(\xi, x) = \frac{G}{2\pi(1-\nu)r^2}\Bigg\{&2\frac{\partial r}{\partial n}[(1-2\nu)\,\delta_{ij}\,r_{,k} + \nu(\delta_{ik}\,r_{,j} + \delta_{jk}\,r_{,i}) - 4r_{,i}\,r_{,j}\,r_{,k}] \\ &+ 2\nu(n_i\,r_{,j}\,r_{,k} + n_j\,r_{,i}\,r_{,k}) + (1-2\nu)(2n_k\,r_{,i}\,r_{,j} + n_j\,\delta_{ik} + n_i\,\delta_{jk}) \\ &- (1-4\nu)\,n_k\,\delta_{ij}\Bigg\}. \end{aligned} \tag{14}$$

In order to obtain an integral equation involving only the variables on the boundary of the body, one can take the limit of Eq. (8) as the point ξ approaches Γ. This limit has to be carefully taken since the boundary integrals become singular at ξ. The resulting equation is [1]

$$c_{ij}(\xi)\,u_j(\xi) + \int_\Gamma p^*_{ij}(\xi, x)\,u_j(x)\,d\Gamma(x) = \int_\Gamma u^*_{ij}(\xi, x)\,p_j(x)\,d\Gamma(x), \quad \xi \in \Gamma \tag{15}$$

where the coefficient c_{ij} is a function of the geometry of Γ at the point ξ and the integral on the left is to be computed in the sense of Cauchy principal value. It is worth mentioning that this coefficient, together with the associated principal value, can be computed by indirect means if one applies Eq. (15) to represent two independent rigid body translations. This procedure is going to be discussed in the next section.

Equation (15) can now be solved, in approximate fashion, by adopting a numerical procedure based on the discretization of Γ into a series of elements which interpolate the variables u_j and p_j. Herein, linear elements have been chosen; therefore, the above equation, after discretization, yields

$$\boldsymbol{c}(\xi_i)\,\boldsymbol{u}(\xi) + \sum_{j=1}^{N} \int_{\Gamma_j} \boldsymbol{p}^*\,\boldsymbol{\Phi}^T\,d\Gamma\,\boldsymbol{u}^{(n)} = \sum_{j=1}^{N} \int_{\Gamma_j} \boldsymbol{u}^*\,\boldsymbol{\Phi}^T\,d\Gamma\,\boldsymbol{p}^{(n)} \tag{16}$$

where ξ_i is a boundary node point, N is the total number of elements Γ_j, the superscript (n) indicates nodal values of the boundary variables and matrix $\boldsymbol{\Phi}$ contains the interpolation functions in the form

$$\boldsymbol{\Phi}^T = [\boldsymbol{I}\,\psi_1 \quad \boldsymbol{I}\,\psi_2] \tag{17}$$

in which $\psi_1 = \frac{1}{2}(1-\eta)$, $\psi_2 = \frac{1}{2}(1+\eta)$, $\boldsymbol{I}$ is the identity matrix of order two and η is a homogeneous coordinate defined in Fig. 1.

Equation (16) can be further written as (summation convention is not implied)

$$\boldsymbol{c}_i\,\boldsymbol{u}_i + \sum_{j=1}^{N} \boldsymbol{h}\,\boldsymbol{u}^{(n)} = \sum_{j=1}^{N} \boldsymbol{g}\,\boldsymbol{p}^{(n)} \tag{18}$$

where matrices $\boldsymbol{h}$ and $\boldsymbol{g}$ are of order 2×4 and can be generally computed by numerical Gaussian integration. The only difficulty is when the singular node ξ_i belongs to the element Γ_j, that case will be dealt with later on.

Henceforth, given a singular node "i", each element "j" will contribute with two matrices of the form (see Fig. 1)

$$\boldsymbol{h} = [\hat{\boldsymbol{h}}_{ij}^1 \;\; \hat{\boldsymbol{h}}_{ij}^2] = \frac{l}{2} \int_{-1}^{1} [\psi_1\,\boldsymbol{p}^* \;\; \psi_2\,\boldsymbol{p}^*]\,d\eta\,,$$

$$\boldsymbol{g} = [\boldsymbol{g}_{ij}^1 \;\; \boldsymbol{g}_{ij}^2] = \frac{l}{2} \int_{-1}^{1} [\psi_1\,\boldsymbol{u}^* \;\; \psi_2\,\boldsymbol{u}^*]\,d\eta \tag{19}$$

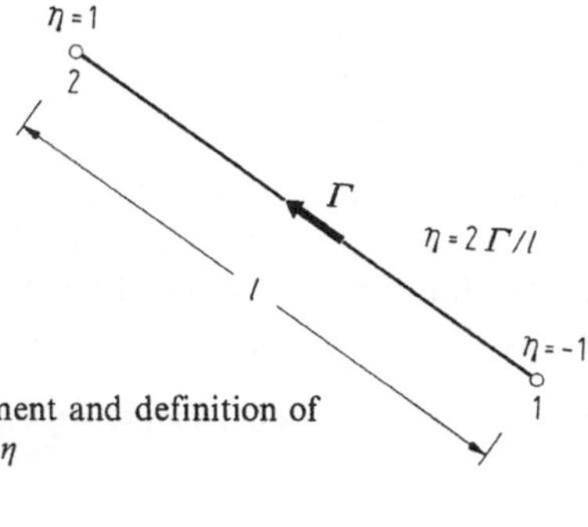

Fig. 1. Linear element and definition of intrinsic coordinate η

and Eq. (18) finally reads (summation not implied)

$$c_i\, u_i + \sum_{k=1}^{M} \hat{h}_{ik}\, u_k = \sum_{k=1}^{M} g_{ik}\, p_k \tag{20}$$

where M represents the total number of nodes. It is worth mentioning that the 2×2 matrices $\hat{h}_{ik}$ and g_{ik} are, except in the double node case (see next section), composed by adding the submatrices $\hat{h}^k_{ij}$ and g^k_{ij} of the two elements connected to node "k".

We can now write a matrix equation, such as Eq. (20), for each of the nodes under consideration. Writing them together, comes

$$\sum_{k=1}^{M} h_{ik}\, u_k = \sum_{k=1}^{M} g_{ik}\, p_k \quad (i = 1, 2, \ldots, M) \tag{21}$$

in which

$$h_{ik} = \begin{cases} \hat{h}_{ik} + c_i & \text{for } i = k \\ \hat{h}_{ik} & \text{for } i \neq k. \end{cases} \tag{22}$$

Equation (21) can than be put in the form

$$HU = GP. \tag{23}$$

Recalling the boundary conditions presented in Eq. (7), one can group the problem unknowns on the left-hand side of Eq. (23) and obtain a system of $2M$ equations ready to be solved by standard methods (such as Gauss elimination). This system can be written as

$$AY = F \tag{24}$$

where A is a fully populated square matrix whose columns originally multiplied the unknowns in Eq. (23), vector Y contains the unknown tractions and displacements and vector F is formed by the product of the prescribed boundary conditions by the corresponding columns of matrices H and G. Note that Eq. (24) can be assembled directly from the element matrices h and g without need to generate first Eq. (23).

Once that all the nodal values of boundary displacements and tractions are calculated, the values of displacements and stresses can be computed at any internal point by simple numerical quadrature (singular integrals are no longer involved). To this end, Eqs. (8) and (12) are used also in discretized form.

Equation (8) then reads

$$u(\xi_i) = \sum_{j=1}^{N} g\, p^{(n)} - \sum_{j=1}^{N} u^{(n)} \tag{25}$$

where $\xi_i \in \Omega$ and matrices h and g have been previously defined.

Similarly, Eq. (12) yields

$$\sigma(\xi_i) = \sum^{N} g'\, p^{(n)} - \sum^{N} h'\, u^{(n)} \tag{26}$$

in which $\boldsymbol{h}'$ and $\boldsymbol{g}'$ are now of order 3×4. These matrices are defined as in Eq. (19) with the fundamental 2×2 matrices $\boldsymbol{u}^*$ and $\boldsymbol{p}^*$ being substituted by their counterparts $'\boldsymbol{u}^*$ and $'\boldsymbol{p}^*$ which are of order 3×2. It should be pointed out that only the three different components of stress have to be computed and the sequence chosen here is σ_{11}, σ_{12} and σ_{22}. The σ_{33} component can be calculated as a function of the other two normal stresses, i.e.

$$\sigma_{33} = \begin{cases} \nu(\sigma_{11} + \sigma_{22}) & \text{for plane strain} \\ 0 & \text{for plane stress.} \end{cases} \tag{27}$$

9.3 Numerical and Computational Procedures

9.3.1 Integration

In the present program all the regular integrals (i.e. those in which $\xi_i \notin \Gamma_j$) are obtained by standard Gaussian quadrature. The number of points employed is a function of the dimensionless distance between the singular node or internal point ξ_i and the element Γ_j under consideration. Therefore, for a typical element

$$\boldsymbol{h} = \sum_{q=1}^{Q} w_q (\boldsymbol{p}^* \boldsymbol{\Phi}^T)_q, \quad \boldsymbol{g} = \sum_{q=1}^{Q} w_q (\boldsymbol{u}^* \boldsymbol{\Phi}^T)_q \tag{28}$$

where Q is the number of integration points and w_q is the weighting factor associated with them. Computation of matrices $\boldsymbol{h}'$ and $\boldsymbol{g}'$ follows the same pattern, in both cases Q is automatically defined by the following criterion

$$\begin{aligned} S \leqq 1.5 &\rightarrow Q = 6 \\ 1.5 < S \leqq 5.5 &\rightarrow Q = 4 \\ 5.5 < S \quad &\rightarrow Q = 2 \end{aligned} \tag{29}$$

in which S is the distance between the centre of the element and the singular node or point ξ_i divided by the length of the element. This criterion was found to be accurate enough without loosing computer efficiency.

For the special case of the singular node being coincident with one of the element end nodes, the coefficients of matrix $\boldsymbol{g}$ can be computed analytically to avoid more sophisticated numerical integration schemes. The resulting expression is

$$g_{ij}^{kn} = C_2 \left\{ C_1 \, l \, [\delta_{kn} + \tfrac{1}{2} - \ln(l)] \, \delta_{ij} + \frac{l_i l_j}{l} \right\}, \quad (i, j, k, n = 1, 2) \tag{30}$$

where the subscripts i and j indicate the position of the coefficient in the 2×2 submatrix k of $\boldsymbol{g}$ and n indicates the singular node. Also (see Fig. 1),

$$\begin{aligned} C_1 &= 3 - 4\nu, \\ C_2 &= 1/[16\pi(1-\nu)\,G], \\ l_i &= x_i(2) - x_i(1). \end{aligned} \tag{31}$$

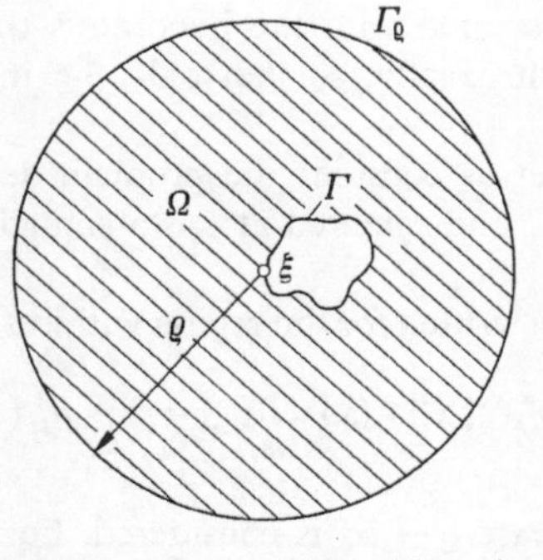

Fig. 2. Infinite region with cavity

The 2×2 submatrix of $\boldsymbol{h}$ which remains regular when the singular node coincides with one of the element nodes can also be computed in closed form. In that case, provided $k \neq n$ one has

$$h_{ij}^{kn} = -(1-\delta_{ij})(\delta_{in}-\delta_{jn})\, C_3\,, \quad k \neq n \tag{32}$$

where the convention adopted for g_{ij}^{kn} is valid and

$$C_3 = (1-2\nu)/[4\pi(1-\nu)]\,. \tag{33}$$

Notice that the singular submatrix $k = n$ of $\boldsymbol{h}$ contributes to the 2×2 submatrix $\boldsymbol{h}_{ii}$ of Eq. (21). Therefore, in addition to the element integrals, the coefficient $c_{ij}(\xi)$ has to be added. These final leading diagonal submatrices of $\boldsymbol{H}$ can be obtained implicitly by imposing to Eq. (23) the condition that rigid body translations result in zero tractions. Therefore, considering finite regions, two independent translations such as $u_i = \delta_{i1}$ and $u_i = \delta_{i2}$ can be applied, resulting in the following relation

$$\sum_{j=1}^{M} \boldsymbol{h}_{ij}\, \boldsymbol{I} = \boldsymbol{0} \quad (i = 1, 2, \ldots, M) \tag{34}$$

where $\boldsymbol{I}$ is the identity matrix of order two.

Equation (34) allows for the computation of the leading diagonal submatrices of $\boldsymbol{H}$ in the form

$$\boldsymbol{h}_{ii} = -\sum_{\substack{j=1 \\ j \neq 2}}^{M} \boldsymbol{h}_{ij} \quad (i = 1, 2, \ldots, M)\,. \tag{35}$$

Expression (35) is valid for finite bodies, for infinite regions (i.e. external problems), however, a further term must be added [1] and this will be discussed in the next section.

9.3.2 External Problems in Infinite Regions

The application of Eq. (15) to solve external problems within an infinite region imply certain assumptions related to the behaviour of the solution at infinity.

These assumptions are associated with the hypothesis that the integrals over the boundary at infinity vanish exactly, so that only the integrals over the internal boundary Γ are considered.

To clarify this matter, let us consider the problem depicted in Fig. 2. Here, a circle of radius ϱ, surface Γ_ϱ and centered at ξ is assumed to enclose the cavity (or cavities) of the problem.

Equation (15) can now be written for the region within Γ and Γ_ϱ as follows

$$c_{ij}(\xi)\, u_j(\xi) + \int_{\Gamma+\Gamma_\varrho} p^*_{ij}(\xi, x)\, u_j(x)\, d\Gamma(x) = \int_{\Gamma+\Gamma_\varrho} u^*_{ij}(\xi, x)\, p_j(x)\, d\Gamma(x)\,. \tag{36}$$

Clearly, if the limiting case $\varrho \to \infty$ is considered, Eq. (36) can be expressed in terms of boundary integrals over Γ alone if

$$\lim_{\varrho\to\infty} \int_{\Gamma_\varrho} [p^*_{ij}(\xi, x)\, u_j(x) - u^*_{ij}(\xi, x)\, p_j(x)]\, d\Gamma(x) = 0 \tag{37}$$

where Eq. (37) represents the so-called regularity conditions at infinity.

Careful examination of the asymptotic behaviour of each term at infinity [1, 3] indicate that unless u_j and p_j present a decay faster then u^*_{ij} and p^*_{ij} as $x \to \infty$, Eq. (37) can only be satisfied if the problem solution at infinity can be exactly simulated by the fundamental solution components multiplied by the resultant of the internal tractions. In that case, the two terms cancel each other as $\varrho \to \infty$. Therefore, the initial assumption that the regularity conditions at infinity are satisfied removes the possibility of free rigid body translations in the solution.

From the above discussion it is seen that since Eq. (37) is violated by rigid body movements, one has to start from Eq. (36) to compute the leading diagonal sub-matrices of $\boldsymbol{H}$.

Henceforth, given a rigid body translation u_j, Eq. (36) yields

$$c_{ij}(\xi)\, u_j + u_j \int_{\Gamma} p^*_{ij}(\xi, x)\, d\Gamma(x) + \lim_{\varrho\to\infty} u_j \int_{\Gamma_\varrho} p^*_{ij}(\xi, x)\, d\Gamma(x) = 0, \quad \xi \in \Gamma\,. \tag{38}$$

The last integral in Eq. (38) can be computed by the condition of equilibrium of the fundamental problem; i.e., since p^*_{ij} corresponds to a positive unit point load applied in "i" direction, one has

$$\lim_{\varrho\to\infty} \int_{\Gamma_\varrho} p^*_{ij}(\xi, x)\, d\Gamma(x) = -\,\delta_{ij} \tag{39}$$

Substituting expression (39) into Eq. (38) for the rigid body translations employed in Eq. (34), and discretizing, gives

$$\boldsymbol{h}_{ii} = \boldsymbol{I} - \sum_{\substack{j=1\\ j\neq i}}^{M} \boldsymbol{h}_{ij} \qquad (i = 1, 2, \ldots, M) \tag{40}$$

in which $\boldsymbol{I}$ is the identity matrix of order two.

Equation (40) substitutes Eq. (35) in the case of external problems in infinite regions.

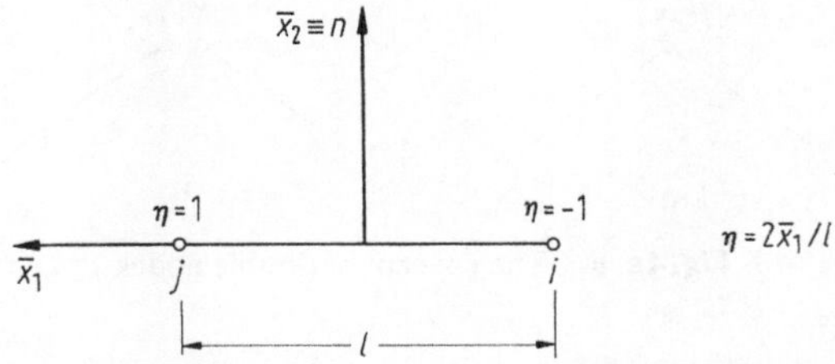

Fig. 3. Linear boundary element and local coordinate system

9.3.3 Stresses on the Boundary

In many practical applications one is interested not only in stresses inside the body and tractions on the surface, but also in the complete stress tensor on the boundary. To this end one can consider a linear boundary element with a local coordinate system tangent to its surface as indicated in Fig. 3. The strain component $\bar{\varepsilon}_{11}$ along the element axis is easily evaluated as

$$\bar{\varepsilon}_{11} = \frac{\partial \bar{u}_1}{\partial \bar{x}_1} = \frac{1}{l}\,({}^j\bar{u}_1 - {}^i\bar{u}_1) \tag{41}$$

where ${}^i\bar{u}_1$ and ${}^j\bar{u}_1$ represent the displacement components at nodes i and j in the direction $\bar{x}_1$ of the local axis.

The traction vector referred to this local system of coordinates gives directly

$$\begin{aligned} \bar{\sigma}_{12} &= p_1 \\ \bar{\sigma}_{22} &= \bar{p}_2\,. \end{aligned} \tag{42}$$

From Hooke's [Eq. (4)] applied for $\bar{\sigma}_{22}$ above one can compute $\bar{\varepsilon}_{22}$ as follows

$$\bar{\varepsilon}_{22} = \frac{1}{1-\nu}\left[(1-2\nu)\frac{\bar{\sigma}_{22}}{2G} - \nu\,\bar{\varepsilon}_{11}\right]. \tag{43}$$

Substitution of Eq. (43) into Eq. (4), now written for $\bar{\sigma}_{11}$, yields

$$\bar{\sigma}_{11} = \frac{1}{1-\nu}\,[2G\,\bar{\varepsilon}_{11} + \nu\,\bar{\sigma}_{22}]\,. \tag{44}$$

Equation (44) together with Eq. (41) and (42) produce the complete stress tensor at any point along the boundary element. These equations were written for plane strain problems, nevertheless they are equally valid for plane stress if one substitute ν by $\bar{\nu}$ as before.

9.3.4 Surface Traction Discontinuities

In order to properly simulate traction discontinuities over the boundary, the concept of double nodes can be employed. This consists of having two boundary

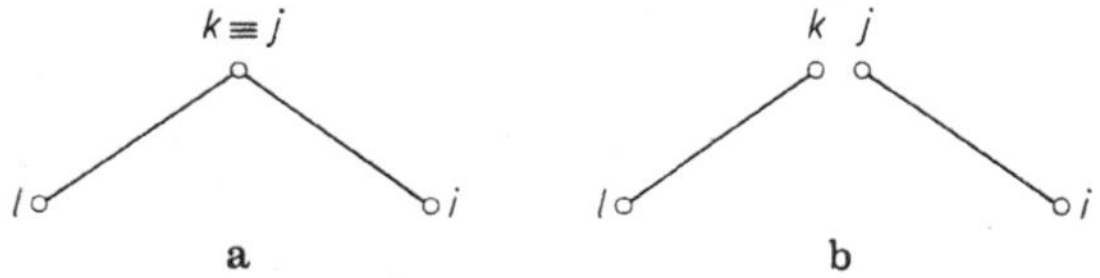

Fig. 4a, b. The concept of double nodes

nodes with exactly the same coordinates without any boundary element in between [4, 5, 6] (see Fig. 4a).

In this procedure the connectivity of the elements is defined as shown in Fig. 4b. Therefore, traction discontinuities can be simulated by assigning different values for the tractions at nodes j and k. It should be noted that here, since the displacement at both nodes is the same, in each direction there are two acceptable combinations of prescribed boundary conditions:

(i) tractions at nodes j and k,
(ii) traction at nodes j and displacement at node k (or vice versa).

The only limitation of this concept is when both nodes have a displacement component, in the same direction, prescribed as a boundary condition. This generates a singular matrix $\boldsymbol{A}$ (such possibility would violate the displacement continuity condition). Nevertheless, an alternative procedure capable of handling this special case has been proposed in Ref. [6] for corner nodes.

9.3.5 Symmetry

An interesting feature of boundary elements is that symmetric bodies under symmetric loads can be analysed without discretization of the symmetry axes. This can be accomplished by an automatic condensation process which integrates over reflected elements and performs the assemble of matrix $\boldsymbol{A}$ already in reduced size. Notice that the matrices $\boldsymbol{h}$ and $\boldsymbol{g}$ of the reflected elements usually need sign changes of some columns before condensation. These changes are required in order to compensate for the contribution of the reflected variables which may change sign from one quadrant to the other.

In addition, instead of actually reflecting the elements, it is usually less cumbersome to reflect the singular node or point and leave the elements in their original position. By doing this the coefficients of the resulting matrices $\boldsymbol{h}$ and $\boldsymbol{g}$ present the same absolute values as the forner procedure, but some signs may be different. Hence, by combining the two sign changes one ends up having to alter the signs of some rows of the matrices. These sign changes are as follows (matrices for internal stresses are included for completeness):

(i) Reflection with respect to plane symmetry perpendicular to y:
$\boldsymbol{h}$ and $\boldsymbol{g}$ change sign of row no. 2
$\boldsymbol{h}'$ and $\boldsymbol{g}'$ change sign of row no. 2

(ii) Reflection with respect to plane of symmetry perpendicular to x:

$\boldsymbol{h}$ and $\boldsymbol{g}$ change sign of row no. 1

$\boldsymbol{h}'$ and $\boldsymbol{g}'$ change sign of row no. 2

(iii) Reflection with respect to both planes of symmetry

$\boldsymbol{h}$ and $\boldsymbol{g}$ change sign of rows no. 1 and 2

$\boldsymbol{h}'$ and $\boldsymbol{g}'$ no change sign is required.

This second procedure has been adopted in the present computer program. It should be noted that in order to compute the leading diagonal submatrices of $\boldsymbol{H}$ by applying rigid body translations, one has to transform matrix $\boldsymbol{h}$ back to the configuration which corresponds to integration over reflected elements with signs unchanged. Therefore, once the assemble of the above modified matrices has been performed, one can modify matrix $\boldsymbol{h}$ again for the rigid body movements. This final change is as follows; case (i) change sign of columns 2 and 4, case (ii) change sign of columns 1 and 3, and finally case (iii) change sign of whole matrix.

9.4 Computer Program

We will now describe a simple FORTRAN computer program for the solution of two dimensional elastostatic problems using linear boundary elements, i.e. elements with linear variations of displacements and tractions.

The program has the following capabilities:

1) Solution of elastostatic plane stress-plane strain problems with isotropic material properties.
2) Computation of surface tractions and displacements plus stresses and displacements at any internal points. In addition, stresses on the boundary are also calculated.
3) Planes of symmetry normal to the X and Y axis can be taken into consideration without need to define boundary elements on these planes (see Fig. 5).
4) The program admits boundary at infinity without any special difficulty (Fig. 6).
5) The code allows multisurfaces.

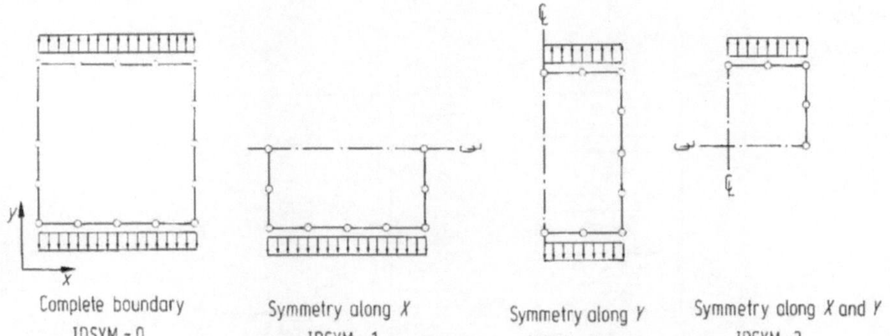

Fig. 5. Symmetry planes

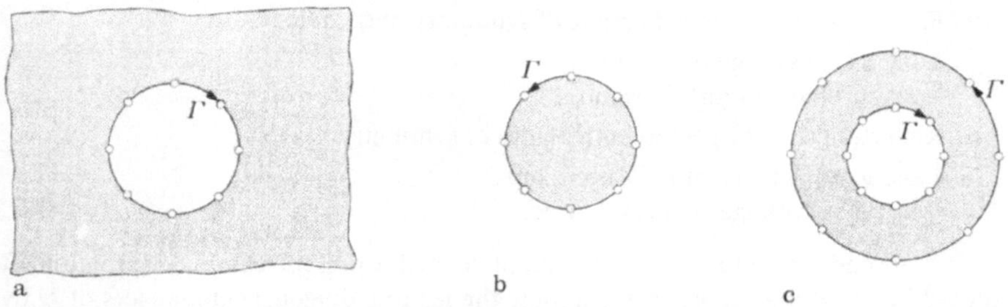

Fig. 6a–c. Definition of connectivity. **a** Unbounded case; connectivity in clockwise sense, INFB = 1. **b** Bounded case; connectivity in anticlockwise direction, INFB = 0. **c** Multisurface case, INFB = 0

6) Traction discontinuities are allowed to occur by using double nodes in the manner described in Sect. 9.3.4. The only limitation is that at least one traction in each direction has to be prescribed in one of the double nodes.

9.4.1 Main Program and Data Structure

The macro flow chart for the boundary element program can be seen in Fig. 7. The main program defines the maximum dimensions of the system of equations (in this case 100 or 50 nodes with 2 unknowns per node). It also allocates chanel 5 for the input (IRE = 5) and 6 for the output (IWR = 6).

The main program calls the following subroutines:

INPUT This routine reads the program input

MATRX Computes matrix $\boldsymbol{A}$ and the right hand side vector $\boldsymbol{F}$. They are called A (100, 100) and XM (100).

SLNPD Solves the system of equations using Gauss elimination. The routine allow for row interchange if required.

OUTPT Outputs the boundary solution including computation and printout of boundary stresses and internal displacement and stresses.

The variables used by the program, together with their meaning are given below:

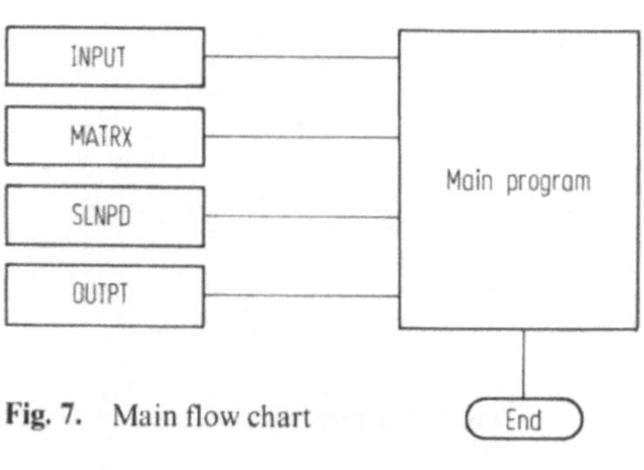

Fig. 7. Main flow chart

NE	Number of elements.
NN	Number of nodes.
NP	Number of internal points.
NN2	Number of nodes multiplied by 2 (NN2 = 2*NN)
NT	Total number of boundary nodes plus internal points (NT = NN + NP).
E	Young's modulus.
PO	Poisson's ratio.
IPL	Index to indicate type of problem (= 1 for plane stress, = 2 for plane strain).
C1 to C11	Constants including material parameters needed for the fundamental solution and other expressions.
IDSYM	Index indicating symmetry (= 0 for no-symmetry case; = 1 for plane of symmetry normal to *Y*; = 2 for plane of symmetry normal to *X*; = 3 when there are two planes of symmetry normal to *Y* and *X*).
XSYM and YSYM	Coordinates *X* and *Y* respectively that define the position of the planes of symmetry.
INFB	Infinite boundary index (= 1 means that there is a boundary at infinity; = 0 if the body is bounded).
IFA and NIF	Parameters needed for symmetry loop.
IFAIL	Index to indicate if matrix ***A*** is singular.

The following arrays are used by the program, the problem dependent ones are indicated with the required dimensions for possible modifications. (Note that herein the listings are presented for NN = 50, NE = 50 and NP = 50).

D(2, 2)	Kronecker delta matrix.
XI(6, 3)	The abscissae of Gaussian integration points.
W(6, 3)	The weights of Gaussian integration points.
IDUP(NN)	Pointer vector which indicates the double node (= 0 no double node; = number of node with the same coordinates).
INC(NE, 2)	Element connectivity.
ISYM(NT)	Vector that indicates if the node or internal point is located on any symmetry plane (= 0 if not; = 1 for plane normal to *Y*; = 2 for plane normal to *X*; = 3 if located at the intersection of the symmetry planes).
C(NE)	Element length vector.
S(NN, 3)	Matrix that will contain the stresses at boundary nodes.
X(NT)	*X* coordinates of boundary nodes and internal points.
Y(NT)	*Y* coordinates of boundary nodes and internal points.
IFIP(NN2)	Boundary condition index vector (= 0 if traction is prescribed; = 1 if displacement is prescribed).
A(NN2, NN2)	System matrix.
P(NN2)	Auxiliary vector to read prescribed tractions and displacements. At the end it contains the nodal tractions.
XM(NN2)	Right hand side vector in system of equations. After subroutine SLNPD is called it contains the boundary solution and at the end the nodal displacements.

```
C.... PROGRAM FOR SOLUTION OF TWO DIMENSIONAL
C.... ELASTOSTATIC PROBLEMS BY THE
C.... BOUNDARY ELEMENT METHOD
C
C.... AUTHOR: J. C. F. TELLES
C
      COMMON /RW/ IRE,IWR
      COMMON /A/ D(2,2),XI(6,3),W(6,3),IDUP(50),INC(50,2),C(50),
     *S(50,3),ISYM(100),X(100),Y(100),IFIP(100),A(100,100),P(100),
     *XM(100)
C
C.... INPUT
C
      IRE=5
      IWR=6
      CALL INPUT(NE,NN,NP,IPL,PO,NN2,NT,C1,C2,
     *C3,C4,C5,C6,C7,C8,C9,C10,C11,IDSYM,XSYM,YSYM,INFB)
C
C.... COMPUTE MATRIX  A  AND INDEPENDENT TERM  XM
C
      CALL MATRX(NE,NN,NN2,NT,C1,C2,C3,C4,C5,C6,C7,C8,
     *C9,C10,C11,PO,IDSYM,XSYM,YSYM,INFB,IFA,NIF)
C
C.... SOLVE SYSTEM OF EQUATIONS
C
      CALL SLNPD(NN2,C8,IFAIL)
      IF(IFAIL.NE.0)GO TO 4
C
C.... OUTPUT RESULTS
C
      CALL OUTPT(NN,NT,NN2,NE,IFA,NIF,C1,C2,C3,C4,C5,C6,
     *C7,C8,C9,C10,C11,PO,XSYM,YSYM)
    4 STOP
      END
```

9.4.2 Subroutine INPUT

The input required by the program is read in this subroutine. The data consists of the following group of cards;

1) *Title Card.* One card containing the title of the problem. The title should start in column 16.
2) *Basic Parameters Card.* One card containing the following information:
   INFB, NE, NN, NP, IPL, IDSYM, E, PO
   FORMAT (I1, I4, 4I5, 2F10.0).
3) *Coordinates of Boundary Nodes and Internal Points.* NT cards indicating for each node or point K (internal points are numbered in sequence to boundary nodes; i.e. NN + 1 ≦ K ≦ NT):
   K, X(K), Y(K), IDUP(K), ISYM(K)
   FORMAT (I5, 2F10.0, 2I5).

Note: In case of two nodes with the same coordinates (double nodes), the first one to be given must have the coordinates *X* and *Y* and IDUP has to be left blank. The second one may have the coordinates left blank but must have IDUP equal to the number K of the former. The program will automatically generate the coordinates and identify the two nodes as double.

4) *Element Connectivity.* NE cards containing the element number K, the first node and the second node in the form:
   K, INC(K, 1), INC(K, 2)
   FORMAT (3I5).

5) *Boundary Values Prescribed.* One card indicating the number of nodes with displacement prescribed (NFIP) and the number of nodes with non-zero tractions prescribed (NDFIP):
   NFIP, NDFIP
   FORMAT (2I5).

Note: The program initially assumes all tractions to be prescribed with zero value in both directions. Therefore only the non-zero prescribed tractions have to be given.

6) *Displacements Prescribed.* NFIP cards containing for each node K;
   K, P(2*K – 1), P(2*K), IFIP(2*K – 1), IFIP(2*K)
   FORMAT (I5, 2F10.0, 2I5).

Note: Nodes located on symmetry planes must not have the displacement in the normal direction to the corresponding plane prescribed. The conditions of zero displacement in this directin is automatically accomplished by the system of equations and consequently this displacement is calculated zero.

7) *Tractions Prescribed.* NDFIP cards indicating for each node K;
   K, P(2*K – 1), P(2*K)
   FORMAT (I5, 2F10.0).

```
      SUBROUTINE INPUT(NE,NN,NP,IPL,PO,NN2,NT,C1,C2,
     *C3,C4,C5,C6,C7,C8,C9,C10,C11,IDSYM,XSYM,YSYM,INFB)
      COMMON /RW/ IRE,IWR
      COMMON /A/ D(2,2),XI(6,3),W(6,3),IDUP(50),INC(50,2),C(50),
     *S(50,3),ISYM(100),X(100),Y(100),IFIP(100),A(100,100),P(100),
     *XM(100)
      WRITE(IWR,1)
    1 FORMAT(//////////,24X,'* * *  B O U N D A R Y  E L E M E N T  M E
     *T H O D  A P P L I E D  T O  * * *',//,24X,'* * *  P L A N E  E L
     *A S T O S T A T I C  P R O B L E M S             * * *',///)
C
C.... TITLE OF PROBLEM
C
      READ(IRE,2)
    2 FORMAT('
      WRITE(IWR,2)
C
C.... GENERAL INFORMATION ABOUT THE PROBLEM
C
      READ(IRE,3)INFB,NE,NN,NP,IPL,IDSYM,E,PO
    3 FORMAT(I1,I4,4I5,2F10.0)
      IF(INFB.EQ.0)GO TO 60
      WRITE(IWR,61)
   61 FORMAT(//,13X,'* INFINITE BOUNDARY *')
   60 WRITE(IWR,4)NE,NN,NP,IPL,IDSYM,E,PO
    4 FORMAT(//,15X,'NO. ELEMENTS =',I5,//,15
     *X,'NO.    NODES =',I5,//,15X,'NO.    POINTS =',I5,//,15X,'PROBL.  T
     *YPE =',I5,//,15X,'SYMME.  TYPE =',I5,///,15X,'MATERIAL PROPERTIES'
     *,//,15X,'E          =',F10.0,
     *//,15X,'POISSON =',F15.5,///,30X,'COORDINATES OF BOUNDARY NODES',
     *//,12X,'NODE',14X,'X',15X,'Y',12X,'DOUBLE',/)
      NN2=NN*2
      NT=NN+NP
C
C.... NODES AND POINTS COORDINATES
C
      DO 5 I=1,NN
      READ(IRE,6)K,X(K),Y(K),IDUP(K),ISYM(K)
    6 FORMAT(I5,2F10.0,2I5)
      IF(IDUP(K).EQ.0)GO TO 5
      J=IDUP(K)
      IDUP(J)=K
```

```
      X(K)=X(J)
      Y(K)=Y(J)
    5 CONTINUE
      DO 63 K=1,NN
      IF(IDUP(K).NE.0)GO TO 62
      WRITE(IWR,7)K,X(K),Y(K)
      GO TO 63
   62 WRITE(IWR,16)K,X(K),Y(K),IDUP(K)
   16 FORMAT(10X,I5,5X,F15.4,1X,F15.4,7X,I5)
   63 CONTINUE
    7 FORMAT(10X,I5,5X,F15.4,1X,F15.4)
      IF(NP.EQ.0)GO TO 9
      WRITE(IWR,8)
    8 FORMAT(//,30X,'COORDINATES OF INTERNAL POINTS',//,11X,'POINT',14X,
     *'X',15X,'Y',/)
      K=NN+1
      READ(IRE,14)(J,X(J),Y(J),ISYM(J),JJ=K,NT)
   14 FORMAT(I5,2F10.0,5X,I5)
      WRITE(IWR,7)(J,X(J),Y(J),J=K,NT)
C
C.... NODES AND POINTS AT SYMMETRY LINES
C
    9 IF(IDSYM.EQ.0)GO TO 49
      WRITE(IWR,42)
   42 FORMAT(//,30X,'BOUNDARY NODES AND INTERNAL POINTS AT SYMMETRY LINE
     *(S)',//,12X,'L. X',12X,'L. Y',/)
      DO 43 K=1,NT
      IF(ISYM(K).EQ.0)GO TO 43
      IZZ=ISYM(K)
      GO TO (44,45,46),IZZ
   44 YSYM=Y(K)
      WRITE(IWR,47)K
   47 FORMAT(10X,I5)
      GO TO 43
   45 XSYM=X(K)
      WRITE(IWR,48)K
   48 FORMAT(26X,I5)
      GO TO 43
   46 WRITE(IWR,50)K,K
   50 FORMAT(10X,I5,11X,I5)
   43 CONTINUE
C
C.... ELEMENT CONNECTIVITY
C
   49 WRITE(IWR,10)
   10 FORMAT(//,30X,'ELEMENT CONNECTIVITY',//,13X,'EL',13X,'N. 1',12X,'N
     *. 2',14X,'L',/)
      DO 11 I=1,NE
      READ(IRE,12)K,INC(K,1),INC(K,2)
   12 FORMAT(3I5)
      II=INC(K,1)
      IF=INC(K,2)
   11 C(K)=SQRT((X(IF)-X(II))**2+(Y(IF)-Y(II))**2)
      WRITE(IWR,13)(I,INC(I,1),INC(I,2),C(I),I=1,NE)
   13 FORMAT(10X,I5,11X,I5,11X,I5,5X,F15.4)
C
C.... CONSTANTS
C
      G=E/(2.*(1.+PO))
      C11=PO
      IF(IPL-1)40,40,41
   40 PO=PO/(1.+PO)
      C11=0.
   41 C2=3.-4.*PO
      C3=1./((1.-PO)*12.56637062)
      C4=1.-2.*PO
      C6=2.*C3*G
      C7=1.-4.*PO
      C1=C3/(2.*G)
      C5=C1/2.
      C8=2.*G/(1.-PO)
      C9=PO/(1.-PO)
      C10=(2.-PO)/(1.-PO)
C
C.... BOUNDARY VALUES PRESCRIBED
C
```

```
      DO 19 I=1,NN2
      P(I)=0.
   19 IFIP(I)=0
      READ(IRE,20)NFIP,NDFIP
   20 FORMAT(2I5)
      WRITE(IWR,21)NFIP,NDFIP
   21 FORMAT(//,15X,'NO. DISPL. PRESC. =',I5,//,15X,'NO. TRACT. PRESC. =
     *',I5,///,15X,'DISPLACEMENTS
     *',//,12X,'NODE',14X,'U',15X,'V',/)
      IF(NFIP.EQ.0)GO TO 22
      DO 23 I=1,NFIP
      READ(IRE,24)K,P(2*K-1),P(2*K),IFIP(2*K-1),IFIP(2*K)
   24 FORMAT(I5,2F10.0,2I5)
      IND=IFIP(2*K-1)+2*IFIP(2*K)
      GO TO (25,26,27),IND
   25 WRITE(IWR,28)K,P(2*K-1)
   28 FORMAT(10X,I5,5X,F15.4)
      GO TO 23
   26 WRITE(IWR,29)K,P(2*K)
   29 FORMAT(10X,I5,21X,F15.4)
      GO TO 23
   27 WRITE(IWR,30)K,P(2*K-1),P(2*K)
   30 FORMAT(10X,I5,5X,F15.4,1X,F15.4)
   23 CONTINUE
   22 IF(NDFIP.EQ.0)GO TO 31
      WRITE(IWR,34)
   34 FORMAT(//,15X,'TRACTIONS',//,12X,'NODE',13X,'PX',14X,'PY',/)
      DO 32 I=1,NDFIP
      READ(IRE,33)K,P(2*K-1),P(2*K)
   33 FORMAT(I5,2F10.0)
   32 WRITE(IWR,30)K,P(2*K-1),P(2*K)
C
C.... INTEGRATION POINTS
C
   31 XI(1,3)=-0.932469514203152
      XI(2,3)=-0.661209386466265
      XI(3,3)=-0.238619186083197
      XI(4,3)=-XI(3,3)
      XI(5,3)=-XI(2,3)
      XI(6,3)=-XI(1,3)
      W(1,3)=0.171324492379170
      W(2,3)=0.360761573048139
      W(3,3)=0.467913934572691
      W(4,3)=W(3,3)
      W(5,3)=W(2,3)
      W(6,3)=W(1,3)
      XI(1,2)=-0.861136311594053
      XI(2,2)=-0.339981043584856
      XI(3,2)=-XI(2,2)
      XI(4,2)=-XI(1,2)
      W(1,2)=0.347854845137454
      W(2,2)=0.652145154862546
      W(3,2)=W(2,2)
      W(4,2)=W(1,2)
      XI(1,1)=-0.577350269189626
      XI(2,1)=-XI(1,1)
      W(1,1)=1.
      W(2,1)=1.
      RETURN
      END
```

9.4.3 Subroutine MATRX

This subroutine computes matrix $\boldsymbol{A}$ and vector $\boldsymbol{F}$ of Eq. (24). The element influence coefficients are computed calling subroutine FUNC.

FUNC: Computes the element influence matrices $\boldsymbol{h}$ and $\boldsymbol{g}$ necessary to form system matrix $\boldsymbol{A}$ and independent term $\boldsymbol{F}$ (now in XM).

Subroutine MATRX generates the system of equations by assembling directly matrix $\boldsymbol{A}$ without going to the global $\boldsymbol{H}$ and $\boldsymbol{G}$ matrices. This is done by considering the prescribed boundary conditions for the node under consideration before assembling. The leading diagonal submatrices corresponding to $\boldsymbol{H}$ are calculated using rigid body translations. Consequently, when the body is unbounded a different type of rigid body consideration needs to be applied [see Eq. (40)].

Symmetry is taken into consideration by instead of reflecting the elements, reflecting the singular nodes. This procedure is very simple and completely general.

```
      SUBROUTINE MATRX(NE,NN,NN2,NT,C1,C2,C3,C4,C5,C6,C7,C8,
     *C9,C10,C11,PO,IDSYM,XSYM,YSYM,INFB,IFA,NIF)
      COMMON /A/ D(2,2),XI(6,3),W(6,3),IDUP(50),INC(50,2),C(50),
     *S(50,3),ISYM(100),X(100),Y(100),IFIP(100),A(100,100),P(100),
     *XM(100)
      COMMON /A4/ H(3,4),G(3,4),HL(3,4),GL(3,4)
C
C.... KRONECKER DELTA
C
      D(1,1)=1.
      D(2,2)=1.
      D(1,2)=0.
      D(2,1)=0.
C
C.... CLEAR ARRAYS
C
      DO 1 I=1,NN2
      XM(I)=0.
      DO 1 J=1,NN2
    1 A(I,J)=0.
C
C.... COMPUTE PARAMETERS FOR SYMMETRY LOOP
C
      IFA=1
      NIF=1
      IF(IDSYM.EQ.1)IFA=2
      IF(IDSYM.NE.2)GO TO 60
      IFA=3
      NIF=2
   60 IF(IDSYM.EQ.3)IFA=4
C
C.... TEST FOR INFINITE BOUNDARY
C
      IF(INFB.EQ.0)GO TO 90
      DO 91 I=1,NN2
      IF(IFIP(I).NE.0)GO TO 92
      A(I,I)=1.
      GO TO 91
   92 XM(I)=-P(I)
   91 CONTINUE
C
C.... SYMMETRY LOOP
C
   90 DO 2 ISY=1,IFA,NIF
C
C.... COMPUTE CHANGE SIGN CONTROLLING PARAMETERS
C
      GO TO (70,71,71,73),ISY
   71 IIS=4-ISY
      IFS=IIS
      GO TO 70
   73 IIS=1
      IFS=2
C
C.... LOOP OVER BOUNDARY NODES
C
   70 DO 2 I=1,NN
      XS=X(I)
      YS=Y(I)
      IF(ISY.EQ.2.OR.ISY.EQ.4)YS=2.*YSYM-YS
      IF(ISY.GE.3)XS=2.*XSYM-XS
C
```

```
C.... GENERATE MATRIX A AND INDEPENDENT TERM XM
C
C.... NOTE: COLUMNS OF  A  THAT CORRESPOND TO  G  ARE MULTIPLIED
C.... BY  C8  TO AVOID ROUNDOFF ERRORS IN THE SOLUTION.
C
      DO 10 J=1,NE
      II=INC(J,1)
      IF=INC(J,2)
      ICOD=1
      IF(ISY.NE.1.AND.ISYM(I).NE.(ISY-1))GO TO 6
      IF(I.EQ.II.OR.I.EQ.IDUP(II))ICOD=2
      IF(I.EQ.IF.OR.I.EQ.IDUP(IF))ICOD=3
    6 CALL FUNC(ICOD,J,C1,C2,C3,C4,C5,C6,C7,PO,II,IF,XS,YS,ISY,IIS,
     *IFS)
      DO 10 K=1,2
      JJ=2*(I-1)+K
      M=0
      DO 10 NX=1,2
      DO 10 NV=1,2
      M=M+1
      IC=2*INC(J,NX)+NV-2
      IF(IFIP(IC).NE.0)GO TO 67
      A(JJ,IC)=A(JJ,IC)+H(K,M)
      XM(JJ)=XM(JJ)+G(K,M)*P(IC)
      GO TO 68
   67 A(JJ,IC)=A(JJ,IC)-G(K,M)*C8
      XM(JJ)=XM(JJ)-H(K,M)*P(IC)
C
C.... COMPUTE REMAINING COEFFICIENTS BY APPLYING RIGID BODY TRANSLATIONS
C
   68 GO TO (61,62,63,64),ISY
   62 IF(NV-2)61,64,61
   63 IF(NV-1)61,64,61
   64 H(K,M)=-H(K,M)
   61 IF(IFIP(JJ+NV-K).NE.0)GO TO 69
      A(JJ,JJ+NV-K)=A(JJ,JJ+NV-K)-H(K,M)
      GO TO 10
   69 XM(JJ)=XM(JJ)+H(K,M)*P(JJ+NV-K)
   10 CONTINUE
    2 CONTINUE
      RETURN
      END
```

9.4.4 Subroutine FUNC

This subroutine computes all the element integrals required for the system of equations, internal displacements and internal stresses. Numerical integrals are performed over non-singular elements by using 2, 4 and 6 Gauss integration points. The number of points is selected according to the relative distance from the singular node or point to the centre of the element under consideration.

For elements with the singularity at one of its extremities the required integrals are computed analytically to obtain more accurate results.

Notice that at the end of the subroutine a symmetry test is included to find if it is necessary to change the sign of some rows of the element submatrices.

```
      SUBROUTINE FUNC(ICOD,JA,C1,C2,C3,C4,C5,C6,C7,PO,II,IF,XS,YS,ISY,
     *IIS,IFS)
C
C.... INTEGRALS OVER BOUNDARY ELEMENTS
C
      COMMON /A/ D(2,2),XI(6,3),W(6,3),IDUP(50),INC(50,2),C(50),
     *S(50,3),ISYM(100),X(100),Y(100),IFIP(100),A(100,100),P(100),
     *XM(100)
      COMMON /A4/ H(3,4),G(3,4),HL(3,4),GL(3,4)
      DIMENSION DXY(2),BN(2),B(2),DR(2),UL(2,2),PL(2,2),ULL(2,2,2),
     *PLL(2,2,2)
      DO 5 KK=1,3
      DO 5 L=1,4
```

```
      GL(KK,L)=0.
      HL(KK,L)=0.
      G(KK,L)=0.
    5 H(KK,L)=0.
      DXY(1)=X(IF)-X(II)
      DXY(2)=Y(IF)-Y(II)
      GO TO (1,2,2,1),ICOD
    1 BN(1)=DXY(2)/C(JA)
      BN(2)=-DXY(1)/C(JA)
C
C.... SELECT NO. INTEGRATION POINTS
C
      SEL=0.5*SQRT((2.*XS-X(II)-X(IF))**2+(2.*YS-Y(II)-Y(IF))**2)/C(JA)
      NPI=4
      IF(SEL.LE.1.5)NPI=6
      IF(SEL.GT.5.5)NPI=2
      INP=NPI/2
C
C.... COMPUTE MATRICES NUMERICALLY
C
      DO 50 KK=1,NPI
      XMXI=0.5*(1.+XI(KK,INP))*DXY(1)+X(II)-XS
      YMYI=0.5*(1.+XI(KK,INP))*DXY(2)+Y(II)-YS
      R=SQRT(XMXI**2+YMYI**2)
      B(1)=-0.25*(XI(KK,INP)-1.)*C(JA)
      B(2)=0.25*(XI(KK,INP)+1.)*C(JA)
      DR(1)=XMXI/R
      DR(2)=YMYI/R
      DRDN=DR(1)*BN(1)+DR(2)*BN(2)
C
C.... COMPUTE MATRICES H AND G
C
      DO 6 I=1,2
      DO 6 J=1,2
      UL(I,J)=-C1*(C2*LOG(R)*D(I,J)-DR(I)*DR(J))
    6 PL(I,J)=-C3*((C4*D(I,J)+2.*DR(I)*DR(J))*DRDN+C4*(DR(J)*BN(I)-DR(I)
     **BN(J)))/R
      DO 7 LA=1,2
      IC=0
      DO 7 LL=1,2
      DO 7 JJ=1,2
      IC=IC+1
      G(LA,IC)=G(LA,IC)+UL(LA,JJ)*B(LL)*W(KK,INP)
    7 H(LA,IC)=H(LA,IC)+PL(LA,JJ)*B(LL)*W(KK,INP)
      IF(ICOD.NE.4)GO TO 50
C
C.... COMPUTE MATRICES HL AND GL (STRESSES AT INTERNAL POINTS)
C
   10 DO 11 I=1,2
      DO 11 J=I,2
      DO 11 K=1,2
      ULL(I,J,K)=C3*(C4*(DR(J)*D(K,I)+DR(I)*D(K,J)-DR(K)*D(I,J))+2.*DR(I
     *)*DR(J)*DR(K))/R
      B1=2.*DRDN*(C4*DR(K)*D(I,J)+PO*(DR(J)*D(I,K)+DR(I)*D(J,K))-4.*DR(I
     *)*DR(J)*DR(K))
      B2=2.*PO*(BN(I)*DR(J)*DR(K)+BN(J)*DR(I)*DR(K))
      B3=C4*(2.*BN(K)*DR(I)*DR(J)+BN(J)*D(I,K)+BN(I)*D(J,K))
   11 PLL(I,J,K)=C6*(B1+B2+B3-C7*BN(K)*D(I,J))/R**2
      IL=0
      DO 12 I=1,2
      DO 12 J=I,2
      IL=IL+1
      IC=0
      DO 12 IAA=1,2
      DO 12 JAA=1,2
      IC=IC+1
      GL(IL,IC)=GL(IL,IC)+B(IAA)*ULL(I,J,JAA)*W(KK,INP)
   12 HL(IL,IC)=HL(IL,IC)+B(IAA)*PLL(I,J,JAA)*W(KK,INP)
   50 CONTINUE
      GO TO 18
C
C.... COMPUTE MATRICES H AND G ANALYTICALLY (BOUNDARY CONSTRAINT EQ.)
C
    2 AL=C5*C2*C(JA)
      AA=AL*(0.5-LOG(C(JA)))
      DO 15 I=1,2
      DO 15 J=1,4
```

```
      IT=(J/2)*2+2-J
      G(I,J)=C5*DXY(I)*DXY(IT)/C(JA)
      IF(IT.EQ.I)G(I,J)=G(I,J)+AA
   15 CONTINUE
      IAA=-2
      IF(ICOD.EQ.3)IAA=0
      G(1,3+IAA)=G(1,3+IAA)+AL
      G(2,4+IAA)=G(2,4+IAA)+AL
      H(1,2-IAA)=C3*C4*(1.+IAA)
      H(2,1-IAA)=-H(1,2-IAA)
C
C.... SYMMETRY TEST
C
   18 IF(ISY.EQ.1)GO TO 8
      DO 24 I=IIS,IFS
      DO 24 J=1,4
      H(I,J)=-H(I,J)
   24 G(I,J)=-G(I,J)
      IF(ICOD.NE.4.OR.ISY.EQ.4)GO TO 8
      DO 25 J=1,4
      HL(2,J)=-HL(2,J)
   25 GL(2,J)=-GL(2,J)
    8 RETURN
      END
```

9.4.5 Subroutine SLNPD

This is a standard subroutine for the solution of non-positive definite matrices using Gauss elimination. If the matrix $\boldsymbol{A}$ has a zero in the diagonal it will interchange rows deciding that the system matrix is singular only when no row interchange will produce a non-zero diagonal coefficient for a given row position.

The final result is a vector with the unknown boundary displacements and tractions and is overwritten in B (NN2) which locally corresponds to XM (NN2) in the main program.

```
      SUBROUTINE SLNPD(N,C8,IFAIL)
C
C.... SOLVE SYSTEM OF EQUATIONS BY GAUSS ELIMINATION PROCESS AND
C.... RETURN RESULT IN VECTOR XM
C
      COMMON /RW/ IRE,IWR
      COMMON /A/ D(2,2),XI(6,3),W(6,3),IDUP(50),INC(50,2),C(50),
     *S(50,3),ISYM(100),X(100),Y(100),IFIP(100),A(100,100),P(100),
     *B(100)
      N1=N-1
      DO 100 K=1,N1
      K1=K+1
      CC=A(K,K)
      IF(ABS(CC))1,1,3
    1 DO 7 J=K1,N
      IF(ABS(A(J,K)))7,7,5
    5 DO 6 L=K,N
      CC=A(K,L)
      A(K,L)=A(J,L)
    6 A(J,L)=CC
      CC=B(K)
      B(K)=B(J)
      B(J)=CC
      CC=A(K,K)
      GO TO 3
    7 CONTINUE
      WRITE(IWR,2)K
    2 FORMAT(/////,20X,'* * * SINGULARITY IN ROW',I5,' * * *')
      IFAIL=1
      GO TO 300
    3 DO 4 J=K1,N
    4 A(K,J)=A(K,J)/CC
      B(K)=B(K)/CC
      DO 10 I=K1,N
```

```
      CC=A(I,K)
      DO 9 J=K1,N
    9 A(I,J)=A(I,J)-CC*A(K,J)
   10 B(I)=B(I)-CC*B(K)
  100 CONTINUE
      IF(ABS(A(N,N)))8,8,101
    8 WRITE(IWR,2)NN
      IFAIL=1
      GO TO 300
  101 B(N)=B(N)/A(N,N)
      DO 200 L=1,N1
      K=N-L
      K1=K+1
      DO 200 J=K1,N
  200 B(K)=B(K)-A(K,J)*B(J)
C
C.... MULTIPLY CALCULATED TRACTIONS BY  C8  TO OBTAIN ACTUAL VALUES
C
      DO 70 I=1,N
      IF(IFIP(I).EQ.0)GO TO 70
      B(I)=B(I)*C8
   70 CONTINUE
      IFAIL=0
  300 RETURN
      END
```

9.4.6 Subroutine OUTPT

This subroutine prints the results for boundary displacements and tractions. In addition, it calculates and prints the boundary stresses and internal displacements and stresses.

The boundary stresses are computed using subroutine FENC. Here, the contribution of adjacent elements to the common boundary nodes is automatically averaged for non-double nodes. The internal displacements and stresses are computed by integrating over the boundary elements using subroutine FUNC.

The total output consists of:

(i) Displacements and tractions at boundary nodes.
(ii) Displacements and stresses at boundary nodes and internal points.

```
      SUBROUTINE OUTPT(NN,NT,NN2,NE,IFA,NIF,C1,C2,C3,C4,C5,C6,
     *C7,C8,C9,C10,C11,PO,XSYM,YSYM)
C
C.... OUTPUT RESULTS
C
      COMMON /RW/ IRE,IWR
      COMMON /A/ D(2,2),XI(6,3),W(6,3),IDUP(50),INC(50,2),C(50),
     *S(50,3),ISYM(100),X(100),Y(100),IFIP(100),A(100,100),P(100),
     *XM(100)
      COMMON /A4/ H(3,4),G(3,4),HL(3,4),GL(3,4)
      DIMENSION U(2),SA(4)
      WRITE(IWR,6)
    6 FORMAT(
     *///,30X,'BOUNDARY DISPLACEMENTS AND TRACTIONS',//,12X,'NODE',14
     *X,'U',15X,'V',14X,'PX',14X,'PY',/)
C
C.... BOUNDARY DISPLACEMENT AND TRACTIONS
C
      DO 8 I=1,NN2
      IF(IFIP(I).EQ.0)GO TO 8
    9 PA=XM(I)
      XM(I)=P(I)
      P(I)=PA
    8 CONTINUE
      WRITE(IWR,11)(I,XM(2*I-1),XM(2*I),P(2*I-1),P(2*I),I=1,NN)
   11 FORMAT(10X,I5,5X,F15.4,1X,F15.4,1X,F15.4,1X,F15.4)
C
```

```
C.... DISPLACEMENTS AND STRESSES AT NODES AND POINTS
C
      WRITE(IWR,12)
   12 FORMAT(//,15X,'DISPLACEMENTS AND STRESSES AT NODES AND INTERNAL PO
     *INTS',//,11X,'NO/PT',14X,'U',15X,'V',14X,'SX',14X,'SXY',13X,'SY',
     *14X,'SZ',/)
C
C.... COMPUTE BOUNDARY STRESSES
C
      DO 14 I=1,NN
      DO 14 J=1,3
   14 S(I,J)=0.
C
C.... LOOP OVER ALL BOUNDARY ELEMENTS
C
      DO 30 I=1,NE
      II=INC(I,1)
      IF=INC(I,2)
      CC1=(Y(IF)-Y(II))/C(I)
      CC2=(X(II)-X(IF))/C(I)
      CALL FENC(C8,C9,C10,CC1,CC2,I)
      DO 30 JP=1,2
      IIF=INC(I,JP)
      XFAC=2.
      IF(IDUP(IIF).NE.0.OR.ISYM(IIF).NE.0)XFAC=1.
      DO 30 IR=1,3
      M=0
      DO 30 IP=1,2
      IO=2*IIF+IP-2
      S(IIF,IR)=S(IIF,IR)+G(IR,IP)*P(IO)/XFAC
      DO 30 JR=1,2
      M=M+1
      IO=2*INC(I,IP)+JR-2
   30 S(IIF,IR)=S(IIF,IR)-H(IR,M)*XM(IO)/XFAC
C
C.... PRINT VALUES ON THE BOUNDARY
C
      DO 13 I=1,NN
      SA(4)=C11*(S(I,1)+S(I,3))
      IF(ISYM(I).NE.0)S(I,2)=0.
   13 WRITE(IWR,15)I,XM(2*I-1),XM(2*I),S(I,1),S(I,2),S(I,3),SA(4)
   15 FORMAT(10X,I5,4X,6(1X,F15.4))
      IF(NN.EQ.NT)GO TO 5
C
C.... COMPUTE INTERNAL VALUES
C
      NNI=NN+1
      ICOD=4
C
C.... LOOP OVER ALL INTERNAL POINTS
C
      DO 16 I=NNI,NT
      U(1)=0.
      U(2)=0.
      DO 17 J=1,3
   17 SA(J)=0.
C
C.... SYMMETRY LOOP
C
      DO 20 ISY=1,IFA,NIF
      XS=X(I)
      YS=Y(I)
      IF(ISY.EQ.2.OR.ISY.EQ.4)YS=2.*YSYM-YS
      IF(ISY.GE.3)XS=2.*XSYM-XS
      GO TO (70,71,71,73),ISY
   71 IIS=4-ISY
      IFS=IIS
      GO TO 70
   73 IIS=1
      IFS=2
C
C.... INTEGRATE OVER THE BOUNDARY
C
   70 DO 20 J=1,NE
      II=INC(J,1)
      IF=INC(J,2)
```

```
      CALL FUNC(ICOD,J,C1,C2,C3,C4,C5,C6,C7,PO,II,IF,XS,YS,ISY,
     *IIS,IFS)
      DO 20 K=1,3
      M=0
      DO 20 NX=1,2
      DO 20 NV=1,2
      M=M+1
      ICA=2*INC(J,NX)+NV-2
      IF(K.LT.3)U(K)=U(K)-H(K,M)*XM(ICA)+G(K,M)*P(ICA)
   20 SA(K)=SA(K)-HL(K,M)*XM(ICA)+GL(K,M)*P(ICA)
      SA(4)=C11*(SA(1)+SA(3))
   16 WRITE(IWR,15)I,U(1),U(2),SA(1),SA(2),SA(3),SA(4)
    5 RETURN
      END
```

9.4.7 Subroutine FENC

This subroutine computes the expressions for stresses at boundary nodes using the formulae presented in Sect. 9.3.3. Notice that the required coordinate transformations are already performed, so that the stresses are calculated with reference to the global system of coordinates.

```
      SUBROUTINE FENC(C8,C9,C10,CC1,CC2,I)
C
C.... EXPRESSIONS FOR STRESSES AT BOUNDARY NODES
C
      COMMON /A/ D(2,2),XI(6,3),W(6,3),IDUP(50),INC(50,2),C(50),
     *S(50,3),ISYM(100),X(100),Y(100),IFIP(100),A(100,100),P(100),
     *XM(100)
      COMMON /A4/ H(3,4),G(3,4),HL(3,4),GL(3,4)
C
C.... MATRIX H
C
      CO=-C8/C(I)
      H(1,1)=CO*CC2**3
      H(1,2)=-CO*CC1*CC2**2
      H(1,3)=-H(1,1)
      H(1,4)=-H(1,2)
      H(2,1)=H(1,2)
      H(2,2)=CO*CC2*CC1**2
      H(2,3)=-H(1,2)
      H(2,4)=-H(2,2)
      H(3,1)=H(2,2)
      H(3,2)=-CO*CC1**3
      H(3,3)=-H(2,2)
      H(3,4)=-H(3,2)
C
C.... MATRIX G
C
      G(1,1)=CC1**3+C10*CC1*CC2**2
      G(1,2)=-CC2*CC1**2+C9*CC2**3
      G(2,1)=CC2**3-C9*CC2*CC1**2
      G(2,2)=CC1**3-C9*CC1*CC2**2
      G(3,1)=-CC1*CC2**2+C9*CC1**3
      G(3,2)=CC2**3+C10*CC2*CC1**2
      RETURN
      END
```

9.4.8 Examples

In order to illustrate the data preparation for the computer program, two simple examples are presented below.

(i) *Square Plate.* This example consists of a square plate under biaxial load. The discretization is presented in Fig. 8 and the corresponding input data is given

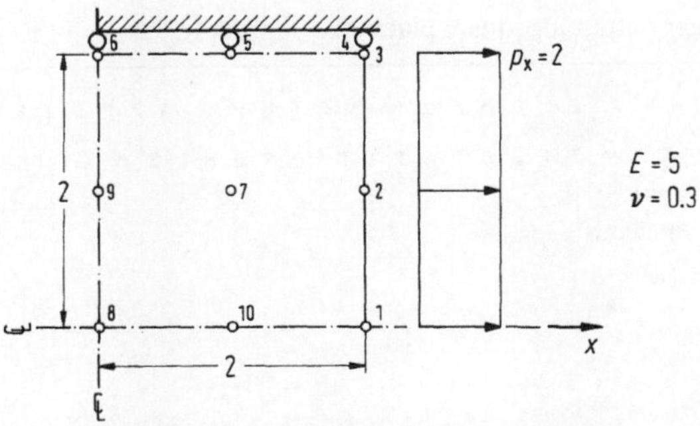

Fig. 8. Definition of square plate problem

Table 1. Data file for square plate

```
             SQUARE PLATE PROBLEM
 4    6    4    1    3    5.         0.3
 1 2.                          1
 2 2.           1.
 3 2.           2.
 4                       3
 5 1.           2.
 6              2.             2
 7 1.           1.
 8                             3
 9              1.             2
10 1.                          1
 1    1    2
 2    2    3
 3    4    5
 4    5    6
 3    3
 4                             1
 5                             1
 6                             1
 1 2.
 2 2.
 3 2.
```

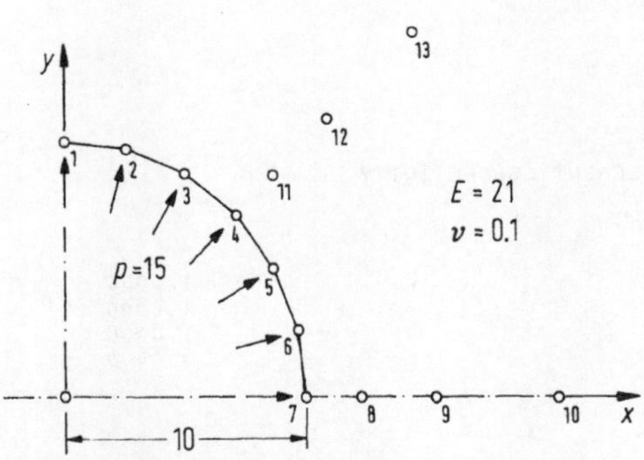

Fig. 9. Definition of cylindrical cavity problem

Table 2. Line printer output for square plate

```
B O U N D A R Y   E L E M E N T   M E T H O D   A P P L I E D   T O

P L A N E   E L A S T O S T A T I C   P R O B L E M S

SQUARE PLATE PROBLEM

NO. ELEMENTS =     4

NO.    NODES =     6

NO.   POINTS =     4

PROBL.  TYPE =     1

SYMME.  TYPE =     3

MATERIAL PROPERTIES

E        =        5.

POISSON =         0.30000

                  COORDINATES OF BOUNDARY NODES

NODE               X                  Y                DOUBLE

  1              2.0000             0.0000
  2              2.0000             1.0000
  3              2.0000             2.0000             4
  4              2.0000             2.0000             3
  5              1.0000             2.0000
  6              0.0000             2.0000

                  COORDINATES OF INTERNAL POINTS

POINT              X                  Y

  7              1.0000             1.0000
  8              0.0000             0.0000
  9              0.0000             1.0000
 10              1.0000             0.0000

                BOUNDARY NODES AND INTERNAL POINTS AT SYMMETRY LINE(S)

L. X             L. Y

  1
                  6
  8               8
                  9
 10

                ELEMENT CONNECTIVITY

EL               N. 1               N. 2               L

 1                1                  2              1.0000
 2                2                  3              1.0000
 3                4                  5              1.0000
 4                5                  6              1.0000

NO. DISPL. PRESC. =    3

NO. TRACT. PRESC. =    3
```

Table 2 (continued)

DISPLACEMENTS

NODE	U	V
4		0.0000
5		0.0000
6		0.0000

TRACTIONS

NODE	PX	PY
1	2.0000	0.0000
2	2.0000	0.0000
3	2.0000	0.0000

BOUNDARY DISPLACEMENTS AND TRACTIONS

NODE	U	V	PX	PY
1	0.7280	0.0000	2.0000	0.0000
2	0.7280	0.0000	2.0000	0.0000
3	0.7280	0.0000	2.0000	0.0000
4	0.7280	0.0000	0.0000	0.6000
5	0.3640	0.0000	0.0000	0.6000
6	0.0000	0.0000	0.0000	0.6000

DISPLACEMENTS AND STRESSES AT NODES AND INTERNAL POINTS

NO/PT	U	V	SX		SY	SZ
1	0.7280	0.0000	2.0000	0.0000	0.6000	0.0000
2	0.7280	0.0000	2.0000	0.0000	0.6000	0.0000
3	0.7280	0.0000	2.0000	0.0000	0.6000	0.0000
4	0.7280	0.0000	2.0000	0.0000	0.6000	0.0000
5	0.3640	0.0000	2.0000	0.0000	0.6000	0.0000
6	0.0000	0.0000	2.0000	0.0000	0.6000	0.0000
7	0.3640	0.0000	2.0000	0.0000	0.6000	0.0000
8	0.0000	0.0000	2.0000	0.0000	0.6000	0.0000
9	0.0000	0.0000	2.0000	0.0000	0.6000	0.0000
10	0.3640	0.0000	2.0000	0.0000	0.6000	0.0000

Table 3. Data file for cylindrical cavity

```
                   CYLINDRICAL CAVITY PROBLEM
1     6     7     6     2     3     21.      0.1
      1                10.              2
      2 2.5882         9.6593
      3 5.             8.6603
      4 7.0711         7.0711
      5 8.6603         5.
      6 9.6593         2.5882
      7 10.                             1
      8 12.                             1
      9 15.                             1
     10 20.                             1
     11 8.4853         8.4853
     12 10.6066        10.6066
     13 14.1421        14.1421
      1     1     2
      2     2     3
      3     3     4
      4     4     5
      5     5     6
      6     6     7
            7
      1                15.
      2 3.8823         14.4889
      3 7.5            12.9904
      4 10.6066        10.6066
      5 12.9904        7.5
      6 14.4889        3.8823
      7 15.
```

Table 4. Line printer output for cylindrical cavity

```
      B O U N D A R Y   E L E M E N T   M E T H O D   A P P L I E D   T O
      P L A N E   E L A S T O S T A T I C   P R O B L E M S

   CYLINDRICAL CAVITY PROBLEM

 * INFINITE BOUNDARY *

   NO. ELEMENTS =    6
   NO.    NODES =    7
   NO.   POINTS =    6
   PROBL.  TYPE =    2
   SYMME.  TYPE =    3

   MATERIAL PROPERTIES
   E        =        21.
   POISSON =          0.10000

                 COORDINATES OF BOUNDARY NODES
 NODE            X                   Y              DOUBLE
   1           0.0000             10.0000
   2           2.5882              9.6593
   3           5.0000              8.6603
   4           7.0711              7.0711
   5           8.6603              5.0000
   6           9.6593              2.5882
   7          10.0000              0.0000

                 COORDINATES OF INTERNAL POINTS
POINT            X                   Y
   8          12.0000              0.0000
   9          15.0000              0.0000
  10          20.0000              0.0000
  11           8.4853              8.4853
  12          10.6066             10.6066
  13          14.1421             14.1421

                 BOUNDARY NODES AND INTERNAL POINTS AT SYMMETRY LINE(S)
 L. X              L. Y

                     1
   7
   8
   9
  10

                 ELEMENT CONNECTIVITY
 EL                N. 1              N. 2              L
  1                 1                 2              2.6105
  2                 2                 3              2.6105
```

Table 4 (continued)

3	3	4	2.6106
4	4	5	2.6106
5	5	6	2.6105
6	6	7	2.6105

NO. DISPL. PRESC. = 0

NO. TRACT. PRESC. = 7

DISPLACEMENTS

NODE	U	V

TRACTIONS

NODE	PX	PY
1	0.0000	15.0000
2	3.8823	14.4889
3	7.5000	12.9904
4	10.6066	10.6066
5	12.9904	7.5000
6	14.4889	3.8823
7	15.0000	0.0000

BOUNDARY DISPLACEMENTS AND TRACTIONS

NODE	U	V	PX	PY
1	0.0000	7.7353	0.0000	15.0000
2	2.0020	7.4717	3.8823	14.4889
3	3.8676	6.6990	7.5000	12.9904
4	5.4697	5.4697	10.6066	10.6066
5	6.6990	3.8676	12.9904	7.5000
6	7.4717	2.0020	14.4889	3.8823
7	7.7353	0.0000	15.0000	0.0000

DISPLACEMENTS AND STRESSES AT NODES AND INTERNAL POINTS

NO/PT	U	V	SX	SXY	SY	SZ
1	0.0000	7.7353	14.7576	0.0000	-14.8737	-0.0116
2	2.0020	7.4717	12.7729	-7.4078	-12.8808	-0.0116
3	3.3676	6.6990	7.3499	-12.8308	-7.4658	-0.0116
4	5.4697	5.4697	-0.0580	-14.8157	-0.0580	-0.0116
5	6.6990	3.8676	-7.4658	-12.8308	7.3500	-0.0116
6	7.4717	2.0020	-12.8888	-7.4078	12.7729	-0.0116
7	7.7353	0.0000	-14.8737	0.0000	14.7576	-0.0116
8	6.4100	0.0000	-10.2220	0.0000	10.2136	-0.0008
9	5.1274	0.0000	-6.5260	0.0000	6.5259	0.0000
10	3.8455	0.0000	-3.6707	0.0000	3.6707	0.0000
11	4.5325	4.5325	-0.0042	-10.2178	-0.0042	-0.0008
12	3.6256	3.6256	0.0000	-6.5259	0.0000	0.0000
13	2.7192	2.7192	0.0000	-3.6707	0.0000	0.0000

in Table 1. In this analysis 4 boundary elements with 6 boundary nodes and 4 internal points are employed.

The complete printout corresponding to the solution of the problem is presented in Table 2. As can be easily checked the analytical solution is obtained.

(ii) *Cylindrical Cavity Problem.* In this example the analysis of a cylindrical cavity under internal pressure is carried out. The problem domain is the infinite medium and consequently, INFB = 1 in this case. The discretization is depicted in Fig. 9 and the input data is presented in Table 3. For the solution presented 6 boundary elements and 7 boundary nodes are used. Additional information is obtained at 6 internal points.

The solution obtained by the program is presented in Table 4. These results are in good agreement with the analytical solution [8], and the maximum error is around 2%.

References

1 Brebbia, C.A., Telles, J.C.F., and Wrobel, L.C., Boundary Element Techniques: Theory and Applications in Engineering. Springer-Verlag, Berlin 1984

2 Love, A.E.H., A Treatise on the Mathematical Theory of Elasticity. Dover, New York 1944

3 Telles, J.C.F., Mansur, W.J., and Wrobel, L.C., On Boundary Elements for External Potential Problems. Mech. Res. Com. **11,** 373–377 (1984)

4 Riccardella, P.C., An Implementation of the Boundary Integral Technique for Planar Problems in Elasticity and Elastoplasticity, Report No. SM-73-10, Dept. Mech. Engng., Carnegie Mellon Univ., Pittsburg 1973

5 Mansur, W.J., Halbritter, A.L., and Telles, J.C.F., The Boundary Element Method: Formulation for Two-Dimensional Elasticity. Commemorative Annals of the 15th Anniversary of COPPE/UFRJ (COPPE/UFRJ ed. in Portuguese). Rio de Janeiro 1979, pp. 1–22

6 Chaudonneret, M., On the Discontinuity of the Stress Vector in the Boundary Integral Equation Method for Elastic Analysis. In: Recent Advances in Boundary Element Methods (C.A. Brebbia, ed.). Pentech-Press, London 1978, pp. 185–194

7 Brebbia, C.A. and Ferrante, A.J., Computational Methods for the Solution of Engineering Problems. Pentech Press, London 1978

8 Timoshenko, S.P. and Goodier, J.N., Theory of Elasticity, 3rd ed. McGraw-Hill, Tokyo 1970

9 Cruse, T.A., Mathematical Foundations of the Boundary Integral Equation Method in Solid Mechanics. Report No. AFOSR-TR-77-1002, Pratt and Whitney Aircraft Group, 1977

10 Brebbia, C.A., The Boundary Element Method for Engineers. Pentech Press, London; Halstead Press, New York 1978

Subject Index